Oscillations and Waves

An Introduction

Second Edition

Oscillations and Waves

An Introduction

Second Edition

Richard Fitzpatrick

CRC Press
Taylor & Francis Group
Boca Raton London New York

CRC Press is an imprint of the
Taylor & Francis Group, an **informa** business

CRC Press
Taylor & Francis Group
6000 Broken Sound Parkway NW, Suite 300
Boca Raton, FL 33487-2742

Printed on acid-free paper
Version Date: 20180628

International Standard Book Number-13: 978-1-138-47971-5 (Paperback)
International Standard Book Number-13: 978-1-138-48035-3 (Hardback)

Library of Congress Cataloging-in-Publication Data

Names: Fitzpatrick, Richard, 1963- author.
Title: Oscillations and waves : an introduction / by Richard Fitzpatrick.
Description: Second edition. | Boca Raton, FL : CRC Press, Taylor & Francis Group, [2018]
Identifiers: LCCN 2018012827| ISBN 9781138480353 (hardback ; alk. paper) | ISBN 1138480355 (hardback ; alk. paper) | ISBN 9781138479715 (pbk. ; alk. paper) | ISBN 1138479713 (pbk. ; alk. paper) | ISBN 9781351063104 (eBook General) | ISBN 1351063103 (eBook General) | ISBN 9781351063098 (eBook Adobe Reader) | ISBN 135106309X (eBook Adobe Reader) | ISBN 9781351063081 (eBook ePub) | ISBN 1351063081 (eBook ePub) | ISBN 9781351063074 (eBook Mobipocket) | ISBN 1351063073 (eBook Mobipocket)
Subjects: LCSH: Wave-motion, Theory of. | Oscillations.
Classification: LCC QA935 .F48 2018 | DDC 531/.32--dc23
LC record available at https://lccn.loc.gov/2018012827

Visit the eResources: https://www.crcpress.com/9781138479715

Visit the Taylor & Francis Web site at
http://www.taylorandfrancis.com

and the CRC Press Web site at
http://www.crcpress.com

Contents

Preface

Oscillations and waves are ubiquitous phenomena in the world around us. An oscillation is defined as a disturbance in a physical system that is repetitive in time. A wave is defined as a disturbance in a continuous, spatially extended, physical system that is both repetitive in time and periodic in space. In general, an oscillation involves a continual back and forth flow between two different energy types. For example, in the case of a pendulum, the two energy types are kinetic and gravitational potential energy. A wave involves similar repetitive energy flows to an oscillation, but, in addition, is capable of transmitting energy (and information) from one place to another. Although sound waves and electromagnetic waves, for example, rely on quite distinct physical mechanisms, they, nevertheless, share many common properties. This is also true of different types of oscillation. It turns out that the common factor linking the various types of wave and oscillation is that they are all described by the same mathematical equations.

The aim of this textbook is to develop a unified mathematical theory of oscillations and waves. Examples are drawn from the physics of discrete mechanical systems; continuous gases, fluids, and elastic solids; electronic circuits; electromagnetic waves; optical systems; and, finally, quantum mechanical systems.

It is assumed that readers of this book possess a basic familiarity with the laws of physics, such as might be obtained from a standard two-semester introductory college-level survey course. Readers are also assumed to be conversant with college-level mathematics up to and including algebra, trigonometry, linear algebra, ordinary differential equations, and partial differential equations.

One unusual feature of this textbook is that the introduction of the conventional complex representation of oscillations and waves is delayed until it becomes absolutely necessary (during the discussion of quantum mechanical waves). The reason for this choice is that, although the complex representation of oscillations and waves greatly facilitates calculations, it is (at least, initially) a significant obstacle to the development of a physical understanding of such phenomena. The author is of the opinion that students should first thoroughly understand how to represent oscillations and waves in terms of regular trigonometric functions before attempting to use the more convenient, but much more abstract, complex representation.

This book only deals with that class of oscillations and waves whose governing differential equations are linear. In most physical systems, this implies a restriction to relatively low amplitude phenomena. The author has resisted the temptation to discuss nonlinear oscillations and waves, mainly because such phenomena require a completely different sort of mathematical analysis to that used to describe linear oscillations and waves, and the main emphasis of this book is the mathematical unity of the subject matter.

Light is ultimately a wave phenomenon. Hence, it is natural that part of this book should be devoted to the study of optics. For the sake of brevity, however, only those aspects of optics that depend crucially on the wave-like nature of light (i.e., wave optics, rather than geometric optics) are discussed in any detail.

This textbook was developed for the "Oscillations and Waves" course that is currently taught at the University of Texas at Austin (UT) immediately following the standard mechanics/heat/sound and electricity/magnetism/light/atomic survey courses. The purpose of the UT waves course is to ease the difficult transition between lower-division physics courses, which mostly rely on algebraic equations, and upper-division courses, which rely almost exclusively on differential equations. Experience at UT indicates that the attrition of physics majors is particularly severe at this transition. On the other hand, experience also suggests that a lower-division waves course—which includes much more interesting applications of physics than the rather pedestrian applications that crop up in the aforementioned survey courses, while not requiring particularly advanced mathematics—is an effective means of converting undecided science students into physics majors.

In the second edition of this book, I have improved the figures, and included many additional exercises at the ends of the chapters. I have also added material on birefringence (in Chapter 7), electromagnetic waves in magnetized plasmas (in Chapter 9), Fresnel diffraction (in Chapter 10), and the applications of wave mechanics (in Chapter 11). Finally, I have developed a large number of Python animations and widgets which I use for illustrative purposes when I teach the Oscillations and Waves course at UT; these are available online at `http://www.crcpress.com/9781138479715`. The Appendices to this book are also available at the same site.

Richard Fitzpatrick
Austin, TX

CHAPTER 1

Simple Harmonic Oscillation

1.1 INTRODUCTION

The aim of this chapter is to investigate a particularly straightforward type of motion known as *simple harmonic oscillation*, and also to introduce the differential equation that governs such motion, which is known as the *simple harmonic oscillator equation*. We shall discover that simple harmonic oscillation always involves a back and forth flow of energy between two different energy types, with the total energy remaining constant in time. We shall also learn that the linear nature of the simple harmonic oscillator equation greatly facilitates its solution. In this chapter, examples are drawn from simple mechanical and electrical systems.

1.2 MASS ON SPRING

Consider a compact mass m that slides over a frictionless horizontal surface. Suppose that the mass is attached to one end of a light horizontal spring whose other end is anchored in an immovable wall. See Figure 1.1. At time t, let $x(t)$ be the extension of the spring; that is, the difference between the spring's actual length and its unstretched length. $x(t)$ can also be used as a coordinate to determine the instantaneous horizontal displacement of the mass.

The equilibrium state of the system corresponds to the situation in which the mass is at rest, and the spring is unextended (i.e., $x = \dot{x} = 0$, where $\dot{} \equiv d/dt$). In this state, zero horizontal force acts on the mass, and so there is no reason for it to start to move. However, if the system is perturbed from its equilibrium state (i.e., if the mass is displaced horizontally, such that the spring becomes extended) then the mass experiences a horizontal force given by *Hooke's law*,

$$f(x) = -k\,x. \tag{1.1}$$

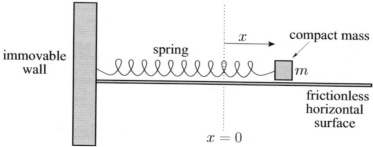

FIGURE 1.1 Mass on a spring.

Here, $k > 0$ is the so-called *force constant* of the spring. The negative sign in the preceding expression indicates that $f(x)$ is a so-called *restoring force* that always acts to return the displacement, x, to its equilibrium value, $x = 0$ (i.e., if the displacement is positive then the force is negative, and vice versa). Note that the magnitude of the restoring force is directly proportional to the displacement of the mass from its equilibrium position (i.e., $|f| \propto x$). Hooke's law only holds for relatively small spring extensions. Hence, the mass's displacement cannot be made too large, otherwise Equation (1.1) ceases to be valid. Incidentally, the motion of this particular dynamical system is representative of the motion of a wide variety of different mechanical systems when they are slightly disturbed from a stable equilibrium state. (See Sections 1.5 and 1.6.)

Newton's second law of motion leads to the following time evolution equation for the system (Fitzpatrick 2012),

$$m \ddot{x} = -k x, \tag{1.2}$$

where $\ddot{} \equiv d^2/dt^2$. This differential equation is known as the *simple harmonic oscillator equation*, and its solution has been known for centuries. The solution can be written

$$x(t) = a \cos(\omega t - \phi), \tag{1.3}$$

where $a > 0$, $\omega > 0$, and ϕ are constants. We can demonstrate that Equation (1.3) is indeed a solution of Equation (1.2) by direct substitution. Plugging the right-hand side of Equation (1.3) into Equation (1.2), and recalling from standard calculus that $d(\cos\theta)/d\theta = -\sin\theta$ and $d(\sin\theta)/d\theta = \cos\theta$ (see Appendix B), so that $\dot{x} = -\omega a \sin(\omega t - \phi)$ and $\ddot{x} = -\omega^2 a \cos(\omega t - \phi)$, where use has been made of the *chain rule* (Riley 1974),

$$\frac{d}{dx}(f[g(x)]) \equiv \frac{df}{dg}\frac{dg}{dx}, \tag{1.4}$$

we obtain

$$-m\omega^2 a \cos(\omega t - \phi) = -k a \cos(\omega t - \phi). \tag{1.5}$$

It follows that Equation (1.3) is the correct solution provided

$$\omega = \sqrt{\frac{k}{m}}. \tag{1.6}$$

Figure 1.2 shows a graph of x versus t derived from Equation (1.3). The type of motion displayed here is called *simple harmonic oscillation*. It can be seen that the displacement x oscillates between $x = -a$ and $x = +a$. This result can be obtained from Equation (1.3) by noting that $-1 \leq \cos\theta \leq +1$. Here, a is termed the *amplitude* of the oscillation. Moreover, the motion is repetitive in time (i.e., it repeats exactly after a certain time period has elapsed). The repetition *period* is

$$T = \frac{2\pi}{\omega}. \tag{1.7}$$

This result can be obtained from Equation (1.3) by noting that $\cos\theta$ is a periodic function of θ with period 2π; that is, $\cos(\theta + 2\pi) \equiv \cos\theta$. It follows that the motion repeats each time ωt increases by 2π. In other words, each time t increases by $2\pi/\omega$. The *frequency* of the motion (i.e., the number of oscillations completed per second) is

$$f = \frac{1}{T} = \frac{\omega}{2\pi}. \tag{1.8}$$

It is apparent that ω is the motion's *angular frequency*; that is, the frequency f converted into radians per second. (The units of f are hertz—otherwise known as cycles per second—whereas the units of ω are radians per second. One cycle per second is equivalent to 2π radians per second.) Finally,

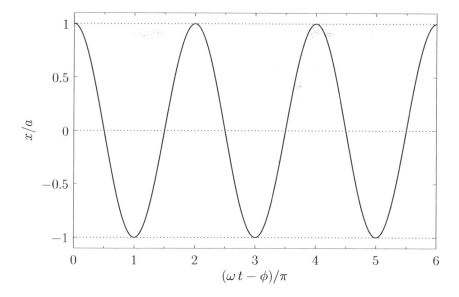

FIGURE 1.2   Simple harmonic oscillation: $x = a \cos(\omega t - \phi)$.

the *phase angle*, ϕ, determines the times at which the oscillation attains its maximum displacement, $x = a$. In fact, because the maxima of $\cos \theta$ occur at $\theta = n\,2\pi$, where n is an arbitrary integer, the times of maximum displacement are

$$t_{\max} = \left(n + \frac{\phi}{2\pi}\right)T. \tag{1.9}$$

Varying the phase angle shifts the pattern of oscillation backward and forward in time. See Figure 1.3.

Table 1.1 lists the displacement, velocity, and acceleration of the mass at various key points on the simple harmonic oscillation cycle. The information contained in this table is derived from Equation (1.3). All of the non-zero values shown in the table represent either the maximum or the minimum value taken by the quantity in question during the oscillation cycle. The variation of the displacement, velocity, and acceleration during the oscillation cycle is illustrated in Figure 1.4.

As we have seen, when a mass on a spring is disturbed it executes simple harmonic oscillation about its equilibrium position. In physical terms, if the mass's initial displacement is positive ($x > 0$) then the force is negative, and pulls the mass toward the equilibrium point ($x = 0$). However, when the mass reaches this point it is moving, and its inertia thus carries it onward, so that it acquires a negative displacement ($x < 0$). The force then becomes positive, and pulls the mass toward the equilibrium point. However, inertia again carries it past this point, and the mass acquires a positive displacement. The motion subsequently repeats itself ad infinitum.

$\omega t - \phi$	0	$\pi/2$	π	$3\pi/2$
x	$+a$	0	$-a$	0
$\dot{x}$	0	$-\omega a$	0	$+\omega a$
$\ddot{x}$	$-\omega^2 a$	0	$+\omega^2 a$	0

TABLE 1.1   Simple harmonic oscillation: $x = a \cos(\omega t - \phi)$.

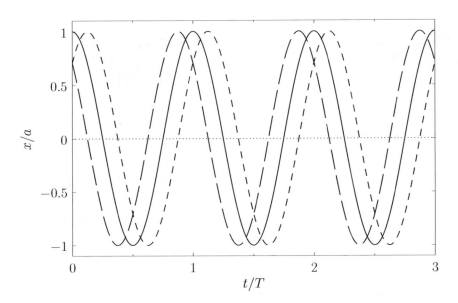

FIGURE 1.3 Simple harmonic oscillation: $x = a \cos(2\pi t/T - \phi)$. The solid, short-dashed, and long-dashed curves correspond to $\phi = 0$, $+\pi/4$, and $-\pi/4$, respectively.

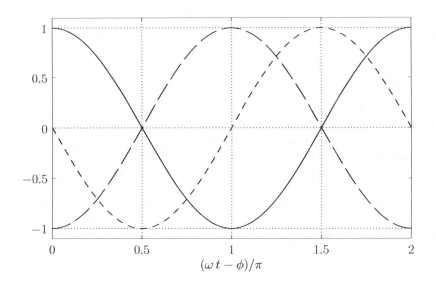

FIGURE 1.4 Simple harmonic oscillation: $x = a \cos(\omega t - \phi)$. The solid, short-dashed, and long-dashed curves show x/a, $\dot{x}/(\omega a)$, and $\ddot{x}/(\omega^2 a)$, respectively.

The standard solution, (1.3), of the simple harmonic oscillator equation, (1.2), contains three constants: the angular frequency, ω; the amplitude, a; and the phase angle, ϕ. The angular frequency is determined by the spring constant, k, and the system inertia, m, via Equation (1.6). It follows that ω is determined by the simple harmonic oscillator equation itself (because both k and m explicitly appear in this equation). On the other hand, the amplitude and phase angle are determined by the *initial conditions*. To be more exact, suppose that the instantaneous displacement and velocity of the mass at $t = 0$ are x_0 and v_0, respectively. It follows from Equation (1.3) that

$$x_0 = x(t = 0) = a \cos \phi, \tag{1.10}$$

$$\frac{v_0}{\omega} = \frac{\dot{x}(t = 0)}{\omega} = a \sin \phi. \tag{1.11}$$

Here, use has been made of the trigonometric identities $\cos(-\theta) \equiv \cos \theta$ and $\sin(-\theta) \equiv -\sin \theta$. (See Appendix B.) Hence, we deduce that

$$a = \sqrt{x_0^2 + \left(\frac{v_0}{\omega}\right)^2}, \tag{1.12}$$

and

$$\phi = \tan^{-1}\left(\frac{v_0}{\omega x_0}\right), \tag{1.13}$$

because $\sin^2 \theta + \cos^2 \theta \equiv 1$ and $\tan \theta \equiv \sin \theta / \cos \theta$. (See Appendix B.)

The kinetic energy of the system, which is the same as the kinetic energy of the mass, is written

$$K = \frac{1}{2} m \dot{x}^2 = \frac{1}{2} m a^2 \omega^2 \sin^2(\omega t - \phi). \tag{1.14}$$

The potential energy of the system, which is the same as the potential energy of the spring, takes the form (Fitzpatrick 2012)

$$U = \frac{1}{2} k x^2 = \frac{1}{2} k a^2 \cos^2(\omega t - \phi). \tag{1.15}$$

Hence, the total energy is

$$E = K + U = \frac{1}{2} k a^2 = \frac{1}{2} m \omega^2 a^2, \tag{1.16}$$

because $m \omega^2 = k$ and $\sin^2 \theta + \cos^2 \theta \equiv 1$. According to the previous expression, the total energy is a constant of the motion, and is proportional to the amplitude squared of the oscillation. Hence, we deduce that the simple harmonic oscillation of a mass on a spring is characterized by a continual back and forth flow of energy between kinetic and potential components. The kinetic energy attains its maximum value, and the potential energy its minimum value, when the displacement is zero (i.e., when $x = 0$). Likewise, the potential energy attains its maximum value, and the kinetic energy its minimum value, when the displacement is maximal (i.e., when $x = \pm a$). The minimum value of K is zero, because the system is instantaneously at rest when the displacement is maximal. The time variation of the kinetic, potential, and total energy of a mass on a spring is illustrated in Figure 1.5.

1.3 SIMPLE HARMONIC OSCILLATOR EQUATION

Suppose that a physical system possessing a single *degree of freedom*—that is, a system whose instantaneous state at time t is fully described by a single dependent variable, $s(t)$—obeys the following time evolution equation [cf., Equation (1.2)],

$$\ddot{s} + \omega^2 s = 0, \tag{1.17}$$

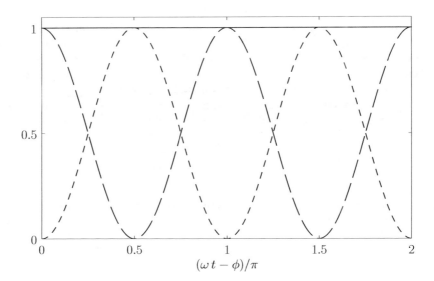

FIGURE 1.5 Simple harmonic oscillation: $x = a \cos(\omega t - \phi)$. The short-dashed, long-dashed, and solid curves show $K(t)/E$, $U(t)/E$, and $[K(t) + U(t)]/E$, respectively.

where $\omega > 0$ is a constant. As we have seen, this differential equation is called the simple harmonic oscillator equation, and has the standard solution

$$s(t) = a \cos(\omega t - \phi), \tag{1.18}$$

where $a > 0$ and ϕ are constants. Moreover, this solution describes a type of oscillation characterized by a constant amplitude, a, and a constant angular frequency, ω. The phase angle, ϕ, determines the times at which the oscillation attains its maximum value. The frequency of the oscillation (in hertz) is $f = \omega/2\pi$, and the period is $T = 2\pi/\omega$. The frequency and period of the oscillation are both determined by the constant ω, which appears in the simple harmonic oscillator equation, whereas the amplitude, a, and phase angle, ϕ, are determined by the initial conditions. [See Equations (1.10)–(1.13).] In fact, a and ϕ are the two arbitrary constants of integration of the second-order ordinary differential equation (1.17). Recall, from standard differential equation theory (Riley 1974), that the most general solution of an nth-order ordinary differential equation (i.e., an equation involving a single independent variable, and a single dependent variable, in which the highest derivative of the dependent with respect to the independent variable is nth-order, and the lowest zeroth-order) involves n arbitrary constants of integration. (Essentially, this is because we have to integrate the equation n times with respect to the independent variable to reduce it to zeroth-order, and so obtain the solution. Furthermore, each integration introduces an arbitrary constant. For example, the integral of $\dot{s} = \alpha$, where α is a known constant, is $s = \alpha t + \beta$, where β is an arbitrary constant.)

Multiplying Equation (1.17) by $\dot{s}$, we obtain

$$\dot{s}\,\ddot{s} + \omega^2\,\dot{s}\,s = 0. \tag{1.19}$$

However, this can also be written

$$\frac{d}{dt}\left(\frac{1}{2}\dot{s}^2\right) + \frac{d}{dt}\left(\frac{1}{2}\omega^2 s^2\right) = 0, \tag{1.20}$$

or

$$\frac{d\mathcal{E}}{dt} = 0, \tag{1.21}$$

where

$$\mathcal{E} = \frac{1}{2}\dot{s}^2 + \frac{1}{2}\omega^2 s^2. \tag{1.22}$$

According to Equation (1.21), $\mathcal{E}$ is a conserved quantity. In other words, it does not vary with time. This quantity is generally proportional to the overall energy of the system. For instance, $\mathcal{E}$ would be the energy divided by the mass in the mass–spring system discussed in Section 1.2. The quantity $\mathcal{E}$ is either zero or positive, because neither of the terms on the right-hand side of Equation (1.22) can be negative.

Let us search for an equilibrium state. Such a state is characterized by $s = $ constant, so that $\dot{s} = \ddot{s} = 0$. It follows from Equation (1.17) that $s = 0$, and from Equation (1.22) that $\mathcal{E} = 0$. We conclude that the system can only remain permanently at rest when $\mathcal{E} = 0$. Conversely, the system can never permanently come to rest when $\mathcal{E} > 0$, and must, therefore, keep moving for ever. Because the equilibrium state is characterized by $s = 0$, we deduce that s represents a kind of "displacement" of the system from this state. It is also apparent, from Equation (1.22), that s attains its maximum value when $\dot{s} = 0$. In fact,

$$s_{max} = \frac{\sqrt{2\mathcal{E}}}{\omega}, \tag{1.23}$$

where $s_{max} = a$ is the amplitude of the oscillation. Likewise, $\dot{s}$ attains its maximum value,

$$\dot{s}_{max} = \sqrt{2\mathcal{E}}, \tag{1.24}$$

when $s = 0$.

The simple harmonic oscillation specified by Equation (1.18) can also be written in the form

$$s(t) = A\,\cos(\omega\,t) + B\,\sin(\omega\,t), \tag{1.25}$$

where $A = a\,\cos\phi$ and $B = a\,\sin\phi$. Here, we have employed the trigonometric identity $\cos(x-y) \equiv \cos x \cos y + \sin x \sin y$. (See Appendix B.) Alternatively, Equation (1.18) can be written

$$s(t) = a\,\sin(\omega\,t - \phi'), \tag{1.26}$$

where $\phi' = \phi - \pi/2$, and use has been made of the trigonometric identity $\cos\theta \equiv \sin(\theta + \pi/2)$. (See Appendix B.) It follows that there are many different ways of representing a simple harmonic oscillation, but they all involve linear combinations of sine and cosine functions whose arguments take the form $\omega\,t+c$, where c is some constant. However, irrespective of its form, a general solution to the simple harmonic oscillator equation must always contain two arbitrary constants. For example, A and B in Equation (1.25), or a and ϕ' in Equation (1.26).

The simple harmonic oscillator equation, (1.17), is a *linear* differential equation, which means that if $s(t)$ is a solution then so is $c\,s(t)$, where c is an arbitrary constant. This can be verified by multiplying the equation by c, and then making use of the fact that $c\,d^2s/dt^2 = d^2(c\,s)/dt^2$. Linear differential equations have the very important and useful property that their solutions are *superposable*. This means that if $s_1(t)$ is a solution to Equation (1.17), so that

$$\ddot{s}_1 = -\omega^2 s_1, \tag{1.27}$$

and $s_2(t)$ is a different solution, so that

$$\ddot{s}_2 = -\omega^2 s_2, \tag{1.28}$$

then $s_1(t) + s_2(t)$ is also a solution. This can be verified by adding the previous two equations, and making use of the fact that $d^2s_1/dt^2 + d^2s_2/dt^2 = d^2(s_1 + s_2)/dt^2$. Furthermore, it can be demonstrated that any linear combination of s_1 and s_2, such as $\alpha\,s_1 + \beta\,s_2$, where α and β are arbitrary constants, is also a solution. It is very helpful to know this fact. For instance, the special solution to the simple harmonic oscillator equation, (1.17), with the simple initial conditions $s(0) = 1$ and $\dot{s}(0) = 0$ can easily be shown to be

$$s_1(t) = \cos(\omega t). \tag{1.29}$$

Likewise, the special solution with the simple initial conditions $s(0) = 0$ and $\dot{s}(0) = 1$ is

$$s_2(t) = \omega^{-1} \sin(\omega t). \tag{1.30}$$

Thus, because the solutions to the simple harmonic oscillator equation are superposable, the solution with the general initial conditions $s(0) = s_0$ and $\dot{s}(0) = \dot{s}_0$ becomes

$$s(t) = s_0 \, s_1(t) + \dot{s}_0 \, s_2(t), \tag{1.31}$$

or

$$s(t) = s_0 \, \cos(\omega t) + \frac{\dot{s}_0}{\omega} \, \sin(\omega t). \tag{1.32}$$

1.4   LC CIRCUIT

Consider an electrical circuit consisting of an inductor, of inductance L, connected in series with a capacitor, of capacitance C. See Figure 1.6. Such a circuit is known as an *LC circuit*, for obvious reasons. Suppose that $I(t)$ is the instantaneous current flowing around the circuit. According to standard electrical circuit theory (Fitzpatrick 2008), the potential drop across the inductor (in the direction of the current flow) is $L\dot{I}$. Again, from standard electrical circuit theory (ibid.), the potential drop across the capacitor is $V = Q/C$, where Q is the charge stored on the capacitor's positive plate. However, because electric charge is a conserved quantity (ibid.), the current flowing around the circuit is equal to the rate at which charge accumulates on the capacitor's positive plate; that is, $I = \dot{Q}$.

According to *Kirchhoff's second circuital law*, the sum of the potential drops across the various components of a closed circuital loop is equal to zero (Grant and Philips 1975). In other words,

$$L\dot{I} + Q/C = 0. \tag{1.33}$$

Dividing by L, and differentiating with respect to t, we obtain

$$\ddot{I} + \omega^2 I = 0, \tag{1.34}$$

where

$$\omega = \frac{1}{\sqrt{LC}}. \tag{1.35}$$

Comparison with Equation (1.17) reveals that Equation (1.34) is a simple harmonic oscillator equation with the associated angular oscillation frequency $\omega = 1/\sqrt{LC}$. We conclude that the current in an LC circuit executes simple harmonic oscillations of the form

$$I(t) = \hat{I} \, \cos(\omega t - \phi), \tag{1.36}$$

where $\hat{I} > 0$ and ϕ are constants.

According to Equation (1.33), the potential drop, $V = Q/C$, across the capacitor is minus that across the inductor, so that $V = -L\dot{I}$, giving

$$V(t) = \sqrt{\frac{L}{C}} \, \hat{I} \, \sin(\omega t - \phi) = \sqrt{\frac{L}{C}} \, \hat{I} \, \cos(\omega t - \phi - \pi/2). \tag{1.37}$$

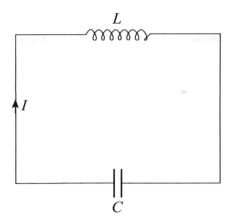

FIGURE 1.6   An LC circuit.

Here, use has been made of the trigonometric identity $\sin\theta \equiv \cos(\theta - \pi/2)$. (See Appendix B.) It follows that the voltage in an LC circuit oscillates at the same frequency as the current, but with a phase shift of $\pi/2$ radians. In other words, the voltage is maximal when the current is zero, and vice versa. The amplitude of the voltage oscillation is that of the current oscillation multiplied by $\sqrt{L/C}$. Thus, we can also write

$$V(t) = \sqrt{\frac{L}{C}}\, I\!\left(t - \frac{T}{4}\right),\tag{1.38}$$

where $T = 2\pi/\omega$ is the period of the oscillation.

Comparing with Equation (1.22), we deduce that

$$\mathcal{E} = \frac{1}{2}\,\dot{I}^2 + \frac{1}{2}\,\omega^2\,I^2\tag{1.39}$$

is a conserved quantity. However, $\omega^2 = 1/LC$, and $\dot{I} = -V/L$. Thus, multiplying the preceding expression by $C\,L^2$, we obtain

$$E = \frac{1}{2}\,C\,V^2 + \frac{1}{2}\,L\,I^2.\tag{1.40}$$

The first and second terms on the right-hand side of the preceding expression can be recognized as the instantaneous energies stored in the capacitor and the inductor, respectively (Fitzpatrick 2008). The former energy is stored in the electric field generated when the capacitor is charged, whereas the latter is stored in the magnetic field induced when current flows through the inductor. It follows that the quantity E in Equation (1.40) is the total energy of the circuit, and that this energy is a conserved quantity. The oscillations of an LC circuit can, thus, be understood as a cyclic interchange between electric energy stored in the capacitor, and magnetic energy stored in the inductor.

Suppose that at $t = 0$ the capacitor is charged to a voltage V_0, and there is zero current flowing through the inductor. In other words, the initial state is one in which all of the circuit energy resides in the capacitor. The initial conditions are $V(0) = -L\,\dot{I}(0) = V_0$ and $I(0) = 0$. In this case, it can be shown that the current evolves in time as

$$I(t) = -\frac{V_0}{\sqrt{L/C}}\,\sin(\omega t).\tag{1.41}$$

Suppose that at $t = 0$ the capacitor is fully discharged, and there is a current I_0 flowing through the inductor. In other words, the initial state is one in which all of the circuit energy resides in

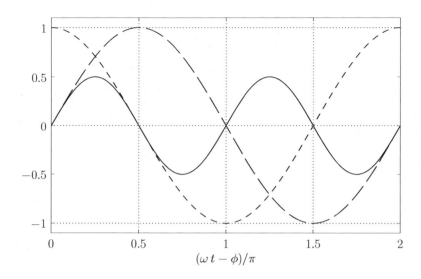

FIGURE 1.7 Simple harmonic oscillation: $I = \hat{I} \cos(\omega t - \phi)$, $V = \hat{V} \sin(\omega t - \phi)$, $P = I V$. The short-dashed, long-dashed, and solid curves show $I(t)/\hat{I}$, $V(t)/\hat{V}$, and $P(t)/(\hat{I} \hat{V})$, respectively.

the inductor. The initial conditions are $V(0) = -L \dot{I}(0) = 0$ and $I(0) = I_0$. In this case, it can be demonstrated that the current evolves in time as

$$I(t) = I_0 \cos(\omega t). \tag{1.42}$$

Suppose, finally, that at $t = 0$ the capacitor is charged to a voltage V_0, and the current flowing through the inductor is I_0. Because the solutions of the simple harmonic oscillator equation are superposable, it follows that the current evolves in time as

$$I(t) = -\frac{V_0}{\sqrt{L/C}} \sin(\omega t) + I_0 \cos(\omega t). \tag{1.43}$$

Furthermore, according to Equation (1.38), the voltage evolves in time as

$$V(t) = -V_0 \sin(\omega t - \pi/2) + \sqrt{\frac{L}{C}} I_0 \cos(\omega t - \pi/2), \tag{1.44}$$

or

$$V(t) = V_0 \cos(\omega t) + \sqrt{\frac{L}{C}} I_0 \sin(\omega t). \tag{1.45}$$

Here, use has been made of the trigonometric identities $\sin(\theta - \pi/2) \equiv -\cos\theta$ and $\cos(\theta - \pi/2) \equiv \sin\theta$. (See Appendix B.)

The instantaneous electrical power absorption by the capacitor, which can be shown to be minus the instantaneous power absorption by the inductor, is (Fitzpatrick 2008)

$$P(t) = I(t) V(t) = I_0 V_0 \cos(2\omega t) + \frac{1}{2}\left(I_0^2 \sqrt{\frac{L}{C}} - \frac{V_0^2}{\sqrt{L/C}} \right) \sin(2\omega t), \tag{1.46}$$

where use has been made of Equations (1.43) and (1.45), as well as the trigonometric identities

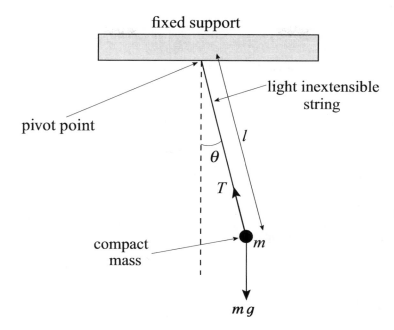

FIGURE 1.8 A simple pendulum.

$\cos(2\,\theta) \equiv \cos^2\theta - \sin^2\theta$ and $\sin(2\,\theta) \equiv 2\,\sin\theta\,\cos\theta$. (See Appendix B.) Hence, the average power absorption during a cycle of the oscillation,

$$\langle P \rangle \equiv \frac{1}{T} \int_0^T P(t)\,dt, \tag{1.47}$$

is zero, because it is readily demonstrated that $\langle \cos(2\,\omega\,t) \rangle = \langle \sin(2\,\omega\,t) \rangle = 0$. In other words, any energy that the capacitor absorbs from the circuit during one half of the oscillation cycle is returned to the circuit, without loss, during the other. The same goes for the inductor. The time variation of the current, I, flowing into the capacitor, the voltage drop, V, across the capacitor, and the power, P, absorbed by the capacitor, during a cycle of the oscillation is illustrated in Figure 1.7.

1.5 SIMPLE PENDULUM

Consider a compact mass m suspended from a light inextensible string of length l, such that the mass is free to swing from side to side in a vertical plane, as shown in Figure 1.8. This setup is known as a *simple pendulum*. Let θ be the angle subtended between the string and the downward vertical. The stable equilibrium state of the system corresponds to the situation in which the mass is stationary, and hangs vertically down (i.e., $\theta = \dot{\theta} = 0$). The angular equation of motion of the pendulum is (Fowles and Cassiday 2005)

$$I\,\ddot{\theta} = \tau, \tag{1.48}$$

where I is the moment of inertia of the mass, and τ the torque acting about the suspension point. For the case in hand, given that the mass is essentially a point particle, and is situated a distance l from the axis of rotation (i.e., from the suspension point), it follows that $I = m\,l^2$ (ibid.).

The two forces acting on the mass are the downward gravitational force, $m\,g$, where g is the acceleration due to gravity, and the tension, T, in the string. However, the tension makes no contribution to the torque, because its line of action passes through the suspension point. From elementary trigonometry, the line of action of the gravitational force passes a perpendicular distance $l \sin \theta$ from the suspension point. Hence, the magnitude of the gravitational torque is $m g l \sin \theta$. Moreover, the gravitational torque is a *restoring torque*; that is, if the mass is displaced slightly from its equilibrium position (i.e., $\theta = 0$) then the gravitational torque acts to push the mass back toward that position. Thus, we can write

$$\tau = -m g l \sin \theta. \tag{1.49}$$

Combining the previous two equations, we obtain the following angular equation of motion of the pendulum,

$$l \ddot{\theta} + g \sin \theta = 0. \tag{1.50}$$

Unlike all of the other time evolution equations that we have examined, so far, in this chapter, the preceding equation is nonlinear [because $\sin(c\, \theta) \neq c \sin \theta$, where c is an arbitrary constant], which means that it is generally very difficult to solve.

Suppose, however, that the system does not stray very far from its equilibrium position ($\theta = 0$). If this is the case then we can expand $\sin \theta$ in a Taylor series about $\theta = 0$. (See Appendix B.) We obtain

$$\sin \theta = \theta - \frac{\theta^3}{6} + \frac{\theta^5}{120} + O(\theta^7). \tag{1.51}$$

If $|\theta|$ is sufficiently small then the series is dominated by its first term, and we can write $\sin \theta \simeq \theta$. This is known as the *small-angle approximation*. Making use of this approximation, the equation of motion (1.50) simplifies to

$$\ddot{\theta} + \omega^2 \theta \simeq 0, \tag{1.52}$$

where

$$\omega = \sqrt{\frac{g}{l}}. \tag{1.53}$$

Equation (1.52) is the simple harmonic oscillator equation. Hence, we can immediately write its solution in the form

$$\theta(t) = \hat{\theta} \cos(\omega t - \phi), \tag{1.54}$$

where $\hat{\theta} > 0$ and ϕ are constants. We conclude that the pendulum swings back and forth at a fixed angular frequency, ω, that depends on l and g, but is independent of the amplitude, $\hat{\theta}$, of the motion. This result only holds as long as the small-angle approximation remains valid. It turns out that $\sin \theta \simeq \theta$ is a reasonably good approximation provided $|\theta| \lesssim 6°$. Hence, the period of a simple pendulum is only amplitude independent when the (angular) amplitude of its motion is less than about $6°$.

1.6   COMPOUND PENDULUM

Consider an extended laminar object of mass M with a hole drilled though it. Suppose that the body is suspended from a fixed horizontal peg, which passes through the hole, such that it is free to swing from side to side in a vertical plane, as shown in Figure 1.9. This setup is known as a *compound pendulum*.

Let P be the pivot point, and let C be the body's center of mass, which is located a distance d from the pivot. Let θ be the angle subtended between the downward vertical (which passes through point P) and the line PC (which is assumed to be confined to a vertical plane). The equilibrium state of the compound pendulum corresponds to the case in which the center of mass lies vertically below the pivot point; that is, $\theta = 0$. The angular equation of motion of the pendulum is

$$I \ddot{\theta} = \tau, \tag{1.55}$$

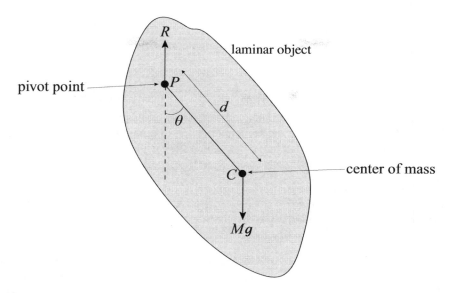

FIGURE 1.9   A compound pendulum.

where I is the moment of inertia about the pivot, and τ is the torque. Using similar arguments to those employed for the case of the simple pendulum (recalling that the weight of the pendulum acts at its center of mass), we can write

$$\tau = -M g d \sin \theta. \tag{1.56}$$

The reaction, R, at the peg does not contribute to the torque, because its line of action passes through the pivot. Combining the previous two equations, we obtain the following angular equation of motion of the pendulum,

$$I \ddot{\theta} = -M g d \sin \theta. \tag{1.57}$$

Finally, adopting the small-angle approximation, $\sin \theta \simeq \theta$, we arrive at the simple harmonic oscillator equation,

$$I \ddot{\theta} \simeq -M g d \theta. \tag{1.58}$$

By analogy with our previous solutions of such equations, the angular frequency of small amplitude oscillations of a compound pendulum is given by

$$\omega = \sqrt{\frac{M g d}{I}}. \tag{1.59}$$

It is helpful to define the so-called *effective length* of the pendulum,

$$L = \frac{I}{M d}. \tag{1.60}$$

Equation (1.59) reduces to

$$\omega = \sqrt{\frac{g}{L}}, \tag{1.61}$$

which is identical in form to the corresponding expression for a simple pendulum. (See Section 1.5.) We conclude that a compound pendulum behaves like a simple pendulum whose length is equal to the effective length.

EXERCISES

1.1 A mass stands on a platform that executes simple harmonic oscillation in a vertical direction at a frequency of 5 Hz. Show that the mass loses contact with the platform when the displacement exceeds 10^{-2} m. [From Pain 1999.]

1.2 A small body rests on a horizontal diaphragm of a loudspeaker that is supplied with an alternating current of constant amplitude but variable frequency. If the diaphragm executes simple harmonic oscillation in the vertical direction of amplitude $10\,\mu$m, at all frequencies, find the greatest frequency (in hertz) for which the small body stays in contact with the diaphragm.

1.3 A mass m is attached to the mid-point of a stretched string of negligible mass, length $2\,l$, and uniform tension T. Let $x \ll l$ be the transverse displacement of the mass from its equilibrium position. Show that the displacement executes simple harmonic oscillation at the angular frequency $\omega = \sqrt{2\,T/l\,m}$.

1.4 Two light springs have spring constants k_1 and k_2, respectively, and are used in a vertical orientation to support an object of mass m. Show that the angular frequency of small amplitude oscillations about the equilibrium state is $[(k_1 + k_2)/m]^{1/2}$ if the springs are connected in parallel, and $[k_1 k_2/(k_1 + k_2) m]^{1/2}$ if the springs are connected in series.

1.5 A mass M is suspended at the end of a uniform spring of unstretched length l and spring constant k. If the mass of the spring is m and the velocity of an element dy of its length is proportional to its distance y from the fixed end of the spring, show that the kinetic energy of this element is

$$\frac{1}{2}\left(\frac{m}{l}\,dy\right)\left(\frac{y}{l}\,v\right)^2,$$

where v is the velocity of the suspended mass. Hence, by integrating over the length of the spring, show that its total kinetic energy is $(1/6)\,m\,v^2$. Finally, deduce, from energy conservation arguments, that the angular oscillation frequency of the system is given by

$$\omega = \sqrt{\frac{k}{M + m/3}}.$$

[From Pain 1999.]

1.6 A body of uniform cross-sectional area A and mass density ρ floats in a liquid of density ρ_0 (where $\rho < \rho_0$), and at equilibrium displaces a volume V. Making use of *Archimedes' principle* (that the buoyancy force acting on a partially submerged body is equal to the weight of the displaced liquid), show that the period of small amplitude oscillations about the equilibrium position is

$$T = 2\pi \sqrt{\frac{V}{g\,A}}.$$

1.7 A U-tube of constant cross-sectional area A consists of a horizontal section connected at either end to two vertical sections. Suppose that the tube is filled with an incompressible liquid of mass density ρ. Let the total length of the liquid column be l. (Where l exceeds the length of the horizontal section.) Suppose that the surface of the liquid in one of the vertical sections is initially displaced (vertically) a small distance h from its equilibrium position. Show that the surface displacement subsequently executes simple harmonic oscillation at the angular frequency $\omega = \sqrt{2\,g/l}$, where g is the acceleration due to gravity.

1.8 A particle of mass m slides in a frictionless semi-circular depression in the ground of radius R. Find the angular frequency of small amplitude oscillations about the particle's equilibrium position, assuming that the oscillations are essentially one-dimensional, so that the particle passes through the lowest point of the depression during each oscillation cycle.

1.9 (a) Imagine a straight tunnel passing through the center of the Earth, which is regarded as a sphere of radius R and uniform mass density. A particle is dropped into the tunnel from the surface. Show that the particle undergoes simple harmonic oscillation at the angular frequency $\omega = \sqrt{g/R}$, where g is the gravitational acceleration at Earth's surface. (Hint: The gravitational acceleration at a point inside a spherically symmetric mass distribution is the same as if all of the mass interior to the point were concentrated at the center, and all of the mass exterior to the point were neglected.) Estimate how long it takes the particle to reach the other end of the tunnel.

   (b) Assuming that the tunnel is smooth (i.e., ignoring friction), show that motion is simple harmonic even if the tunnel does not pass through the center of the Earth, and that the travel time from one end of the tunnel to the other is the same as before.

   [From Ingard 1988.]

1.10 A particle executing simple harmonic oscillation in one dimension has speeds u and v at displacements a and b, respectively, from its equilibrium position.

   (a) Show that the period of the motion can be written

$$T = 2\pi \left(\frac{b^2 - a^2}{u^2 - v^2} \right)^{1/2}.$$

   (b) Show that the amplitude of the motion can be written

$$A = \left(\frac{u^2 b^2 - v^2 a^2}{u^2 - v^2} \right)^{1/2}.$$

1.11 If a thin wire is twisted through an angle θ then a restoring torque $\tau = -k\theta$ develops, where $k > 0$ is known as the *torsional force constant*. Consider a so-called *torsional pendulum*, which consists of a horizontal disk of mass M, and moment of inertia I, suspended at its center from a thin vertical wire of negligible mass and length l, whose other end is attached to a fixed support. The disk is free to rotate about a vertical axis passing through the suspension point, but such rotation twists the wire. Find the frequency of torsional oscillations of the disk about its equilibrium position.

1.12 A circular hoop of diameter d hangs on a nail. What is the period of its small amplitude oscillations? [From French 1971.]

1.13 A compound pendulum consists of a uniform bar of length l that pivots about one of its ends. Show that the pendulum has the same period of oscillation as a simple pendulum of length $(2/3)\,l$.

1.14 A compound pendulum consists of a uniform circular disk of radius r that is free to turn about a horizontal axis perpendicular to its plane. Find the position of the axis for which the periodic time is a minimum.

1.15 A laminar object of mass M has a moment of inertia I_0 about a perpendicular axis passing through its center of mass. Suppose that the object is converted into a compound pendulum by suspending it about a horizontal axis perpendicular to its plane. Show that the minimum effective length of the pendulum occurs when the distance of the suspension point from the center of gravity is equal to the radius of gyration, $k = (I_0/M)^{1/2}$.

1.16 A uniform disk of mass M and radius a rolls without slipping over a rough horizontal surface. Suppose that a small mass m is attached to the edge of the disk. Show that the angular frequency of small amplitude oscillations of the disk about its equilibrium position is

$$\omega = \sqrt{\frac{2}{3}\frac{m}{M}\frac{g}{a}}.$$

1.17 A body hung at the end of a light vertical spring stretches the spring statically to twice its original length. The system can be set into motion either as a simple pendulum or as a mass-spring oscillator. Determine the ratio between the periods of these motions. (In the pendulum mode of motion, assume the length of the spring to be constant.) [From Ingard 1988.]

1.18 Show that the average speed of a particle executing simple harmonic oscillation is $2/\pi$ times the maximum speed. [From Ingard 1988.]

1.19 A particle of mass m executes one-dimensional simple harmonic oscillation such that its instantaneous x coordinate is

$$x(t) = a\,\cos(\omega t - \phi).$$

   (a) Find the average values of x, x^2, $\dot{x}$, and $\dot{x}^2$ over a single cycle of the oscillation.

   (b) Find the average values of the kinetic and potential energies of the particle over a single cycle of the oscillation.

1.20 A particle executes two-dimensional simple harmonic oscillation such that its instantaneous coordinates in the x-y plane are

$$x(t) = a\,\cos(\omega t),$$
$$y(t) = a\,\cos(\omega t - \phi).$$

Describe the motion when (a) $\phi = 0$, (b) $\phi = \pi/4$, and (c) $\phi = \pi/2$. In each case, plot the trajectory of the particle in the x-y plane.

1.21 An LC circuit is such that at $t = 0$ the capacitor is uncharged and a current I_0 flows through the inductor. Find an expression for the charge Q stored on the positive plate of the capacitor as a function of time.

1.22 (a) A simple pendulum of mass m and length l is such that $\theta(0) = 0$ and $\dot{\theta}(0) = \omega_0$. Find the subsequent motion, $\theta(t)$, assuming that its amplitude remains small.

   (b) Suppose, instead, that $\theta(0) = \theta_0$ and $\dot{\theta}(0) = 0$. Find the subsequent motion.

   (c) Suppose, finally, that $\theta(0) = \theta_0$ and $\dot{\theta}(0) = \omega_0$. Find the subsequent motion.

1.23 (a) Demonstrate that

$$E = \frac{1}{2}ml^2\dot{\theta}^2 + mgl(1 - \cos\theta)$$

   is a constant of the motion of a simple pendulum whose time evolution equation is given by Equation (1.50). (Do not make the small-angle approximation.)

(b) Show that the amplitude of the motion, $\hat{\theta}$, can be written

$$\hat{\theta} = 2 \sin^{-1}\left(\frac{E}{2mgl}\right)^{1/2}.$$

(c) Demonstrate that the period of the motion is determined by

$$\frac{T}{T_0} = \frac{1}{\pi} \int_0^{\hat{\theta}} \frac{d\theta}{\sqrt{\sin^2(\hat{\theta}/2) - \sin^2(\theta/2)}},$$

where T_0 is the period of small-angle oscillations.

(d) Making use of the substitution $\sin u = \sin(\theta/2)/\sin(\hat{\theta}/2)$, show that the previous expression transforms to

$$\frac{T}{T_0} = \frac{2}{\pi} \int_0^{\pi/2} \frac{du}{\sqrt{1 - \sin^2(\hat{\theta}/2) \sin^2 u}}.$$

Hence, deduce that

$$\frac{T}{T_0} = 1 + \frac{\hat{\theta}^2}{16} + O(\hat{\theta}^4).$$

CHAPTER 2

Damped and Driven Harmonic Oscillation

2.1 INTRODUCTION

In the previous chapter, we encountered a number of energy conserving physical systems that exhibit simple harmonic oscillation about a stable equilibrium state. One of the main features of such oscillation is that, once excited, it never dies away. However, the majority of the oscillatory systems that we encounter in everyday life suffer some sort of irreversible energy loss while they are in motion, due, for instance, to frictional or viscous heat generation. We would therefore expect oscillations excited in such systems eventually to be damped away. The aim of this chapter is to examine so-called *damped harmonic oscillation*, and also to introduce the differential equation that governs such motion, which is known as the *damped harmonic oscillator equation*. In addition, we shall examine the phenomenon of *resonance* in periodically driven, damped, oscillating systems. In this chapter, examples are again drawn from simple mechanical and electrical systems.

2.2 DAMPED HARMONIC OSCILLATION

Consider the mass–spring system discussed in Section 1.2. Suppose that, as it slides over the horizontal surface, the mass is subject to a frictional damping force that opposes its motion, and is directly proportional to its instantaneous velocity. It follows that the net force acting on the mass when its instantaneous displacement is $x(t)$ takes the form

$$f(x, \dot{x}) = -k\,x - m\,v\,\dot{x}, \tag{2.1}$$

where $m > 0$ is the mass, $k > 0$ the spring force constant, and $v > 0$ a constant (with the dimensions of angular frequency) that parameterizes the strength of the damping. The time evolution equation of the system thus becomes [cf., Equation (1.2)]

$$\ddot{x} + v\,\dot{x} + \omega_0^2\,x = 0, \tag{2.2}$$

where $\omega_0 = \sqrt{k/m}$ is the undamped oscillation frequency [cf., Equation (1.6)]. We shall refer to the preceding equation as the *damped harmonic oscillator equation*.

It is worth discussing the two forces that appear on the right-hand side of Equation (2.1) in more detail. The first is the restoring force that develops when a mechanical system in a *stable* equilibrium state is *slightly* disturbed from that state. We can suppose, without loss of generality, that the equilibrium state corresponds to $x = 0$. Suppose that $f(x)$ is the restoring force. It follows that $f(0) = 0$ and $df(0)/dx < 0$, otherwise $x = 0$ is not a stable equilibrium point. Taylor expanding about $x = 0$ (Riley 1974), we obtain

$$f(x) \simeq -k x + \alpha x^2 + O\left(x^3\right), \tag{2.3}$$

where $k = -df(0)/dx$, and $\alpha = 2 d^2 f(0)/dx^2$. It follows that, provided the magnitude of the displacement from the equilibrium point, $|x|$, remains sufficiently small, we can always approximate $f(x)$ by the first non-zero term in its Taylor expansion; that is, $f(x) \simeq -k x$. Thus, the first term that appears on the right-hand side of Equation (2.1) is exact (as long as the amplitude of the motion remains sufficiently small). The second term, on the other hand, which represents the damping force, is completely phenomenological. In other words, damping forces that arise in nature are not necessarily directly proportional to the instantaneous velocity of the system. For example, the damping force that arises from air resistance is usually directly proportional to the square of the instantaneous velocity (Batchelor 2000). In fact, the only reason that we have chosen a damping force that is directly proportional to the instantaneous velocity is that it leads to a linear equation of motion.

Let us search for a solution to Equation (2.2) of the form

$$x(t) = a e^{-\gamma t} \cos(\omega_1 t - \phi), \tag{2.4}$$

where $a > 0$, $\gamma > 0$, $\omega_1 > 0$, and ϕ are all constants. By analogy with the discussion in Section 1.2, we can interpret the preceding solution as a periodic oscillation, of fixed angular frequency ω_1, and phase angle ϕ, whose amplitude decays exponentially in time as $a(t) = a \exp(-\gamma t)$. It can be demonstrated that

$$\dot{x} = -\gamma a e^{-\gamma t} \cos(\omega_1 t - \phi) - \omega_1 a e^{-\gamma t} \sin(\omega_1 t - \phi), \tag{2.5}$$

$$\ddot{x} = \left(\gamma^2 - \omega_1^2\right) a e^{-\gamma t} \cos(\omega_1 t - \phi) + 2 \gamma \omega_1 a e^{-\gamma t} \sin(\omega_1 t - \phi). \tag{2.6}$$

Hence, collecting similar terms, Equation (2.2) becomes

$$0 = \left[\left(\gamma^2 - \omega_1^2\right) - v \gamma + \omega_0^2\right] a e^{-\gamma t} \cos(\omega_1 t - \phi)$$
$$+ \left(2 \gamma \omega_1 - v \omega_1\right) a e^{-\gamma t} \sin(\omega_1 t - \phi). \tag{2.7}$$

The only way that the preceding equation can be satisfied at all times is if the (constant) coefficients of $\exp(-\gamma t) \cos(\omega_1 t - \phi)$ and $\exp(-\gamma t) \sin(\omega_1 t - \phi)$ separately equate to zero, so that

$$\left(\gamma^2 - \omega_1^2\right) - v \gamma + \omega_0^2 = 0, \tag{2.8}$$

$$2 \gamma \omega_1 - v \omega_1 = 0. \tag{2.9}$$

These equations can be solved to give

$$\gamma = v/2, \tag{2.10}$$

and

$$\omega_1 = \left(\omega_0^2 - v^2/4\right)^{1/2}. \tag{2.11}$$

Thus, the solution to the damped harmonic oscillator equation is written

$$x(t) = a e^{-vt/2} \cos\left(\omega_1 t - \phi\right), \tag{2.12}$$

assuming that $v < 2 \omega_0$ (because $\omega_1^2 = \omega_0^2 - v^2/4$ cannot be negative). We conclude that the effect of a relatively small amount of damping, parameterized by the *damping constant*, v, on a system that exhibits simple harmonic oscillation about a stable equilibrium state is to reduce the angular

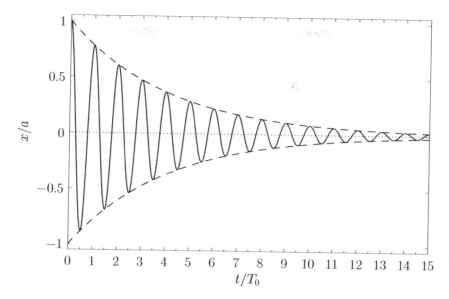

FIGURE 2.1 Damped harmonic oscillation. Here, $T_0 = 2\pi/\omega_0$, $\nu T_0 = 0.5$, and $\phi = 0$. The solid line shows $x(t)/a$, whereas the dashed lines show $\pm a(t)/a$.

frequency of the oscillation from its undamped value ω_0 to $\omega_1 = (\omega_0^2 - \nu^2/4)^{1/2}$, and to cause the amplitude of the oscillation to decay exponentially in time at the rate $\nu/2$. This modified type of oscillation, which we shall refer to as *damped harmonic oscillation*, is illustrated in Figure 2.1. Incidentally, if the damping is sufficiently large that $\nu \geq 2\,\omega_0$ (which we shall assume not to be the case) then the system does not oscillate at all, and any motion simply decays away exponentially in time. (See Exercise 2.7.)

Although the angular frequency, ω_1, and decay rate, $\nu/2$, of the damped harmonic oscillation specified in Equation (2.12) are determined by the constants appearing in the damped harmonic oscillator equation, (2.2), the initial amplitude, a, and the phase angle, ϕ, of the oscillation are determined by the initial conditions. In fact, if $x(0) = x_0$ and $\dot{x}(0) = v_0$ then it follows from Equation (2.12) that

$$x_0 = a \cos\phi, \tag{2.13}$$

$$v_0 = -\frac{\nu}{2} a \cos\phi + \omega_1 a \sin\phi = -\frac{\nu x_0}{2} + \omega_1 a \sin\phi, \tag{2.14}$$

which implies that

$$a \cos\phi = x_0, \tag{2.15}$$

$$a \sin\phi = \frac{v_0 + \nu x_0/2}{\omega_1}, \tag{2.16}$$

giving

$$a = \left[x_0^2 + \left(\frac{v_0 + \nu x_0/2}{\omega_1} \right)^2 \right]^{1/2}, \tag{2.17}$$

$$\phi = \tan^{-1}\left(\frac{v_0 + \nu x_0/2}{\omega_1 x_0} \right). \tag{2.18}$$

The damped harmonic oscillator equation is a *linear* differential equation. In other words, if $x(t)$ is a solution then so is $c\,x(t)$, where c is an arbitrary constant. It follows that the solutions of this equation are superposable, so that if $x_1(t)$ and $x_2(t)$ are two solutions corresponding to different initial conditions then $\alpha\,x_1(t) + \beta\,x_2(t)$ is a third solution, where α and β are arbitrary constants.

Multiplying the damped harmonic oscillator equation, (2.2), by $\dot{x}$, we obtain

$$\dot{x}\,\ddot{x} + \nu\,\dot{x}^2 + \omega_0^2\,\dot{x}\,x = 0, \tag{2.19}$$

which can be rearranged to give

$$\frac{dE}{dt} = -m\,\nu\,\dot{x}^2, \tag{2.20}$$

where

$$E = \frac{1}{2}\,m\,\dot{x}^2 + \frac{1}{2}\,k\,x^2 \tag{2.21}$$

is the total energy of the system; that is, the sum of the kinetic and potential energies. Because the right-hand side of (2.20) cannot be positive, and is only zero when the system is stationary, the total energy is not a conserved quantity, but instead decays monotonically in time due to the action of the damping. The net rate at which the force (2.1) does work on the mass is

$$P = f\,\dot{x} = -k\,\dot{x}\,x - m\,\nu\,\dot{x}^2. \tag{2.22}$$

The spring force (i.e., the first term on the right-hand side) does negative work on the mass (i.e., it reduces the system kinetic energy) when $\dot{x}$ and x are of the same sign, and does positive work when they are of the opposite sign. It can easily be demonstrated that, on average, the spring force does no net work on the mass during an oscillation cycle. The damping force, on the other hand, (i.e., the second term on the right-hand side) always does negative work on the mass, and, therefore, always acts to reduce the system kinetic energy.

2.3 QUALITY FACTOR

The energy loss rate of a weakly damped (i.e., $\nu \ll 2\,\omega_0$) harmonic oscillator is conveniently characterized in terms of a parameter, Q_f, which is known as the *quality factor*. (Note that the standard symbol for the quality factor is Q. We are only using Q_f here to avoid confusion with electrical charge.) This quantity is defined to be 2π *times the energy stored in the oscillator, divided by the energy lost in a single oscillation period.* If the oscillator is weakly damped then the energy lost per period is relatively small, and Q_f is therefore much larger than unity. Roughly speaking, Q_f is the number of oscillations that the oscillator typically completes, after being set in motion, before its amplitude decays to a negligible value. (To be more exact, Q_f is the number of oscillations that the oscillator completes before its amplitude decays to about 4% of its original value. See Exercise 2.6.) For instance, the quality factor for the damped oscillation shown in Figure 2.1 is 12.6. Let us find an expression for Q_f.

As we have seen, the motion of a weakly damped harmonic oscillator is specified by [see Equation (2.12)]

$$x = a\,e^{-\nu t/2}\,\cos(\omega_1\,t - \phi). \tag{2.23}$$

It follows that

$$\dot{x} = -\frac{a\,\nu}{2}\,e^{-\nu t/2}\,\cos(\omega_1\,t - \phi) - a\,\omega_1\,e^{-\nu t/2}\,\sin(\omega_1\,t - \phi). \tag{2.24}$$

Thus, making use of Equation (2.20), the energy lost during a single oscillation period is

$$\Delta E = - \int_{\phi/\omega_1}^{(2\pi+\phi)/\omega_1} \frac{dE}{dt} \, dt$$

$$= m \, v \, a^2 \int_{\phi/\omega_1}^{(2\pi+\phi)/\omega_1} e^{-\nu t} \left[\frac{\nu}{2} \cos(\omega_1 t - \phi) + \omega_1 \sin(\omega_1 t - \phi) \right]^2 \, dt. \tag{2.25}$$

In the weakly damped limit, $\nu \ll 2\,\omega_0$, the exponential factor is approximately unity in the interval $t = \phi/\omega_1$ to $(2\pi + \phi)/\omega_1$, so that

$$\Delta E \simeq \frac{m \, v \, a^2}{\omega_1} \int_0^{2\pi} \left(\frac{\nu^2}{4} \cos^2 \theta + \nu \, \omega_1 \cos \theta \sin \theta + \omega_1^2 \sin^2 \theta \right) d\theta, \tag{2.26}$$

where $\theta = \omega_1 t - \phi$. Thus,

$$\Delta E \simeq \frac{\pi \, m \, v \, a^2}{\omega_1} \left(\nu^2/4 + \omega_1^2 \right) = \pi \, m \, \omega_0^2 \, a^2 \left(\frac{\nu}{\omega_1} \right), \tag{2.27}$$

because, as is readily demonstrated,

$$\int_0^{2\pi} \cos^2 \theta \, d\theta = \int_0^{2\pi} \sin^2 \theta \, d\theta = \pi, \tag{2.28}$$

$$\int_0^{2\pi} \cos \theta \sin \theta \, d\theta = 0. \tag{2.29}$$

The energy stored in the oscillator (at $t = 0$) is [cf., Equation (1.16)]

$$E = \frac{1}{2} m \, \omega_0^2 \, a^2. \tag{2.30}$$

Hence, we obtain

$$Q_f = 2\pi \frac{E}{\Delta E} = \frac{\omega_1}{\nu} \simeq \frac{\omega_0}{\nu}. \tag{2.31}$$

2.4 LCR CIRCUIT

Consider an electrical circuit consisting of an inductor, of inductance L, connected in series with a capacitor, of capacitance C, and a resistor, of resistance R. See Figure 2.2. Such a circuit is known as an *LCR circuit*, for obvious reasons. Suppose that $I(t)$ is the instantaneous current flowing around the circuit. As we saw in Section 1.4, the potential drops across the inductor and the capacitor are $L\dot{I}$ and Q/C, respectively. Here, Q is the charge on the capacitor's positive plate, and $I = \dot{Q}$. Moreover, from *Ohm's law*, the potential drop across the resistor is $V = I R$ (Fitzpatrick 2008). Kirchhoff's second circuital law states that the sum of the potential drops across the various components of a closed circuit loop is zero. It follows that

$$L\dot{I} + R I + Q/C = 0. \tag{2.32}$$

Dividing by L, and differentiating with respect to time, we obtain

$$\ddot{I} + \nu \dot{I} + \omega_0^2 I = 0, \tag{2.33}$$

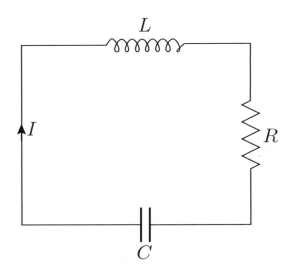

FIGURE 2.2 An LCR circuit.

where

$$\omega_0 = \frac{1}{\sqrt{LC}}, \tag{2.34}$$

$$\nu = \frac{R}{L}. \tag{2.35}$$

Comparison with Equation (2.2) reveals that Equation (2.33) is a damped harmonic oscillator equation. Thus, provided that the resistance is not too high (i.e., provided $\nu < 2\,\omega_0$, which is equivalent to $R < 2\,\sqrt{L/C}$), the current in the circuit executes damped harmonic oscillations of the form [cf., Equation (2.12)]

$$I(t) = \hat{I}\,e^{-\nu t/2}\,\cos(\omega_1 t - \phi), \tag{2.36}$$

where $\hat{I}$ and ϕ are constants, and $\omega_1 = (\omega_0^2 - \nu^2/4)^{1/2}$. We conclude that if a small amount of resistance is introduced into an LC circuit then the characteristic oscillations in the current damp away exponentially at a rate proportional to the resistance.

Multiplying Equation (2.32) by I, and making use of the fact that $I = \dot{Q}$, we obtain

$$L\dot{I}\,I + R\,I^2 + \dot{Q}\,Q/C = 0, \tag{2.37}$$

which can be rearranged to give

$$\frac{dE}{dt} = -R\,I^2, \tag{2.38}$$

where

$$E = \frac{1}{2}L\,I^2 + \frac{1}{2}\frac{Q^2}{C}. \tag{2.39}$$

Here, E is the circuit energy; that is, the sum of the energies stored in the inductor and the capacitor. Moreover, according to Equation (2.38), the circuit energy decays in time due to the power $R\,I^2$ dissipated via *Joule heating* in the resistor (Fitzpatrick 2008). Of course, the dissipated power is always positive. In other words, the circuit never gains energy from the resistor.

Finally, a comparison of Equations (2.31), (2.34), and (2.35) reveals that the quality factor of an LCR circuit is

$$Q_f = \frac{\sqrt{L/C}}{R}. \tag{2.40}$$

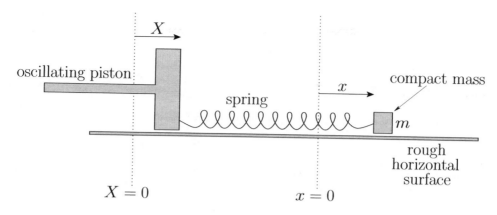

FIGURE 2.3 A driven oscillatory system.

2.5 DRIVEN DAMPED HARMONIC OSCILLATION

We saw earlier, in Section 2.2, that if a damped mechanical oscillator is set into motion then the oscillations eventually die away due to frictional energy losses. In fact, the only way of maintaining the amplitude of a damped oscillator is to continuously feed energy into the system in such a manner as to offset the frictional losses. A steady (i.e., constant amplitude) oscillation of this type is called *driven damped harmonic oscillation*. Consider a modified version of the mass–spring system investigated in Section 2.2 in which one end of the spring is attached to the mass, and the other to an oscillating piston. See Figure 2.3. Let $x(t)$ be the horizontal displacement of the mass, and $X(t)$ the horizontal displacement of the piston. The extension of the spring is thus $x(t) - X(t)$, assuming that the spring is unstretched when $x = X = 0$. Thus, the horizontal force acting on the mass can be written [cf., Equation (2.1)]

$$f = -k\,(x - X) - m\,v\,\dot{x}.\tag{2.41}$$

The equation of motion of the system then becomes [cf., Equation (2.2)]

$$\ddot{x} + v\,\dot{x} + \omega_0^2\,x = \omega_0^2\,X,\tag{2.42}$$

where $v > 0$ is the damping constant, and $\omega_0 > 0$ the undamped oscillation frequency. Suppose, finally, that the piston executes simple harmonic oscillation of angular frequency $\omega > 0$ and amplitude $\hat{X} > 0$, so that the time evolution equation of the system takes the form

$$\ddot{x} + v\,\dot{x} + \omega_0^2\,x = \omega_0^2\,\hat{X}\,\cos(\omega\,t).\tag{2.43}$$

We shall refer to the preceding equation as the *driven damped harmonic oscillator equation*.

   We would generally expect the periodically driven oscillator shown in Figure 2.3 to eventually settle down to a steady (i.e., constant amplitude) pattern of oscillation, with the same frequency as the piston, in which the frictional energy loss per cycle is exactly matched by the work done by the piston per cycle. (See Exercise 2.9.) This suggests that we should search for a solution to Equation (2.43) of the form

$$x(t) = \hat{x}\,\cos(\omega\,t - \varphi).\tag{2.44}$$

Here, $\hat{x} > 0$ is the amplitude of the driven oscillation, whereas φ is the *phase lag* of this oscillation (with respect to the phase of the piston oscillation). Because

$$\dot{x} = -\omega\,\hat{x}\,\sin(\omega\,t - \varphi),\tag{2.45}$$

$$\ddot{x} = -\omega^2\,\hat{x}\,\cos(\omega\,t - \varphi),\tag{2.46}$$

Equation (2.43) becomes

$$\left(\omega_0^2 - \omega^2\right)\hat{x}\,\cos(\omega t - \varphi) - \nu\omega\,\hat{x}\,\sin(\omega t - \varphi) = \omega_0^2\,\hat{X}\,\cos(\omega t). \tag{2.47}$$

However, $\cos(\omega t - \varphi) \equiv \cos(\omega t)\,\cos\varphi + \sin(\omega t)\,\sin\varphi$ and $\sin(\omega t - \varphi) \equiv \sin(\omega t)\,\cos\varphi - \cos(\omega t)\,\sin\varphi$ (see Appendix B), so we obtain

$$\left[\hat{x}\left(\omega_0^2 - \omega^2\right)\cos\varphi + \hat{x}\,\nu\omega\,\sin\varphi - \omega_0^2\,\hat{X}\right]\cos(\omega t)$$
$$+\hat{x}\left[\left(\omega_0^2 - \omega^2\right)\sin\varphi - \nu\omega\,\cos\varphi\right]\sin(\omega t) = 0, \tag{2.48}$$

where we have collected similar terms. The only way in which the preceding equation can be satisfied at all times is if the (constant) coefficients of $\cos(\omega t)$ and $\sin(\omega t)$ separately equate to zero. In other words, if

$$\hat{x}\left(\omega_0^2 - \omega^2\right)\cos\varphi + \hat{x}\,\nu\omega\,\sin\varphi - \omega_0^2\,\hat{X} = 0, \tag{2.49}$$

$$\left(\omega_0^2 - \omega^2\right)\sin\varphi - \nu\omega\,\cos\varphi = 0. \tag{2.50}$$

These two expressions can be combined to give

$$\hat{x} = \frac{\omega_0^2\,\hat{X}}{\left[\left(\omega_0^2 - \omega^2\right)^2 + \nu^2\,\omega^2\right]^{1/2}}, \tag{2.51}$$

$$\varphi = \tan^{-1}\left(\frac{\nu\omega}{\omega_0^2 - \omega^2}\right). \tag{2.52}$$

This follows because Equation (2.50) yields

$$\tan\varphi = \frac{\nu\omega}{\omega_0^2 - \omega^2}, \tag{2.53}$$

and so

$$\cos\varphi \equiv \frac{1}{\sqrt{1 + \tan^2\varphi}} = \frac{\omega_0^2 - \omega^2}{\left[\left(\omega_0^2 - \omega^2\right)^2 + \nu^2\,\omega^2\right]^{1/2}}, \tag{2.54}$$

$$\sin\varphi \equiv \frac{\tan\varphi}{\sqrt{1 + \tan^2\varphi}} = \frac{\nu\omega}{\left[\left(\omega_0^2 - \omega^2\right)^2 + \nu^2\,\omega^2\right]^{1/2}}. \tag{2.55}$$

(See Appendix B.) Hence, substitution into Equation (2.49) gives Equation (2.51).

Let us investigate the dependence of the amplitude, $\hat{x}$, and phase lag, φ, of the driven oscillation on the driving frequency, ω. This is most easily done graphically. Figures 2.4 and 2.5 show $\hat{x}/\hat{X}$ and φ plotted as functions of ω for various different values of ν/ω_0. It can be seen that as the amount of damping in the system is decreased the amplitude of the response becomes progressively more peaked at the system's natural frequency of oscillation, ω_0. This effect is known as *resonance*, and ω_0 is termed the *resonant frequency*. Thus, a weakly damped oscillator (i.e., $\nu \ll \omega_0$) can be driven to large amplitude by the application of a relatively small amplitude external driving force that oscillates at a frequency close to the resonant frequency. The response of the oscillator is in phase (i.e., $\varphi \simeq 0$) with the external drive for driving frequencies well below the resonant frequency, is in phase quadrature (i.e., $\varphi = \pi/2$) at the resonant frequency, and is in anti-phase (i.e., $\varphi \simeq \pi$) for frequencies well above the resonant frequency.

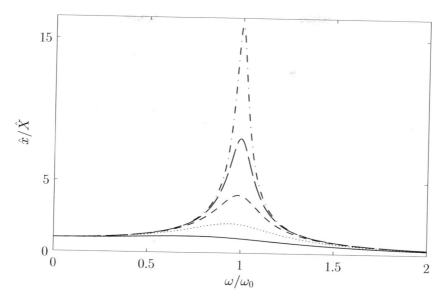

FIGURE 2.4 Driven damped harmonic oscillation. The figure shows the relative amplitude of the motion plotted as a function of the driving frequency for various different values of the damping rate. In fact, $\nu/\omega_0 = 1/Q_f = 1$, 1/2, 1/4, 1/8, and 1/16 correspond to the solid, dotted, short-dashed, long-dashed, and dot-dashed curves, respectively.

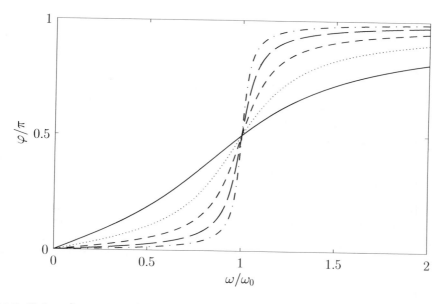

FIGURE 2.5 Driven damped harmonic oscillation. The figure shows the relative phase of the motion plotted as a function of the driving frequency for various different values of the damping rate. In fact, $\nu/\omega_0 = 1/Q_f = 1$, 1/2, 1/4, 1/8, and 1/16 correspond to the solid, dotted, short-dashed, long-dashed, and dot-dashed curves, respectively.

According to Equations (2.31) and (2.51),

$$\frac{\hat{x}(\omega = \omega_0)}{\hat{X}} = \frac{\omega_0}{\nu} = Q_f. \tag{2.56}$$

In other words, if the driving frequency matches the resonant frequency then the ratio of the amplitude of the driven oscillation to that of the piston oscillation is equal to the quality factor, Q_f. Hence, Q_f can be interpreted as the resonant amplification factor of the oscillator. Equations (2.51) and (2.55) imply that, for a weakly damped oscillator (i.e., $\nu \ll \omega_0$) which is close to resonance [i.e., $|\omega - \omega_0| \sim \nu \ll \omega_0$],

$$\frac{\hat{x}(\omega)}{\hat{x}(\omega = \omega_0)} \simeq \sin\varphi \simeq \frac{\nu}{[4(\omega_0 - \omega)^2 + \nu^2]^{1/2}}. \tag{2.57}$$

This follows because $\omega_0^2 - \omega^2 = (\omega_0 + \omega)(\omega_0 - \omega) \simeq 2\omega_0(\omega_0 - \omega)$. Hence, the width of the resonance peak (in angular frequency) is $\Delta\omega = \nu$, where the edges of the peak are defined as the points where the driven amplitude is reduced to $1/\sqrt{2}$ of its maximum value; that is, $\omega = \omega_0 \pm \nu/2$. The phase lag at the low- and high-frequency edges of the peak are $\pi/4$ and $3\pi/4$, respectively. Furthermore, the fractional width of the peak is

$$\frac{\Delta\omega}{\omega_0} = \frac{\nu}{\omega_0} = \frac{1}{Q_f}. \tag{2.58}$$

We conclude that the height and width of the resonance peak of a weakly damped ($Q_f \gg 1$) harmonic oscillator scale as Q_f and Q_f^{-1}, respectively. Thus, the area under the resonance peak stays approximately constant as Q_f varies.

The force exerted on the system by the piston is

$$F(t) = k\hat{X}\cos(\omega t). \tag{2.59}$$

Hence, the instantaneous power absorption from the piston becomes

$$
\begin{aligned}
P(t) &= F(t)\dot{x}(t) \\
&= -k\hat{X}\hat{x}\omega\cos(\omega t)\sin(\omega t - \varphi) \\
&= -k\hat{X}\hat{x}\omega\left[\cos(\omega t)\sin(\omega t)\cos\varphi - \cos^2(\omega t)\sin\varphi\right].
\end{aligned} \tag{2.60}
$$

The average power absorption during an oscillation cycle is

$$\langle P\rangle = \frac{1}{2}k\hat{X}\hat{x}\omega\sin\varphi, \tag{2.61}$$

because $\langle\cos(\omega t)\sin(\omega t)\rangle = 0$ and $\langle\cos^2(\omega t)\rangle = 1/2$. Given that the amplitude of the driven oscillation neither grows nor decays, the average power absorption from the piston during an oscillation cycle must equal the average power dissipation due to friction. (See Exercise 2.9.) Making use of Equations (2.56) and (2.57), the mean power absorption when the driving frequency is close to the resonant frequency is

$$\langle P\rangle \simeq \frac{1}{2}\omega_0 k\hat{X}^2 Q_f\left[\frac{\nu^2}{4(\omega_0 - \omega)^2 + \nu^2}\right]. \tag{2.62}$$

Thus, the maximum power absorption occurs at the resonance (i.e., $\omega = \omega_0$), and the absorption is reduced to half of this maximum value at the edges of the resonance (i.e., $\omega = \omega_0 \pm \nu/2$). Furthermore, the peak power absorption is proportional to the quality factor, Q_f, which means that it is inversely proportional to the damping constant, ν.

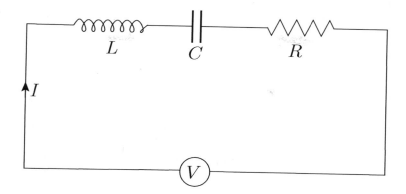

FIGURE 2.6   A driven LCR circuit.

2.6   DRIVEN LCR CIRCUIT

Consider an LCR circuit consisting of an inductor, L, a capacitor, C, and a resistor, R, connected in series with an oscillating emf of voltage $V(t)$. See Figure 2.6. Let $I(t)$ be the instantaneous current flowing around the circuit. According to Kirchhoff's second circuital law, the sum of the potential drops across the various components of a closed circuit loop is equal to zero. Thus, because the potential drop across an emf is minus the associated voltage (Fitzpatrick 2008), we obtain [cf., Equation (2.32)]

$$L\dot{I} + RI + Q/C = V, \tag{2.63}$$

where $\dot{Q} = I$. Suppose that the emf is such that its voltage oscillates sinusoidally at the angular frequency $\omega > 0$, with the peak value $\hat{V} > 0$, so that

$$V(t) = \hat{V}\,\sin(\omega\,t). \tag{2.64}$$

Dividing Equation (2.63) by L, and differentiating with respect to time, we obtain [cf., Equation (2.33)]

$$\ddot{I} + \nu\dot{I} + \omega_0^2\,I = \frac{\omega\,\hat{V}}{L}\,\cos(\omega\,t), \tag{2.65}$$

where $\omega_0 = 1/\sqrt{LC}$ and $\nu = R/L$. Comparison with Equation (2.43) reveals that this is a driven damped harmonic oscillator equation. It follows, by analogy with the analysis contained in the previous section, that the current driven in the circuit by the oscillating emf is of the form

$$I(t) = \hat{I}\,\cos(\omega\,t - \varphi), \tag{2.66}$$

where

$$\hat{I} = \frac{\omega\,\hat{V}/L}{\left[(\omega_0^2 - \omega^2)^2 + \nu^2\,\omega^2\right]^{1/2}}, \tag{2.67}$$

$$\varphi = \tan^{-1}\left(\frac{\nu\,\omega}{\omega_0^2 - \omega^2}\right). \tag{2.68}$$

In the immediate vicinity of the resonance (i.e., $|\omega - \omega_0| \sim v \ll \omega_0$), these expression simplify to

$$\hat{I} \simeq \frac{\hat{V}}{R} \frac{v}{[4\,(\omega_0 - \omega)^2 + v^2]^{1/2}}, \tag{2.69}$$

$$\sin\varphi \simeq \frac{v}{[4\,(\omega - \omega_0)^2 + v^2]^{1/2}}. \tag{2.70}$$

The circuit's mean power absorption from the emf is written

$$\langle P \rangle = \langle I(t)\,V(t) \rangle = \hat{I}\,\hat{V}\,\langle \cos(\omega t - \varphi)\,\sin(\omega t) \rangle = \frac{1}{2}\,\hat{I}\,\hat{V}\,\sin\varphi, \tag{2.71}$$

so that

$$\langle P \rangle \simeq \frac{1}{2}\,\frac{\hat{V}^2}{R}\left[\frac{v^2}{4\,(\omega_0 - \omega)^2 + v^2}\right] \tag{2.72}$$

close to the resonance. It follows that the peak power absorption, $(1/2)\,\hat{V}^2/R$, takes place when the emf oscillates at the resonant frequency, ω_0. Moreover, the power absorption falls to half of this peak value at the edges of the resonant peak; that is, $\omega = \omega_0 \pm v/2$.

LCR circuits can be used as *analog radio tuners*. In this application, the emf represents the analog signal picked up by a radio antenna. According to the previous analysis, the circuit only has a strong response (i.e., it only absorbs significant energy) when the signal oscillates in the angular frequency range $\omega = \omega_0 \pm v/2$, which corresponds to $\omega = 1/\sqrt{LC} \pm R/(2\,L)$. Thus, if the values of L, C, and R are properly chosen then the circuit can be made to strongly absorb the signal from a particular radio station, which has a given carrier frequency and bandwidth (see Section 8.5), while essentially ignoring the signals from other stations with different carrier frequencies. In practice, the values of L and R are fixed, while the value of C is varied (by turning a knob that adjusts the degree of overlap between two sets of parallel semicircular conducting plates) until the signal from the desired radio station is found.

2.7 TRANSIENT OSCILLATOR RESPONSE

The time evolution of the driven mechanical oscillator discussed in Section 2.5 is governed by the driven damped harmonic oscillator equation,

$$\ddot{x} + v\,\dot{x} + \omega_0^2\,x = \omega_0^2\,\hat{X}\,\cos(\omega t). \tag{2.73}$$

Recall that the steady (i.e., constant amplitude) solution to this equation that we found earlier takes the form

$$x_{ta}(t) = \hat{x}\,\cos(\omega t - \varphi), \tag{2.74}$$

where

$$\hat{x} = \frac{\omega_0^2\,\hat{X}}{\left[(\omega_0^2 - \omega^2)^2 + v^2\,\omega^2\right]^{1/2}}, \tag{2.75}$$

$$\varphi = \tan^{-1}\left(\frac{v\,\omega}{\omega_0^2 - \omega^2}\right). \tag{2.76}$$

Equation (2.73) is a second-order ordinary differential equation, which means that its general solution should contain two arbitrary constants (Riley 1974). However, Equation (2.74) contains no arbitrary constants. It follows that the right-hand side of (2.74) cannot be the most general solution

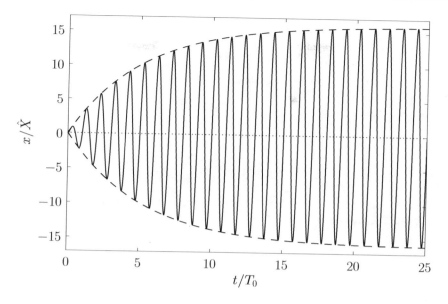

FIGURE 2.7 Resonant response of a driven damped harmonic oscillator. Here, $T_0 = 2\pi/\omega_0$, ω/ω_0, and $v/\omega_0 = 1/16$. The solid curve shows $x(t)/\hat{X}$, and the dashed curves show $\pm a(t)/\hat{X}$.

to the driven damped harmonic oscillator equation, (2.73). However, as is easily demonstrated, if we add any solution of the undriven damped harmonic oscillator equation,

$$\ddot{x} + v\,\dot{x} + \omega_0^2\,x = 0, \tag{2.77}$$

to the right-hand side of (2.74) then the result is still a solution to Equation (2.73). [Essentially, this is because Equation (2.73) is linear.] From Section 2.2, the most general solution to the preceding equation can be written

$$x_{tr}(t) = A\,e^{-v t/2}\,\cos(\omega_1\,t) + B\,e^{-v t/2}\,\sin(\omega_1\,t), \tag{2.78}$$

where $\omega_1 = (\omega_0^2 - v^2/4)^{1/2}$, and A and B are arbitrary constants. [In terms of the standard solution, (2.12), $A = a\,\cos\phi$ and $B = a\,\sin\phi$.] Thus, a more general solution to Equation (2.73) is

$$x(t) = x_{ta}(t) + x_{tr}(t)$$
$$= \hat{x}\,\cos(\omega\,t - \varphi) + A\,e^{-v t/2}\,\cos(\omega_1\,t) + B\,e^{-v t/2}\,\sin(\omega_1\,t). \tag{2.79}$$

In fact, because the preceding solution contains two arbitrary constants, we can be sure that it is the most general solution. The arbitrary constants, A and B, are determined by the initial conditions. Thus, the most general solution to the driven damped harmonic oscillator equation, (2.73), consists of two parts: first, the solution (2.74), which oscillates at the driving frequency ω with a constant amplitude, and which is independent of the initial conditions; second, the solution (2.78), which oscillates at the natural frequency ω_1 with an amplitude that decays exponentially in time, and which depends on the initial conditions. The former is termed the *time asymptotic solution*, because if we wait long enough then it becomes dominant. The latter is called the *transient solution*, because if we wait long enough then it decays away.

Suppose, for the sake of argument, that the system is initially in its equilibrium state. In other words, $x(0) = \dot{x}(0) = 0$. It follows from Equation (2.79) that

$$x(0) = \hat{x} \cos \varphi + A = 0, \tag{2.80}$$

$$\dot{x}(0) = \hat{x} \omega \sin \varphi - \frac{\nu}{2} A + \omega_1 B = 0. \tag{2.81}$$

The preceding equations can be solved to give

$$A = -\hat{x} \cos \varphi, \tag{2.82}$$

$$B = -\hat{x} \left[\frac{\omega \sin \varphi + (\nu/2) \cos \varphi}{\omega_1} \right]. \tag{2.83}$$

According to the analysis in Section 2.5, for driving frequencies close to the resonant frequency (i.e., $|\omega - \omega_0| \sim \nu$), we can write

$$\hat{x} \simeq \frac{\omega_0 \hat{X}}{[4 (\omega_0 - \omega)^2 + \nu^2]^{1/2}}, \tag{2.84}$$

$$\sin \varphi \simeq \frac{\nu}{[4 (\omega_0 - \omega)^2 + \nu^2]^{1/2}}, \tag{2.85}$$

$$\cos \varphi \simeq \frac{2 (\omega_0 - \omega)}{[4 (\omega_0 - \omega)^2 + \nu^2]^{1/2}}. \tag{2.86}$$

Hence, in this case, the solution (2.79), combined with Equations (2.82)–(2.86), reduces to

$$x(t) \simeq \hat{X} \left[\frac{2 \omega_0 (\omega_0 - \omega)}{4 (\omega_0 - \omega)^2 + \nu^2} \right] \left[\cos(\omega t) - e^{-\nu t/2} \cos(\omega_0 t) \right]$$

$$+ \hat{X} \left[\frac{\omega_0 \nu}{4 (\omega_0 - \omega)^2 + \nu^2} \right] \left[\sin(\omega t) - e^{-\nu t/2} \sin(\omega_0 t) \right]. \tag{2.87}$$

There are a number of interesting cases that are worth discussing. Consider, first, the situation in which the driving frequency is equal to the resonant frequency; that is, $\omega = \omega_0$. In this case, Equation (2.87) reduces to

$$x(t) = \hat{X} Q_f \left(1 - e^{-\nu t/2} \right) \sin(\omega_0 t), \tag{2.88}$$

because $Q_f = \omega_0/\nu$. Thus, the driven response oscillates at the resonant frequency, ω_0, because both the time asymptotic and transient solutions oscillate at this frequency. However, the amplitude of the oscillation grows monotonically as $a(t) = \hat{X} Q_f \left(1 - e^{-\nu t/2} \right)$, and so takes a time of order ν^{-1} (i.e., a time of order Q_f oscillation periods) to attain its final value $\hat{X} Q_f$, which is, of course, larger that the driving amplitude by the resonant amplification factor (or quality factor), Q_f. This behavior is illustrated in Figure 2.7.

Consider the situation in which there is no damping, so that $\nu = 0$. In this case, Equation (2.87) yields

$$x(t) = \hat{X} \left(\frac{\omega_0}{\omega_0 - \omega} \right) \sin \left[\frac{(\omega_0 - \omega) t}{2} \right] \sin \left[\frac{(\omega_0 + \omega) t}{2} \right], \tag{2.89}$$

where use has been made of the trigonometry identity $\cos a - \cos b \equiv -2 \sin[(a + b)/2] \sin[(a - b)/2]$. (See Appendix B.) It can be seen that the driven response oscillates relatively rapidly at the "sum frequency" $(\omega_0 + \omega)/2$ with an amplitude $a(t) = \hat{X} [\omega_0/(\omega_0 - \omega)] \sin[(\omega_0 - \omega)/t]$ that modulates relatively slowly at the "difference frequency" $(\omega_0 - \omega)/2$. (Recall, that we are assuming that ω is close to ω_0.) This behavior is illustrated in Figure 2.8. The amplitude modulations shown in Figure 2.8 are called *beats*, and are produced whenever two sinusoidal oscillations of similar

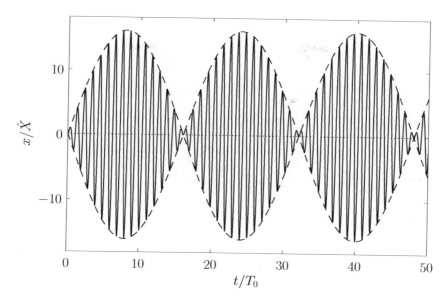

FIGURE 2.8 Off-resonant response of a driven undamped harmonic oscillator. Here, $T_0 = 2\pi/\omega_0$, $\omega/\omega_0 = 15/16$, and $\nu/\omega_0 = 0$. The solid curve shows $x(t)/\hat{X}$, and the dashed curves show $\pm a(t)/\hat{X}$.

amplitude, and slightly different frequency, are superposed. In this case, the two oscillations are the time asymptotic solution, which oscillates at the driving frequency, ω, and the transient solution, which oscillates at the resonant frequency, ω_0. The beats modulate at the difference frequency, $(\omega_0 - \omega)/2$. In the limit $\omega \to \omega_0$, Equation (2.89) yields

$$x(t) = \frac{\hat{X}}{2} \omega_0 t \sin(\omega_0 t), \qquad (2.90)$$

because $\sin x \simeq x$ when $|x| \ll 1$. (See Appendix B.) Thus, the resonant response of a driven un-damped oscillator is an oscillation at the resonant frequency whose amplitude, $a(t) = (\hat{X}/2)\omega_0 t$, increases linearly in time. In this case, the period of the beats has effectively become infinite.

Finally, Figure 2.9 illustrates the non-resonant response of a driven damped harmonic oscillator, obtained from Equation (2.87). It can be seen that the driven response grows, showing some initial evidence of beat modulation, but eventually settles down to a steady pattern of oscillation. This behavior occurs because the transient solution, which is needed to produce beats, initially grows, but then damps away, leaving behind the constant amplitude time asymptotic solution.

EXERCISES

2.1 Show that the period between successive zeros of a damped harmonic oscillator is constant, and is half the period between successive maxima.

2.2 Show that the ratio of two successive maxima in the displacement of a damped harmonic oscillator is constant. [From Fowles and Cassiday 2005.]

2.3 If the amplitude of a damped harmonic oscillator decreases to $1/e$ of its initial value after $n \gg 1$ periods show that the ratio of the period of oscillation to the period of the oscillation with no damping is approximately

$$1 + \frac{1}{8\pi^2 n^2}.$$

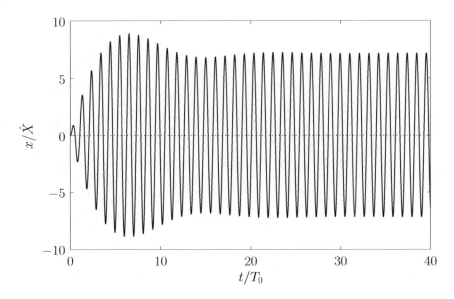

FIGURE 2.9 Off-resonant response of a driven damped harmonic oscillator. Here, $T_0 = 2\pi/\omega_0$, $\omega/\omega_0 = 15/16$, and $\nu/\omega_0 = 1/16$.

[From Fowles and Cassiday 2005.]

2.4 Many oscillatory systems are subject to damping effects that are not exactly analogous to the linear frictional damping considered in Section 2.2. Nevertheless, such systems typically exhibit an exponential decrease in their average stored energy of the form $\langle E \rangle = E_0 \exp(-\nu t)$. It is possible to define an effective quality factor for such oscillators as $Q_f = \omega_0/\nu$, where ω_0 is the natural angular oscillation frequency.

For example, when the note "middle C" on a piano is struck its oscillation energy decreases to one half of its initial value in about 1 second. The frequency of middle C is 256 Hz. What is the effective Q_f of the system? [Modified from French 1971.]

2.5 According to classical electromagnetic theory, an accelerated electron radiates energy at the rate $K e^2 a^2/c^3$, where $K = 6 \times 10^9\,\mathrm{N\,m^2/C^2}$, e is the charge on an electron, a the instantaneous acceleration, and c the velocity of light in vacuum (Fitzpatrick 2008).

   (a) If an electron were oscillating in a straight line with displacement $x(t) = A\,\sin(2\pi f t)$, how much energy would it radiate away during a single cycle?

   (b) What is the effective Q_f of this oscillator?

   (c) How many periods of oscillation would elapse before the energy of the oscillation was reduced to half of its initial value?

   (d) Substituting a typical optical frequency (e.g., for green light) for f, give numerical estimates for the Q_f and half-life of the radiating system.

2.6 Show that, on average, the energy of a damped harmonic oscillator of quality factor $Q_f \gg 1$ decays by a factor $e^{-2\pi} \simeq 1.9 \times 10^{-3}$ during Q_f oscillation cycles. By what factor does the amplitude decay in the same time interval?

2.7 Demonstrate that in the limit $\nu \rightarrow 2\,\omega_0$ the solution to the damped harmonic oscillator equation becomes

$$x(t) = (x_0 + [v_0 + \omega_0\,x_0]\,t)\,e^{-\omega_0 t},$$

where $x_0 = x(0)$ and $v_0 = \dot{x}(0)$.

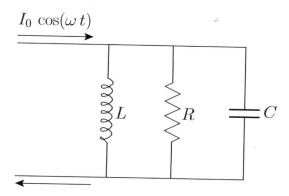

$I_0 \cos(\omega t)$

FIGURE 2.10 Figure for Exercise 2.8.

2.8 (a) What are the resonant angular frequency and quality factor of the circuit shown in Figure 2.10?

(b) What is the average power absorbed at resonance?

2.9 The power input $\langle P \rangle$ required to maintain a constant amplitude oscillation in a driven damped harmonic oscillator can be calculated by recognizing that this power is minus the average rate that work is done by the damping force, $-m\,\nu\,\dot{x}$.

(a) Using $x = \hat{x}\,\cos(\omega t - \varphi)$, show that the average rate that the damping force does work is $-m\,\nu\,\omega^2\,\hat{x}^2/2$.

(b) Substitute the value of $\hat{x}$ at an arbitrary driving frequency and, hence, obtain an expression for $\langle P \rangle$.

(c) Demonstrate that this expression yields Equation (2.62) in the limit that the driving frequency is close to the resonant frequency.

2.10 The equation $m\,\ddot{x} + k\,x = F_0\,\sin(\omega t)$ governs the motion of an undamped harmonic oscillator driven by a sinusoidal force of angular frequency ω.

(a) Show that the "time asymptotic" solution (i.e., the solution that would be time asymptotic were a small amount of damping added to the system) is

$$x = \frac{F_0\,\sin(\omega t)}{m\,(\omega_0^2 - \omega^2)},$$

where $\omega_0 = \sqrt{k/m}$. Sketch the behavior of x versus t for $\omega < \omega_0$ and $\omega > \omega_0$.

(b) Demonstrate that if $x = \dot{x} = 0$ at $t = 0$ then the general solution is

$$x = \frac{F_0}{m\,(\omega_0^2 - \omega^2)}\left[\sin(\omega t) - \frac{\omega}{\omega_0}\,\sin(\omega_0 t)\right].$$

(c) Show, finally, that if ω is close to the resonant frequency ω_0 then

$$x \simeq \frac{F_0}{2\,m\,\omega_0^2}\,[\sin(\omega_0\,t) - \omega_0\,t\,\cos(\omega_0\,t)]\,.$$

Sketch the behavior of x versus t.

[Modified from Pain 1999.]

CHAPTER 3

Coupled Oscillations

3.1 INTRODUCTION

In Chapter 1, we investigated single-degree-of-freedom mechanical and electrical systems that exhibit simple harmonic oscillation. The aim of this chapter is to extend this analysis to deal with analogous multiple-degree-of-freedom systems. Of course, a single-degree-of-freedom system is one that only requires a single coordinate to specify its instantaneous state, whereas a multiple-degree-of-freedom system requires multiple coordinates.

3.2 TWO SPRING-COUPLED MASSES

Consider a mechanical system consisting of two identical masses m that are free to slide over a frictionless horizontal surface. Suppose that the masses are attached to one another, and to two immovable walls, by means of three identical light horizontal springs of spring constant k, as shown in Figure 3.1. The instantaneous state of the system is conveniently specified by the displacements of the left and right masses, $x_1(t)$ and $x_2(t)$, respectively. The extensions of the left, middle, and right springs are x_1, $x_2 - x_1$, and $-x_2$, respectively, assuming that $x_1 = x_2 = 0$ corresponds to the equilibrium configuration in which the springs are all unextended. The equations of motion of the two masses are thus

$$m\,\ddot{x}_1 = -k\,x_1 + k\,(x_2 - x_1), \tag{3.1}$$

$$m\,\ddot{x}_2 = -k\,(x_2 - x_1) + k\,(-x_2). \tag{3.2}$$

Here, we have made use of the fact that a mass attached to the left end of a spring of extension x and spring constant k experiences a horizontal force $+k\,x$, whereas a mass attached to the right end of the same spring experiences an equal and opposite force $-k\,x$.

Equations (3.1)–(3.2) can be rewritten in the form

$$\ddot{x}_1 = -2\,\omega_0^2\,x_1 + \omega_0^2\,x_2, \tag{3.3}$$

$$\ddot{x}_2 = \omega_0^2\,x_1 - 2\,\omega_0^2\,x_2, \tag{3.4}$$

where $\omega_0 = \sqrt{k/m}$. Let us search for a solution in which the two masses oscillate *in phase* with one another at the *same* angular frequency, ω. In other words,

$$x_1(t) = \hat{x}_1\,\cos(\omega\,t - \phi), \tag{3.5}$$

$$x_2(t) = \hat{x}_2\,\cos(\omega\,t - \phi), \tag{3.6}$$

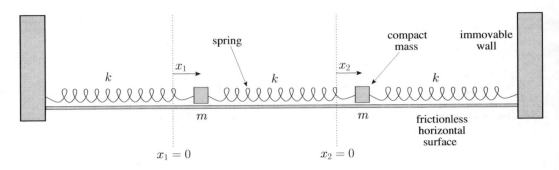

FIGURE 3.1   Two-degree-of-freedom mass-spring system.

where $\hat{x}_1$, $\hat{x}_2$, and ϕ are constants. Equations (3.3) and (3.4) yield

$$-\omega^2\,\hat{x}_1\,\cos(\omega t - \phi) = \left(-2\,\omega_0^2\,\hat{x}_1 + \omega_0^2\,\hat{x}_2\right)\cos(\omega t - \phi), \tag{3.7}$$

$$-\omega^2\,\hat{x}_2\,\cos(\omega t - \phi) = \left(\omega_0^2\,\hat{x}_1 - 2\,\omega_0^2\,\hat{x}_2\right)\cos(\omega t - \phi), \tag{3.8}$$

or (after cancelling the common factor $\cos(\omega t - \phi)$, dividing through by ω_0, and rearranging),

$$(\hat{\omega}^2 - 2)\,\hat{x}_1 + \hat{x}_2 = 0, \tag{3.9}$$

$$\hat{x}_1 + (\hat{\omega}^2 - 2)\,\hat{x}_2 = 0, \tag{3.10}$$

where $\hat{\omega} = \omega/\omega_0$. Note that, by searching for a solution of the form (3.5)–(3.6), we have effectively converted the system of two coupled linear differential equations (3.3)–(3.4) into the much simpler system of two coupled linear algebraic equations (3.9)–(3.10). The latter equations have the trivial solutions $\hat{x}_1 = \hat{x}_2 = 0$, but also yield

$$\frac{\hat{x}_1}{\hat{x}_2} = -\frac{1}{(\hat{\omega}^2 - 2)} = -(\hat{\omega}^2 - 2). \tag{3.11}$$

Hence, the condition for a nontrivial solution is

$$(\hat{\omega}^2 - 2)(\hat{\omega}^2 - 2) - 1 = 0. \tag{3.12}$$

In fact, if we write Equations (3.9)–(3.10) in the form of a homogenous (i.e., with a null right-hand side) 2×2 matrix equation, so that

$$\begin{pmatrix} \hat{\omega}^2 - 2, & 1 \\ 1, & \hat{\omega}^2 - 2 \end{pmatrix} \begin{pmatrix} \hat{x}_1 \\ \hat{x}_2 \end{pmatrix} = \begin{pmatrix} 0 \\ 0 \end{pmatrix}, \tag{3.13}$$

then it is apparent that the criterion (3.12) can also be obtained by setting the determinant of the associated matrix to zero (Riley 1974).

Equation (3.12) can be rewritten

$$\hat{\omega}^4 - 4\,\hat{\omega}^2 + 3 = (\hat{\omega}^2 - 1)(\hat{\omega}^2 - 3) = 0. \tag{3.14}$$

It follows that

$$\hat{\omega} = 1 \text{ or } \sqrt{3}. \tag{3.15}$$

Here, we have neglected the two negative frequency roots of (3.14)—that is, $\hat{\omega} = -1$ and

$\hat{\omega} = -\sqrt{3}$—because a negative frequency oscillation is equivalent to an oscillation with an equal and opposite positive frequency, and an equal and opposite phase. In other words, $\cos(\omega t - \phi) \equiv \cos(-\omega t + \phi)$. It is thus apparent that the dynamical system pictured in Figure 3.1 has two unique frequencies of oscillation: namely, $\omega = \omega_0$ and $\omega = \sqrt{3}\,\omega_0$. These are called the *normal frequencies* of the system. Because the system possesses two degrees of freedom (i.e., two independent coordinates are needed to specify its instantaneous configuration), it is not entirely surprising that it possesses two normal frequencies. In fact, it is a general rule that a dynamical system with N degrees of freedom possesses N normal frequencies.

The patterns of motion associated with the two normal frequencies can be deduced from Equation (3.11). Thus, for $\omega = \omega_0$ (i.e., $\hat{\omega} = 1$), we get $\hat{x}_1 = \hat{x}_2$, so that

$$x_1(t) = \hat{\eta}_1 \, \cos(\omega_0 t - \phi_1), \tag{3.16}$$

$$x_2(t) = \hat{\eta}_1 \, \cos(\omega_0 t - \phi_1), \tag{3.17}$$

where $\hat{\eta}_1$ and ϕ_1 are arbitrary constants. This first pattern of motion corresponds to the two masses executing simple harmonic oscillation with the same amplitude and phase. Such an oscillation does not stretch the middle spring. On the other hand, for $\omega = \sqrt{3}\,\omega_0$ (i.e., $\hat{\omega} = \sqrt{3}$), we get $\hat{x}_1 = -\hat{x}_2$, so that

$$x_1(t) = \hat{\eta}_2 \, \cos\left(\sqrt{3}\,\omega_0 t - \phi_2\right), \tag{3.18}$$

$$x_2(t) = -\hat{\eta}_2 \, \cos\left(\sqrt{3}\,\omega_0 t - \phi_2\right), \tag{3.19}$$

where $\hat{\eta}_2$ and ϕ_2 are arbitrary constants. This second pattern of motion corresponds to the two masses executing simple harmonic oscillation with the same amplitude but in anti-phase; that is, with a phase shift of π radians. Such oscillations do stretch the middle spring, implying that the restoring force associated with similar amplitude displacements is greater for the second pattern of motion than for the first. This accounts for the higher oscillation frequency in the second case. (The inertia is the same in both cases, so the oscillation frequency is proportional to the square root of the restoring force associated with similar amplitude displacements.) The two distinctive patterns of motion that we have found are called the *normal modes of oscillation* of the system. Incidentally, it is a general rule that a dynamical system possessing N degrees of freedom has N unique normal modes of oscillation.

The most general motion of the system is a linear combination of the two normal modes. This immediately follows because Equations (3.1) and (3.2) are linear equations. [In other words, if $x_1(t)$ and $x_2(t)$ are solutions then so are $\alpha\, x_1(t)$ and $\alpha\, x_2(t)$, where α is an arbitrary constant.] Thus, we can write

$$x_1(t) = \hat{\eta}_1 \, \cos(\omega_0 t - \phi_1) + \hat{\eta}_2 \, \cos\left(\sqrt{3}\,\omega_0 t - \phi_2\right), \tag{3.20}$$

$$x_2(t) = \hat{\eta}_1 \, \cos(\omega_0 t - \phi_1) - \hat{\eta}_2 \, \cos\left(\sqrt{3}\,\omega_0 t - \phi_2\right). \tag{3.21}$$

We can be sure that this represents the most general solution to Equations (3.1) and (3.2) because it contains four arbitrary constants; namely, $\hat{\eta}_1$, ϕ_1, $\hat{\eta}_2$, and ϕ_2. (In general, we expect the solution of a second-order ordinary differential equation to contain two arbitrary constants. It, thus, follows that the solution of a system of two coupled, second-order, ordinary differential equations should contain four arbitrary constants.) These constants are determined by the initial conditions.

For instance, suppose that $x_1 = a$, $\dot{x}_1 = 0$, $x_2 = 0$, and $\dot{x}_2 = 0$ at $t = 0$. It follows, from

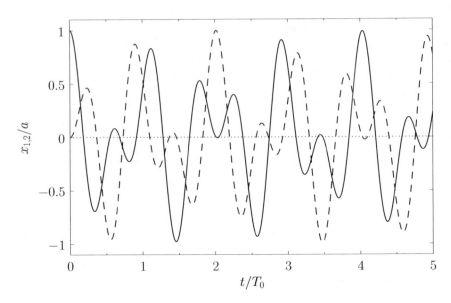

FIGURE 3.2   Time evolution of the physical coordinates of the two-degree-of-freedom mass-spring system pictured in Figure 3.1. The initial conditions are $x_1 = a$, $\dot{x}_1 = x_2 = \dot{x}_2 = 0$. The solid curve corresponds to x_1, and the dashed curve to x_2. Here, $T_0 = 2\pi/\omega_0$.

Equations (3.20) and (3.21), that

$$a = \hat{\eta}_1 \cos \phi_1 + \hat{\eta}_2 \cos \phi_2, \tag{3.22}$$

$$0 = \hat{\eta}_1 \sin \phi_1 + \sqrt{3}\,\hat{\eta}_2 \sin \phi_2, \tag{3.23}$$

$$0 = \hat{\eta}_1 \cos \phi_1 - \hat{\eta}_2 \cos \phi_2, \tag{3.24}$$

$$0 = \hat{\eta}_1 \sin \phi_1 - \sqrt{3}\,\hat{\eta}_2 \sin \phi_2, \tag{3.25}$$

which implies that $\phi_1 = \phi_2 = 0$ and $\hat{\eta}_1 = \hat{\eta}_2 = a/2$. Thus, the system evolves in time as

$$x_1(t) = \frac{a}{2} \cos(\omega_0 t) + \frac{a}{2} \cos(\sqrt{3}\,\omega_0 t) = a \cos(\omega_- t) \cos(\omega_+ t), \tag{3.26}$$

$$x_2(t) = \frac{a}{2} \cos(\omega_0 t) - \frac{a}{2} \cos(\sqrt{3}\,\omega_0 t) = a \sin(\omega_- t) \sin(\omega_+ t), \tag{3.27}$$

where $\omega_\pm = [(\sqrt{3} \pm 1)/2]]\,\omega_0$, and use has been made of the trigonometric identities $\cos a + \cos b \equiv 2 \cos[(a+b)/2] \cos[(a-b)/2]$ and $\cos a - \cos b \equiv -2 \sin[(a+b)/2] \sin[(a-b)/2]$. (See Appendix B.) This evolution is illustrated in Figure 3.2.

Finally, let us define the so-called *normal coordinates*, which (in the present case) take the form

$$\eta_1(t) = [x_1(t) + x_2(t)]/2, \tag{3.28}$$

$$\eta_2(t) = [x_1(t) - x_2(t)]/2. \tag{3.29}$$

It follows from Equations (3.20) and (3.21) that, in the presence of both normal modes,

$$\eta_1(t) = \hat{\eta}_1 \cos(\omega_0 t - \phi_1), \tag{3.30}$$

$$\eta_2(t) = \hat{\eta}_2 \cos\left(\sqrt{3}\,\omega_0 t - \phi_2\right). \tag{3.31}$$

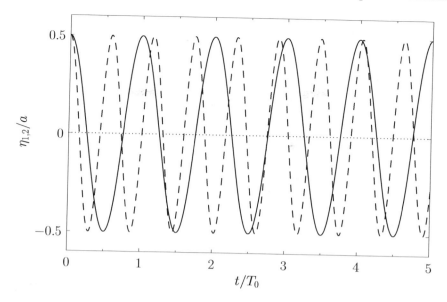

FIGURE 3.3 Time evolution of the normal coordinates of the two-degree-of-freedom mass-spring system pictured in Figure 3.1. The initial conditions are $x_1 = a$, $\dot{x}_1 = x_2 = \dot{x}_2 = 0$. The solid curve corresponds to η_1, and the dashed curve to η_2. Here, $T_0 = 2\pi/\omega_0$.

Thus, in general, the two normal coordinates oscillate sinusoidally with unique frequencies, unlike the physical coordinates, $x_1(t)$ and $x_2(t)$. See Figures 3.2 and 3.3. This suggests that the equations of motion of the system should look particularly simple when expressed in terms of the normal coordinates. In fact, it can be seen that the sum of Equations (3.3) and (3.4) reduces to

$$\ddot{\eta}_1 = -\omega_0^2\,\eta_1, \tag{3.32}$$

whereas the difference gives

$$\ddot{\eta}_2 = -3\,\omega_0^2\,\eta_2. \tag{3.33}$$

Thus, when expressed in terms of the normal coordinates, the equations of motion of the system reduce to two uncoupled simple harmonic oscillator equations. The most general solution to Equation (3.32) is (3.30), whereas the most general solution to Equation (3.33) is (3.31). Hence, if we can guess the normal coordinates of a coupled oscillatory system then the determination of the normal modes of oscillation is considerably simplified.

Note that, in general, the normal coordinates of a multiple-degree-of-freedom oscillatory system are linear combinations of the physical coordinates. By definition, the equations of motion of the system become decoupled from one another when expressed in terms of the normal coordinates. In fact, each equation of motion takes the form of a simple harmonic oscillator equation whose characteristic frequency is one of the normal frequencies. Consequently, each normal coordinate executes simple harmonic oscillation at its associated normal frequency.

3.3 TWO COUPLED LC CIRCUITS

Consider the LC circuit pictured in Figure 3.4. Let $I_1(t)$, $I_2(t)$, and $I_3(t)$ be the currents flowing in the three legs of the circuit, which meet at junctions A and B. According to *Kirchhoff's first circuital law*, the net current flowing into each junction is zero (Grant and Phillips 1975). It follows

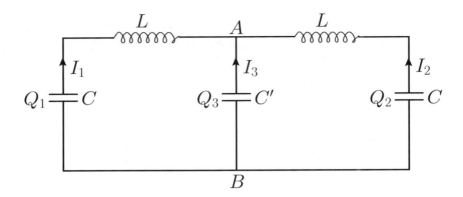

FIGURE 3.4   A two-degree-of-freedom LC circuit.

that $I_3 = -(I_1 + I_2)$. We deduce that this is a two-degree-of-freedom system whose instantaneous configuration is specified by the two independent variables $I_1(t)$ and $I_2(t)$. It follows that there are two independent normal modes of oscillation. The potential drops across the left, middle, and right legs of the circuit are $Q_1/C + L\dot{I}_1$, Q_3/C', and $Q_2/C + L\dot{I}_2$, respectively, where $\dot{Q}_1 = I_1$, $\dot{Q}_2 = I_2$, and $Q_3 = -(Q_1 + Q_2)$. However, because the three legs are connected in parallel with one another, the potential drops must all be equal, so that

$$Q_1/C + L\dot{I}_1 = Q_3/C' = -(Q_1 + Q_2)/C', \tag{3.34}$$

$$Q_2/C + L\dot{I}_2 = Q_3/C' = -(Q_1 + Q_2)/C'. \tag{3.35}$$

Differentiating with respect to t, dividing by L, and rearranging, we obtain the coupled time evolution equations of the system:

$$\ddot{I}_1 + \omega_0^2(1+\alpha)I_1 + \omega_0^2\,\alpha\,I_2 = 0, \tag{3.36}$$

$$\ddot{I}_2 + \omega_0^2(1+\alpha)I_2 + \omega_0^2\,\alpha\,I_1 = 0, \tag{3.37}$$

where $\omega_0 = 1/\sqrt{LC}$ and $\alpha = C/C'$.

We can solve the problem in a systematic manner by searching for a normal mode of the form

$$I_1(t) = \hat{I}_1\,\cos(\omega t - \phi), \tag{3.38}$$

$$I_2(t) = \hat{I}_2\,\cos(\omega t - \phi). \tag{3.39}$$

Substitution into the time evolution equations (3.36) and (3.37) yields the homogeneous matrix equation

$$\begin{pmatrix} \hat{\omega}^2 - (1+\alpha), & -\alpha \\ -\alpha, & \hat{\omega}^2 - (1+\alpha) \end{pmatrix}\begin{pmatrix} \hat{I}_1 \\ \hat{I}_2 \end{pmatrix} = \begin{pmatrix} 0 \\ 0 \end{pmatrix}, \tag{3.40}$$

where $\hat{\omega} = \omega/\omega_0$. The normal frequencies are determined by setting the determinant of the matrix to zero. This gives

$$\left[\hat{\omega}^2 - (1+\alpha)\right]^2 - \alpha^2 = 0, \tag{3.41}$$

or

$$\hat{\omega}^4 - 2(1+\alpha)\hat{\omega}^2 + 1 + 2\alpha = \left(\hat{\omega}^2 - 1\right)\left(\hat{\omega}^2 - [1+2\alpha]\right) = 0. \tag{3.42}$$

The roots of the preceding equation are $\hat{\omega} = 1$ and $\hat{\omega} = (1+2\alpha)^{1/2}$. (Again, we have neglected the

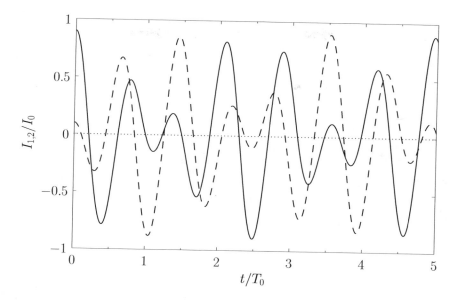

FIGURE 3.5 Time evolution of the physical coordinates of the two-degree-of-freedom LC circuit pictured in Figure 3.4. The initial conditions are $I_1 = 0.9\,I_0$, $I_2 = 0.1\,I_0$, $\dot{I}_1 = \dot{I}_2 = 0$. The solid curve corresponds to I_1, and the dashed curve to I_2. Here, $\alpha = 0.5$, and $T_0 = 2\pi/\omega_0$.

negative frequency roots, because they generate the same patterns of motion as the corresponding positive frequency roots.) Hence, the two normal frequencies are ω_0 and $(1 + 2\,\alpha)^{1/2}\,\omega_0$. The characteristic patterns of motion associated with the normal modes can be calculated from the first row of the matrix equation (3.40), which can be rearranged to give

$$\frac{\hat{I}_1}{\hat{I}_2} = \frac{\alpha}{\hat{\omega}^2 - (1 + \alpha)}. \tag{3.43}$$

It follows that $\hat{I}_1 = -\hat{I}_2$ for the normal mode with $\hat{\omega} = 1$, and $\hat{I}_1 = \hat{I}_2$ for the normal mode with $\hat{\omega} = (1 + 2\,\alpha)^{1/2}$. The most general solution, thus, takes the form

$$I_1(t) = \hat{\eta}_1\,\cos(\omega_1\,t - \phi_1) + \hat{\eta}_2\,\cos(\omega_0\,t - \phi_2), \tag{3.44}$$

$$I_2(t) = \hat{\eta}_1\,\cos(\omega_1\,t - \phi_1) - \hat{\eta}_2\,\cos(\omega_0\,t - \phi_2), \tag{3.45}$$

where $\hat{\eta}_1$ and ϕ_1 are the amplitude and phase of the higher frequency normal mode, whereas $\hat{\eta}_2$ and ϕ_2 are the amplitude and phase of the lower frequency mode.

It is fairly easy to guess that the normal coordinates of the system are

$$\eta_1 = (I_1 + I_2)/2, \tag{3.46}$$

$$\eta_2 = (I_1 - I_2)/2. \tag{3.47}$$

Forming the sum and difference of Equations (3.36) and (3.37), we obtain the evolution equations for the two independent normal modes of oscillation,

$$\ddot{\eta}_1 + (1 + 2\,\alpha)\,\omega_0^2\,\eta_1 = 0, \tag{3.48}$$

$$\ddot{\eta}_2 + \omega_0^2\,\eta_2 = 0. \tag{3.49}$$

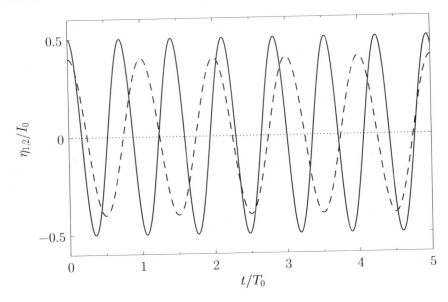

FIGURE 3.6 Time evolution of the normal coordinates of the two-degree-of-freedom LC circuit pictured in Figure 3.4. The initial conditions are $I_1 = 0.9\,I_0$, $I_2 = 0.1\,I_0$, $\dot{I}_1 = \dot{I}_2 = 0$. The solid curve corresponds to η_1, and the dashed curve to η_2. Here, $\alpha = 0.5$, and $T_0 = 2\pi/\omega_0$.

(We can be sure that we have correctly guessed the normal coordinates because the previous two equations do not couple to one another.) These equations can readily be solved to give

$$\eta_1(t) = \hat{\eta}_1 \, \cos(\omega_1\, t - \phi_1), \tag{3.50}$$

$$\eta_2(t) = \hat{\eta}_2 \, \cos(\omega_0\, t - \phi_2), \tag{3.51}$$

where $\omega_1 = (1 + 2\,\alpha)^{1/2}\,\omega_0$. Here, $\hat{\eta}_1$, ϕ_1, $\hat{\eta}_2$, and ϕ_2 are arbitrary constants. Note that the previous two equations, when combined with Equations (3.46) and (3.47) (which imply that $I_1 = \eta_1 + \eta_2$ and $I_2 = \eta_1 - \eta_2$), are equivalent to our previous solution, (3.44) and (3.45).

As an example, suppose that $\alpha = 0.5$. Furthermore, let $I_1 = 0.9\,I_0$, $I_2 = 0.1\,I_0$, and $\dot{I}_1 = \dot{I}_2 = 0$, at $t = 0$, where I_0 is an arbitrary constant. The time evolution of the system is illustrated in Figures 3.5 and 3.6. Note that the normal coordinates oscillate sinusoidally, whereas the time evolution of the physical coordinates is more complicated.

3.4 THREE SPRING-COUPLED MASSES

Consider a generalized version of the mechanical system discussed in Section 3.2 that consists of three identical masses m which slide over a frictionless horizontal surface, and are connected by identical light horizontal springs of spring constant k. As before, the outermost masses are attached to immovable walls by springs of spring constant k. The instantaneous configuration of the system is specified by the horizontal displacements of the three masses from their equilibrium positions; namely, $x_1(t)$, $x_2(t)$, and $x_3(t)$. This is manifestly a three-degree-of-freedom system. We, therefore, expect it to possess three independent normal modes of oscillation. Equations (3.1)–(3.2) generalize to give

$$m\ddot{x}_1 = -k\,x_1 + k\,(x_2 - x_1),$$ (3.52)

$$m\ddot{x}_2 = -k\,(x_2 - x_1) + k\,(x_3 - x_2),$$ (3.53)

$$m\ddot{x}_3 = -k\,(x_3 - x_2) + k\,(-x_3).$$ (3.54)

These equations can be rewritten

$$\ddot{x}_1 = -2\,\omega_0^2\,x_1 + \omega_0^2\,x_2,$$ (3.55)

$$\ddot{x}_2 = \omega_0^2\,x_1 - 2\,\omega_0^2\,x_2 + \omega_0^2\,x_3,$$ (3.56)

$$\ddot{x}_3 = \omega_0^2\,x_2 - 2\,\omega_0^2\,x_3,$$ (3.57)

where $\omega_0 = \sqrt{k/m}$. Let us search for a normal mode solution of the form

$$x_1(t) = \hat{x}_1\,\cos(\omega t - \phi),$$ (3.58)

$$x_2(t) = \hat{x}_2\,\cos(\omega t - \phi),$$ (3.59)

$$x_3(t) = \hat{x}_3\,\cos(\omega t - \phi).$$ (3.60)

Equations (3.55)–(3.60) can be combined to give the 3×3 homogeneous matrix equation

$$\begin{pmatrix} \hat{\omega}^2 - 2, & 1, & 0 \\ 1, & \hat{\omega}^2 - 2, & 1 \\ 0, & 1, & \hat{\omega}^2 - 2 \end{pmatrix} \begin{pmatrix} \hat{x}_1 \\ \hat{x}_2 \\ \hat{x}_3 \end{pmatrix} = \begin{pmatrix} 0 \\ 0 \\ 0 \end{pmatrix},$$ (3.61)

where $\hat{\omega} = \omega/\omega_0$. The normal frequencies are determined by setting the determinant of the matrix to zero; that is,

$$(\hat{\omega}^2 - 2)\left[(\hat{\omega}^2 - 2)^2 - 1\right] - (\hat{\omega}^2 - 2) = 0,$$ (3.62)

or

$$(\hat{\omega}^2 - 2)\left[\hat{\omega}^2 - 2 - \sqrt{2}\right]\left[\hat{\omega}^2 - 2 + \sqrt{2}\right] = 0.$$ (3.63)

Thus, the normal frequencies are $\hat{\omega} = \sqrt{2}\left(1 - 1/\sqrt{2}\right)^{1/2}$, $\sqrt{2}$, and $\sqrt{2}\left(1 + 1/\sqrt{2}\right)^{1/2}$. According to the first and third rows of Equation (3.61),

$$\hat{x}_1 : \hat{x}_2 : \hat{x}_3 :: 1 : 2 - \hat{\omega}^2 : 1,$$ (3.64)

provided $\hat{\omega}^2 \neq 2$. According to the second row,

$$\hat{x}_1 : \hat{x}_2 : \hat{x}_3 :: -1 : 0 : 1$$ (3.65)

when $\hat{\omega}^2 = 2$. Incidentally, we can only determine the ratios of $\hat{x}_1$, $\hat{x}_2$, and $\hat{x}_3$, rather than the absolute values of these quantities. In other words, only the direction of the vector $\hat{\mathbf{x}} = (\hat{x}_1, \hat{x}_2, \hat{x}_3)$ is well-defined. [This follows because the most general solution, (3.69), is undetermined to an arbitrary multiplicative constant. That is, if $\mathbf{x}(t) = (x_1(t), x_2(t), x_3(t))$ is a solution to the dynamical equations (3.55)–(3.57) then so is $\alpha\,\mathbf{x}(t)$, where α is an arbitrary constant. This, in turn, follows because the dynamical equations are linear.] Let us arbitrarily set the magnitude of $\hat{\mathbf{x}}$ to unity. It follows that the normal mode associated with the normal frequency $\hat{\omega}_1 = \sqrt{2}\left(1 - 1/\sqrt{2}\right)^{1/2}$ is

$$\hat{\mathbf{x}}_1 = \left(\frac{1}{2}, \frac{1}{\sqrt{2}}, \frac{1}{2}\right).$$ (3.66)

Likewise, the normal mode associated with the normal frequency $\hat{\omega}_2 = \sqrt{2}$ is

$$\hat{\mathbf{x}}_2 = \left(-\frac{1}{\sqrt{2}}, 0, \frac{1}{\sqrt{2}}\right). \tag{3.67}$$

Finally, the normal mode associated with the normal frequency $\hat{\omega}_3 = \sqrt{2}\left(1 + 1/\sqrt{2}\right)^{1/2}$ is

$$\hat{\mathbf{x}}_3 = \left(\frac{1}{2}, -\frac{1}{\sqrt{2}}, \frac{1}{2}\right). \tag{3.68}$$

Note that the vectors $\hat{\mathbf{x}}_1$, $\hat{\mathbf{x}}_2$, and $\hat{\mathbf{x}}_3$ are mutually perpendicular. In other words, they are normal vectors. (Hence, the name "normal" mode.)

Let $\mathbf{x} = (x_1, x_2, x_2)$. It follows that the most general solution to the problem is

$$\mathbf{x}(t) = \eta_1(t)\,\hat{\mathbf{x}}_1 + \eta_2(t)\,\hat{\mathbf{x}}_2 + \eta_3(t)\,\hat{\mathbf{x}}_3, \tag{3.69}$$

where

$$\eta_1(t) = \hat{\eta}_1 \cos(\omega_1 t - \phi_1), \tag{3.70}$$

$$\eta_2(t) = \hat{\eta}_2 \cos(\omega_2 t - \phi_2), \tag{3.71}$$

$$\eta_3(t) = \hat{\eta}_3 \cos(\omega_3 t - \phi_3). \tag{3.72}$$

Here, $\hat{\eta}_{1,2,3}$ and $\phi_{1,2,3}$ are arbitrary constants. Incidentally, we need to introduce the arbitrary amplitudes $\hat{\eta}_{1,2,3}$ to make up for the fact that we set the magnitudes of the vectors $\hat{\mathbf{x}}_{1,2,3}$ to unity. Equation (3.69) yields

$$\begin{pmatrix} x_1 \\ x_2 \\ x_3 \end{pmatrix} = \begin{pmatrix} 1/2, & -1/\sqrt{2}, & 1/2 \\ 1/\sqrt{2}, & 0, & -1/\sqrt{2} \\ 1/2, & 1/\sqrt{2}, & 1/2 \end{pmatrix} \begin{pmatrix} \eta_1 \\ \eta_2 \\ \eta_3 \end{pmatrix} \tag{3.73}$$

The preceding equation can be inverted by noting that $\eta_1 = \hat{\mathbf{x}}_1 \cdot \hat{\mathbf{x}}$, et cetera, because $\hat{\mathbf{x}}_1$, $\hat{\mathbf{x}}_2$, and $\hat{\mathbf{x}}_3$ are mutually perpendicular unit vectors. Thus, we obtain

$$\begin{pmatrix} \eta_1 \\ \eta_2 \\ \eta_3 \end{pmatrix} = \begin{pmatrix} 1/2, & 1/\sqrt{2}, & 1/2 \\ -1/\sqrt{2}, & 0, & 1/\sqrt{2} \\ 1/2, & -1/\sqrt{2}, & 1/2 \end{pmatrix} \begin{pmatrix} x_1 \\ x_2 \\ x_3 \end{pmatrix} \tag{3.74}$$

This equation determines the three normal coordinates, η_1, η_2, η_3, in terms of the three physical coordinates, x_1, x_2, x_3. In general, the normal coordinates are undetermined to arbitrary multiplicative constants.

EXERCISES

3.1 A particle of mass m is attached to a rigid support by means of a spring of spring constant k. At equilibrium, the spring hangs vertically downward. An identical oscillator is added to this system, the spring of the former being attached to the mass of the latter. Calculate the normal frequencies for one-dimensional vertical oscillations about the equilibrium state, and describe the associated normal modes.

3.2 Consider a mass-spring system of the general form shown in Figure 3.1 in which the two masses are of mass m, the two outer springs have spring constant k, and the middle spring has spring constant k'. Find the normal frequencies and normal modes in terms of $\omega_0 = \sqrt{k/m}$ and $\alpha = k'/k$.

3.3 Consider a mass-spring system of the general form shown in Figure 3.1 in which the two masses are of mass m, the two leftmost springs have spring constant k, and the rightmost spring is absent. Find the normal frequencies and normal modes in terms of $\omega_0 = \sqrt{k/m}$.

3.4 Consider a mass-spring system of the general form shown in Figure 3.1 in which the springs all have spring constant k, and the left and right masses are of mass m and m', respectively. Find the normal frequencies and normal modes in terms of $\omega_0 = \sqrt{k/m}$ and $\alpha = m'/m$.

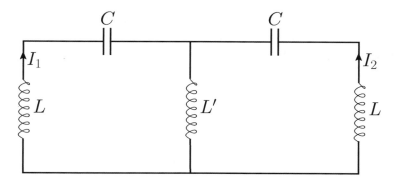

FIGURE 3.7   Figure for Exercise 3.5.

3.5 Find the normal frequencies and normal modes of the coupled LC circuit shown in Figure 3.7 in terms of $\omega_0 = 1/\sqrt{LC}$ and $\alpha = L'/L$.

3.6 Consider two simple pendula with the same length, l, but different bob masses, m_1 and m_2. Suppose that the pendula are connected by a spring of spring constant k. Let the spring be unextended when the two bobs are in their equilibrium positions.

   (a) Demonstrate that the equations of motion of the system (for small amplitude oscillations) are

$$m_1 \ddot{\theta}_1 = -m_1 \frac{g}{l} \theta_1 + k(\theta_2 - \theta_1),$$

$$m_2 \ddot{\theta}_2 = -m_2 \frac{g}{l} \theta_2 + k(\theta_1 - \theta_2),$$

   where θ_1 and θ_2 are the angular displacements of the respective pendula from their equilibrium positions.

   (b) Show that the normal coordinates are $\eta_1 = (m_1 \theta_1 + m_2 \theta_2)/(m_1 + m_2)$ and $\eta_2 = \theta_1 - \theta_2$. Find the normal frequencies.

   (c) Find a superposition of the two modes such that at $t = 0$ the two pendula are stationary, with $\theta_1 = \theta_0$, and $\theta_2 = 0$.

3.7 Two masses, m_1 and m_2, slide over a horizontal frictionless surface, and are connected via a spring of force constant k. Mass m_1 is acted on by a horizontal force $F_0 \cos(\omega t)$. In the absence of the second mass, this force causes the first mass to execute simple harmonic motion of amplitude $F_0/(m_1 \, \omega^2)$. Find an appropriate choice of the combination of values m_2 and k that reduces the oscillation amplitude of m_1 as much as possible. What is the oscillation amplitude of m_2 in this case?

3.8 A linear triatomic molecule (e.g., carbon dioxide) consists of a central atom of mass M flanked by two identical atoms of mass m. The atomic bonds are represented as springs of spring constant k. Find the molecule's normal frequencies and modes of linear oscillation.

3.9 Consider the mass-spring system discussed in Section 3.2.

(a) Show that, when written in terms of the physical coordinates, the total energy of the system takes the form

$$E = m \left[\frac{1}{2} \left(\dot{x}_1^2 + \dot{x}_2^2 \right) + \omega_0^2 \left(x_1^2 - x_1 \, x_2 + x_2^2 \right) \right].$$

(b) Furthermore, show that the total energy takes the form

$$E = m \left[\left(\dot{\eta}_1^2 + \dot{\eta}_2^2 \right) + \omega_0^2 \left(\eta_1^2 + 3 \, \eta_2^2 \right) \right]$$

when expressed in terms of the normal coordinates.

(c) Hence, deduce that

$$E = m \left(\mathcal{E}_1 + \mathcal{E}_2 \right),$$

$$\mathcal{E}_1 = \dot{\eta}_1^2 + \omega_0^2 \, \eta_1^2,$$

$$\mathcal{E}_2 = \dot{\eta}_2^2 + 3 \, \omega_0^2 \, \eta_2^2,$$

$$\frac{d\mathcal{E}_1}{dt} = 0,$$

$$\frac{d\mathcal{E}_2}{dt} = 0.$$

Here, $\mathcal{E}_1$ and $\mathcal{E}_2$ are the separately conserved energies per unit masses of the first and second normal modes, respectively.

CHAPTER 4

Transverse Standing Waves

4.1 INTRODUCTION

In Chapter 3, we investigated few-degree-of-freedom mechanical and electrical systems that exhibit simple harmonic oscillation about a stable equilibrium state. In this chapter, we shall extend this investigation to deal with many-degree-of-freedom mechanical systems, made up of a number of *identical* coupled single-degree-of-freedom systems, that likewise exhibit simple harmonic oscillation about a stable equilibrium state. We shall find that, in the limit as the number of degrees of freedom tends to infinity, such systems morph into physically continuous, uniform, mechanical systems that exhibit *standing wave* oscillations. A standing wave is a disturbance in a physically continuous mechanical system that is periodic in space as well as in time, but which does not propagate; that is, both the *nodes*, where the amplitude of the oscillation is zero, and the *anti-nodes*, where the amplitude of the oscillation is maximal, are stationary. In this chapter, we shall restrict our investigation to *transverse waves*; that is, waves in which the direction of oscillation is perpendicular to the direction along which the phase of the waves varies sinusoidally.

4.2 NORMAL MODES OF BEADED STRING

Consider a mechanical system consisting of a taut string that is stretched between two immovable walls. Suppose that N identical beads of mass m are attached to the string in such a manner that they cannot slide along it. Let the beads be equally spaced a distance a apart, and let the distance between the first and the last beads and the neighboring walls also be a. See Figure 4.1. Consider transverse oscillations of the string; that is, oscillations in which the string moves in a direction perpendicular to its length. It is assumed that the inertia of the string is negligible with respect to that of the beads. It follows that the sections of the string between neighboring beads, and between the outermost beads and the walls, are straight. (Otherwise, there would be a net tension force acting on the sections, and they would consequently suffer an infinite acceleration.) In fact, we expect the instantaneous configuration of the string to be a set of continuous straight-line segments of varying inclinations, as shown in the figure. Finally, assuming that the transverse displacement of the string is relatively small, it is reasonable to suppose that each section of the string possesses the same tension, T. [See Equation (4.11).]

It is convenient to introduce a Cartesian coordinate system such that x measures distance along the string from the left wall, and y measures the transverse displacement of the string from its equilibrium position. See Figure 4.1. Thus, when the string is in its equilibrium position it runs along the x-axis. We can define

$$x_i = i\, a, \tag{4.1}$$

where $i = 1, 2, \cdots, N$. Here, x_1 is the x-coordinate of the closest bead to the left wall, x_2 the x-

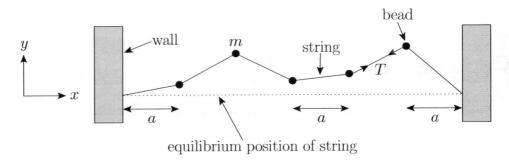

FIGURE 4.1 A beaded string.

coordinate of the second-closest bead, et cetera. The x-coordinates of the beads are assumed to remain constant during their transverse oscillations. We can also define $x_0 = 0$ and $x_{N+1} = (N+1)a$ as the x-coordinates of the left and right ends of the string, respectively. Let the transverse displacement of the ith bead be $y_i(t)$, for $i = 1, N$. Because each displacement can vary independently, we are dealing with an N degree of freedom system. We would, therefore, expect such a system to possess N unique normal modes of oscillation.

Consider the section of the string lying between the $i - 1$th and $i + 1$th beads, as shown in Figure 4.2. Here, $x_{i-1} = x_i - a$, x_i, and $x_{i+1} = x_i + a$ are the distances of the $i - 1$th, ith, and $i + 1$th beads, respectively, from the left wall, whereas y_{i-1}, y_i, and y_{i+1} are the corresponding transverse displacements of these beads. The two sections of the string that are attached to the ith bead subtend angles θ_i and θ_{i+1} with the x-axis, as illustrated in the figure. Standard trigonometry reveals that

$$\tan \theta_i = \frac{y_i - y_{i-1}}{x_i - x_{i-1}} = \frac{y_i - y_{i-1}}{a}, \tag{4.2}$$

and

$$\tan \theta_{i+1} = \frac{y_{i+1} - y_i}{a}. \tag{4.3}$$

However, if the transverse displacement of the string is relatively small—that is, if $|y_i| \ll a$ for all i—which we shall assume to be the case, then θ_i and θ_{i+1} are both small angles. Thus, we can use the small-angle approximation $\tan \theta \simeq \theta$. (See Appendix B.) It follows that

$$\theta_i \simeq \frac{y_i - y_{i-1}}{a}, \tag{4.4}$$

$$\theta_{i+1} \simeq \frac{y_{i+1} - y_i}{a}. \tag{4.5}$$

Let us find the transverse equation of motion of the ith bead. This bead is subject to two forces: namely, the tensions in the sections of the string to the left and to the right of it. (Incidentally, we are neglecting any gravitational forces acting on the beads, compared to the tension forces.) These tensions are of magnitude T, and are directed parallel to the associated string sections, as shown in Figure 4.2. Thus, the transverse (i.e., y-directed) components of these two tensions are $-T \sin \theta_i$ and $T \sin \theta_{i+1}$, respectively. Hence, the transverse equation of motion of the ith bead becomes

$$m \ddot{y}_i = -T \sin \theta_i + T \sin \theta_{i+1}. \tag{4.6}$$

However, because θ_i and θ_{i+1} are both small angles, we can employ the small-angle approximation $\sin \theta \simeq \theta$. It follows that

$$\ddot{y}_i \simeq \frac{T}{m} (\theta_{i+1} - \theta_i). \tag{4.7}$$

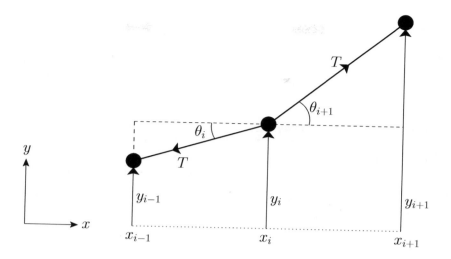

FIGURE 4.2   A short section of a beaded string.

Finally, making use of Equations (4.4) and (4.5), we obtain

$$\ddot{y}_i = \omega_0^2 \, (y_{i-1} - 2\,y_i + y_{i+1}),$$

(4.8)

where $\omega_0 = \sqrt{T/m\,a}$. Because there is nothing special about the ith bead, we deduce that the preceding equation of motion applies to all N beads; that is, it is valid for $i = 1, N$. Of course, the first ($i = 1$) and last ($i = N$) beads are special cases, because there is no bead corresponding to $i = 0$ or $i = N + 1$. In fact, $i = 0$ and $i = N + 1$ correspond to the left and right ends of the string, respectively. However, Equation (4.8) still applies to the first and last beads, as long as we set $y_0 = 0$ and $y_{N+1} = 0$. What we are effectively demanding is that the two ends of the string, which are attached to the left and right walls, must both have zero transverse displacement.

Incidentally, we can prove that the tensions in the two sections of the string shown in Figure 4.2 must be equal by considering the longitudinal equation of motion of the ith bead. This equation takes the form

$$m\,\ddot{x}_i = -T_i \, \cos\theta_i + T_{i+1} \, \cos\theta_{i+1},$$

(4.9)

where T_i and T_{i+1} are the, supposedly different, tensions in the sections of the string to the immediate left and right of the ith bead, respectively. We are assuming that the motion of the beads is purely transverse; that is, all of the x_i are constant in time. Thus, it follows from the preceding equation that

$$T_i \, \cos\theta_i = T_{i+1} \, \cos\theta_{i+1}.$$

(4.10)

However, if the transverse displacement of the string is such that all of the θ_i are small then we can make use of the small-angle approximation $\cos\theta_i \simeq 1$. (See Appendix B.) Hence, we obtain

$$T_i \simeq T_{i+1}.$$

(4.11)

A straightforward extension of this argument reveals that the tension is the same in all sections of the string.

Let us search for a normal mode solution to Equation (4.8) that takes the form

$$y_i(t) = A \, \sin(k\,x_i) \, \cos(\omega\,t - \phi),$$

(4.12)

where $A > 0$, $k > 0$, $\omega > 0$, and ϕ are constants. This particular type of solution is such that all of the beads execute transverse simple harmonic oscillations that are in phase with one another. See Figure 4.4. Moreover, the oscillations have an amplitude $A \sin(k\,x_i)$ that varies sinusoidally along the length of the string (i.e., in the x-direction). The pattern of oscillations is thus periodic in space. The spatial repetition period, which is generally termed the *wavelength*, is $\lambda = 2\pi/k$. [This follows from Equation (4.12) because $\sin\theta$ is a periodic function with period 2π: i.e., $\sin(\theta + 2\pi) \equiv \sin\theta$.] The constant k, which determines the wavelength, is usually referred to as the *wavenumber*. Thus, a small wavenumber corresponds to a long wavelength, and vice versa. The type of solution specified in Equation (4.12) is generally known as a *standing wave*. It is a wave because it is periodic in both space and time. (An oscillation is periodic in time only.) It is a standing wave, rather than a traveling wave, because the points of maximum and minimum amplitude oscillation are stationary (in x). See Figure 4.4.

Substituting Equation (4.12) into Equation (4.8), we obtain

$$-\omega^2 A \, \sin(k\,x_i)\,\cos(\omega t - \phi) = \omega_0^2 A\,[\sin(k\,x_{i-1}) - 2\,\sin(k\,x_i)$$
$$+ \sin(k\,x_{i+1})]\,\cos(\omega t - \phi), \tag{4.13}$$

which yields

$$-\omega^2 \sin(k\,x_i) = \omega_0^2\,(\sin[k\,(x_i - a)] - 2\,\sin(k\,x_i) + \sin[k\,(x_i + a)]). \tag{4.14}$$

However, because $\sin(a + b) \equiv \sin a \cos b + \cos a \sin b$ (see Appendix B), we get

$$-\omega^2 \sin(k\,x_i) = \omega_0^2\,[\cos(k\,a) - 2 + \cos(k\,a)]\,\sin(k\,x_i), \tag{4.15}$$

which gives

$$\omega = 2\,\omega_0 \sin(k\,a/2), \tag{4.16}$$

where use has been made of the trigonometric identity $1 - \cos\theta \equiv 2\,\sin^2(\theta/2)$. (See Appendix B.) An expression, such as Equation (4.16), that determines the angular frequency, ω, of a wave in terms of its wavenumber, k, is generally termed a *dispersion relation*.

The solution (4.12) is only physical provided $y_0 = y_{N+1} = 0$. In other words, provided the two ends of the string remain stationary. The first constraint is automatically satisfied, because $x_0 = 0$. See Equation (4.1). The second constraint implies that

$$\sin(k\,x_{N+1}) = \sin[(N + 1)\,k\,a] = 0. \tag{4.17}$$

This condition can only be satisfied if

$$k = \frac{n}{N + 1}\frac{\pi}{a}, \tag{4.18}$$

where the integer n is known as a *mode number*. A small mode number translates to a small wavenumber, and, hence, to a long wavelength, and vice versa. We conclude that the possible wavenumbers, k, of the normal modes of the system are quantized such that they are integer multiples of $\pi/[(N+1)\,a]$. Thus, the nth normal mode is associated with the characteristic pattern of bead displacements

$$y_{n,i}(t) = A_n \sin\left(\frac{n\,i}{N + 1}\pi\right)\cos(\omega_n t - \phi_n), \tag{4.19}$$

where

$$\omega_n = 2\,\omega_0 \sin\left(\frac{n}{N + 1}\frac{\pi}{2}\right). \tag{4.20}$$

Here, the integer $i = 1, N$ indexes the beads, whereas the mode number n indexes the normal modes.

Furthermore, A_n and ϕ_n are arbitrary constants determined by the initial conditions. Of course, A_n is the peak amplitude of the nth normal mode, whereas ϕ_n is the associated phase angle.

How many unique normal modes does the system possess? At first sight, it might seem that there are an infinite number of normal modes, corresponding to the infinite number of possible values that the integer n can take. However, this is not the case. For instance, if $n = 0$ or $n = N + 1$ then all of the $y_{n,i}$ are zero. Clearly, these cases are not real normal modes. Moreover, it can be demonstrated that

$$\omega_{-n} = -\omega_n, \tag{4.21}$$

$$y_{-n,i}(t) = y_{n,i}(t), \tag{4.22}$$

provided $A_{-n} = -A_n$ and $\phi_{-n} = -\phi_n$, as well as

$$\omega_{N+1+n} = \omega_{N+1-n}, \tag{4.23}$$

$$y_{N+1+n,i}(t) = y_{N+1-n,i}(t), \tag{4.24}$$

provided $A_{N+1+n} = -A_{N+1-n}$ and $\phi_{N+1+n} = \phi_{N+1-n}$. We, thus, conclude that only those normal modes that have n in the range 1 to N correspond to unique modes. Modes with n values lying outside this range are either null modes, or modes that are identical to other modes with n values lying within the prescribed range. It follows that there are N unique normal modes of the form (4.19). Hence, given that we are dealing with an N degree of freedom system, which we would expect to only possess N unique normal modes, we can be sure that we have found all the normal modes.

Figure 4.3 illustrates the spatial variation of the normal modes of a beaded string possessing eight beads. That is, an $N = 8$ system. It can be seen that the low-n—that is, long-wavelength—modes cause the string to oscillate in a fairly smoothly-varying (in x) sine wave pattern. On the other hand, the high-n—that is, short-wavelength—modes cause the string to oscillate in a rapidly-varying zig-zag pattern that bears little resemblance to a sine wave. The crucial distinction between the two different types of mode is that the wavelength of the oscillation (in the x-direction) is much larger than the bead spacing in the former case, while it is similar to the bead spacing in the latter. For instance, $\lambda = 18\,a$ for the $n = 1$ mode, $\lambda = 9\,a$ for the $n = 2$ mode, but $\lambda = 2.25\,a$ for the $n = 8$ mode.

Figure 4.5 displays the temporal variation of the $n = 2$ normal mode of an $N = 8$ beaded string. It can be seen that the beads oscillate in phase with one another. In other words, they all attain their maximal transverse displacements, and pass through zero displacement, simultaneously. Moreover, the mid-way point of the string always remains stationary. Such a point is known as a *node*. The $n = 1$ normal mode has two nodes (counting the stationary points at each end of the string as nodes), the $n = 2$ mode has three nodes, the $n = 3$ mode four nodes, et cetera. In fact, the existence of nodes is one of the distinguishing features of a standing wave.

Figure 4.5 shows the normal frequencies of an $N = 8$ beaded string plotted as a function of the normalized wavenumber. Recall that, for an $N = 8$ system, the relationship between the normalized wavenumber, $k\,a$, and the mode number, n, is $k\,a = (n/9)\,\pi$. It can be seen that the angular frequency increases as the wavenumber increases, which implies that shorter wavelength modes have higher oscillation frequencies. The dependence of the angular frequency on the normalized wavenumber, $k\,a$, is approximately linear when $k\,a \ll 1$. Indeed, it can be seen from Equation (4.20) that if $k\,a \ll 1$ then the small-angle approximation $\sin\theta \simeq \theta$ yields a linear dispersion relation of the form

$$\omega_n \simeq (k\,a)\,\omega_0 = \left(\frac{n}{N+1}\right)\pi\,\omega_0. \tag{4.25}$$

We, thus, conclude that those normal modes of a uniformly-beaded string whose wavelengths greatly exceed the bead spacing (i.e., modes with $k\,a \ll 1$) have approximately linear dispersion

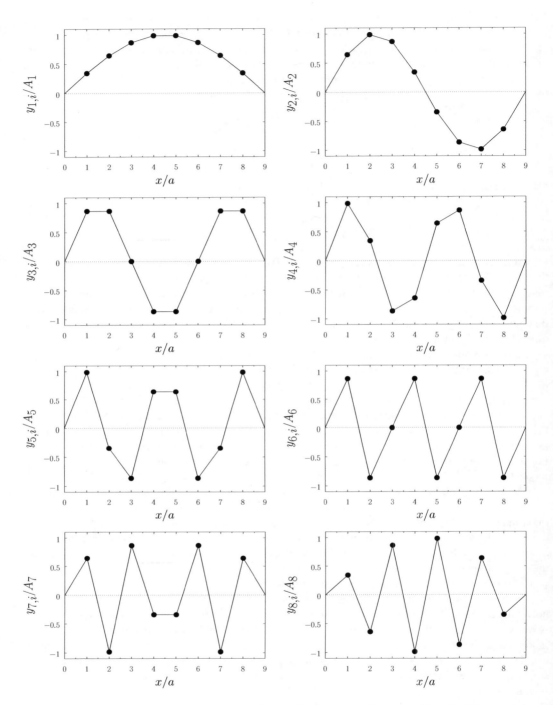

FIGURE 4.3 Normal modes of a beaded string with eight equally-spaced beads. The modes are all shown at the instances in time when they attain their maximum amplitudes: in other words, at $\omega_n t - \phi_n = 0$.

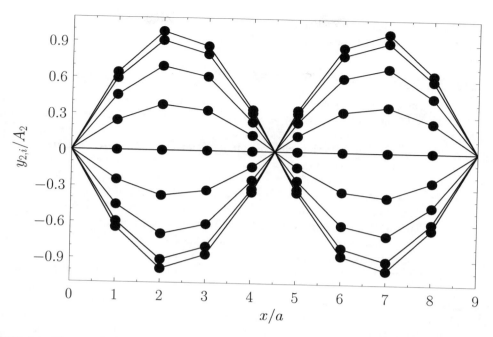

FIGURE 4.4 Time evolution of the $n = 2$ normal mode of a beaded string with eight equally-spaced beads. The mode is shown at $\omega_2 t - \phi_2 = 0, \pi/8, \pi/4, 3\pi/8, \pi/2, 5\pi/8, 3\pi/2, 7\pi/8$ and π.

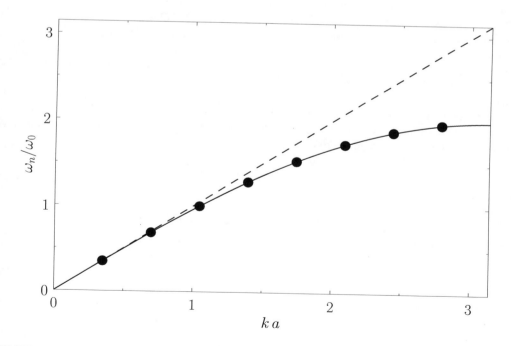

FIGURE 4.5 Normal frequencies of a beaded string with eight equally-spaced beads.

relations in which their angular frequencies are directly proportional to their mode numbers. However, it is evident from the figure that this linear relationship breaks down as $k\,a \to 1$, and the mode wavelength consequently becomes similar to the bead spacing.

4.3 NORMAL MODES OF UNIFORM STRING

Consider a uniformly-beaded string in the limit in which the number of beads, N, becomes increasingly large, while the spacing, a, and the individual mass, m, of the beads become increasingly small. Let the limit be taken in such a manner that the length, $l = (N + 1)\,a$, and the average mass per unit length, $\rho = m/a$, of the string both remain constant. As N increases, and becomes very large, such a string will more and more closely approximate a uniform string of length l and mass per unit length ρ. What can we guess about the normal modes of a uniform string on the basis of the analysis contained in the preceding section? First of all, we would guess that a uniform string is an infinite degree of freedom system, with an infinite number of unique normal modes of oscillation. This follows because a uniform string is the $N \to \infty$ limit of a beaded string, and a beaded string possesses N unique normal modes. Next, we would guess that the normal modes of a uniform string exhibit smooth sinusoidal spatial variation in the x-direction, and that the angular frequency of the modes is directly proportional to their wavenumber. These last two conclusions follow because all of the normal modes of a beaded string are characterized by $k\,a \ll 1$, in the limit in which the spacing between the beads becomes zero. Let us see whether these guesses are correct.

Consider the transverse oscillations of a uniform string of length l and mass per unit length ρ that is stretched between two immovable walls. It is again convenient to define a Cartesian coordinate system in which x measures distance along the string from the left wall, and y measures the transverse displacement of the string. Thus, the instantaneous state of the system at time t is determined by the function $y(x, t)$ for $0 \le x \le l$. This function consists of an infinite number of different y values, corresponding to the infinite number of different x values in the range 0 to l. Moreover, all of these y values are free to vary independently of one another. It follows that we are indeed dealing with a dynamical system possessing an infinite number of degrees of freedom.

Let us try to reuse some of the analysis of the previous section. We can reinterpret $y_i(t)$ as $y(x, t)$, $y_{i-1}(t)$ as $y(x - \delta x, t)$, and $y_{i+1}(t)$ as $y(x + \delta x, t)$, assuming that $x_i = x$ and $a = \delta x$. Moreover, $\ddot{y}_i(t)$ becomes $\partial^2 y(x, t)/\partial t^2$; namely, a second derivative of $y(x, t)$ with respect to t at constant x. Finally, $\omega_0^2 = T/(m\,a)$, where T is the tension in the string, can be rewritten as $T/[\rho\,(\delta x)^2]$, because $\rho = m/\delta x$. Incidentally, we are again assuming that the transverse displacement of the string remains sufficiently small that the tension is approximately constant in x. Thus, the equation of motion of the beaded string, (4.8), transforms into

$$\frac{\partial^2 y(x, t)}{\partial t^2} = \frac{T}{\rho}\left[\frac{y(x - \delta x, t) - 2\,y(x, t) + y(x + \delta x, t)}{(\delta x)^2}\right]. \tag{4.26}$$

However, Taylor expanding $y(x + \delta x, t)$ in x at constant t (see Appendix B), we obtain

$$y(x + \delta x, t) = y(x, t) + \frac{\partial y(x, t)}{\partial x}\,\delta x + \frac{1}{2}\frac{\partial^2 y(x, t)}{\partial x^2}\,(\delta x)^2 + O\!\left(\delta x^3\right). \tag{4.27}$$

Likewise,

$$y(x - \delta x, t) = y(x, t) - \frac{\partial y(x, t)}{\partial x}\,\delta x + \frac{1}{2}\frac{\partial^2 y(x, t)}{\partial x^2}\,(\delta x)^2 + O\!\left(\delta x^3\right). \tag{4.28}$$

It follows that

$$\left[\frac{y(x - \delta x, t) - 2\,y(x, t) + y(x + \delta x, t)}{(\delta x)^2}\right] = \frac{\partial^2 y(x, t)}{\partial x^2} + O(\delta x). \tag{4.29}$$

Thus, in the limit that $\delta x \to 0$, Equation (4.26) reduces to

$$\frac{\partial^2 y}{\partial t^2} = v^2 \frac{\partial^2 y}{\partial x^2},$$

(4.30)

where

$$v = \sqrt{\frac{T}{\rho}}$$

(4.31)

is a quantity having the dimensions of velocity. Equation (4.30), which is the transverse equation of motion of the string, is an example of a very famous partial differential equation known as the *wave equation*. The quantity v turns out to be the propagation velocity of transverse waves along the string. (See Section 6.2.)

By analogy with Equation (4.12), let us search for a solution of the wave equation of the form

$$y(x, t) = A \, \sin(k\,x) \, \cos(\omega\,t - \phi),$$

(4.32)

where $A > 0$, $k > 0$, $\omega > 0$, and ϕ are constants. We can interpret such a solution as a standing wave of wavenumber k, wavelength $\lambda = 2\pi/k$, angular frequency ω, peak amplitude A, and phase angle ϕ. Substitution of the preceding expression into Equation (4.30) yields the dispersion relation [cf., Equation (4.16)]

$$\omega = k\,v.$$

(4.33)

The standing wave solution (4.32) is subject to the physical constraint that the two ends of the string, which are both attached to immovable walls, remain stationary. This leads directly to the spatial boundary conditions

$$y(0, t) = 0,$$

(4.34)

$$y(l, t) = 0.$$

(4.35)

It can be seen that the solution (4.32) automatically satisfies the first boundary condition. However, the second boundary condition is only satisfied when $\sin(k\,l) = 0$, which immediately implies that

$$k = n\,\frac{\pi}{l},$$

(4.36)

where the mode number, n, is an integer. We, thus, conclude that the possible normal modes of a taut string, of length l and fixed ends, have wavenumbers which are quantized such that they are integer multiples of π/l. Moreover, this quantization is a direct consequence of the imposition of the physical boundary conditions at the two ends of the string.

It follows, from the previous analysis, that the nth normal mode of the string is associated with the pattern of motion

$$y_n(x, t) = A_n \, \sin\left(n\,\pi\,\frac{x}{l}\right) \cos(\omega_n\,t - \phi_n),$$

(4.37)

where

$$\omega_n = n\,\frac{\pi\,v}{l}.$$

(4.38)

Here, A_n and ϕ_n are constants that are determined by the initial conditions. (See Section 4.4.) How many unique normal modes are there? The choice $n = 0$ yields $y_0(x, t) = 0$ for all x and t, so this is not a real normal mode. Moreover,

$$\omega_{-n} = -\omega_n,$$

(4.39)

$$y_{-n}(x, t) = y_n(x, t),$$

(4.40)

provided that $A_{-n} = -A_n$ and $\phi_{-n} = -\phi_n$. We conclude that modes with negative mode numbers give rise to the same patterns of motion as modes with corresponding positive mode numbers. However, modes with different positive mode numbers correspond to unique patterns of motion that oscillate at unique frequencies. It follows that the string possesses an infinite number of normal modes, corresponding to the mode numbers $n = 1, 2, 3$, et cetera. Recall that we are dealing with an infinite-degree-of-freedom system, which we would expect to possess an infinite number of unique normal modes. The fact that we have actually obtained an infinite number of such modes suggests that we have found all of the normal modes.

Figure 4.6 illustrates the spatial variation of the first eight normal modes of a uniform string with fixed ends. It can be seen that the modes are all smoothly-varying sine waves. The low-mode-number (i.e., long-wavelength) modes are actually quite similar in form to the corresponding normal modes of a uniformly-beaded string. (See Figure 4.3.) However, the high-mode-number modes are substantially different. We conclude that the normal modes of a beaded string are similar to those of a uniform string, with the same length and mass per unit length, provided that the wavelength of the mode is much larger than the spacing between the beads.

Figure 4.7 illustrates the temporal variation of the $n = 4$ normal mode of a uniform string. All points on the string attain their maximal transverse displacements, and pass through zero displacement, simultaneously. Moreover, the $n = 4$ mode possesses five nodes (i.e., points where the string remains stationary). Two of these are located at the ends of the string, and three in the middle. In fact, according to Equation (4.37), the nodes correspond to points where $\sin[n\,(x/l)\,\pi] = 0$. Hence, the nodes are located at

$$x_{n,j} = \left(\frac{j}{n}\right)l, \tag{4.41}$$

where j is an integer lying in the range 0 to n. Here, n indexes the normal mode, and j the node. Thus, the $j = 0$ node lies at the left end of the string, the $j = 1$ node is the next node to the right, et cetera. It is apparent, from the preceding formula, that the nth normal mode has $n + 1$ nodes that are uniformly spaced a distance l/n apart.

Finally, Figure 4.8 shows the first eight normal frequencies of a uniform string with fixed ends, plotted as a function of the mode number. It can be seen that the angular frequency of oscillation increases linearly with the mode number. Recall that the low-mode-number (i.e., long-wavelength) normal modes of a beaded string also exhibit a linear relationship between normal frequency and mode number of the form [see Equation (4.25)]

$$\omega_n = \frac{n\,\pi}{N+1}\,\omega_0 = \frac{n\,\pi}{N+1}\left(\frac{T}{m\,a}\right)^{1/2}. \tag{4.42}$$

However, $m = \rho\,a$ and $l = (N+1)\,a$, so we obtain

$$\omega_n = \frac{n\,\pi}{l}\left(\frac{T}{\rho}\right)^{1/2} = n\,\frac{\pi\,v}{l}, \tag{4.43}$$

which is identical to Equation (4.38). We, thus, conclude that the normal frequencies of a uniformly-beaded string are similar to those of a uniform string, with the same length and mass per unit length, as long as the wavelength of the associated normal mode is much larger than the spacing between the beads.

4.4 GENERAL TIME EVOLUTION OF UNIFORM STRING

In the preceding section, we found the normal modes of a uniform string of length l, both ends of which are attached to immovable walls. These modes are spatially-periodic solutions of the wave equation, (4.30), that oscillate at unique frequencies and satisfy the spatial boundary conditions

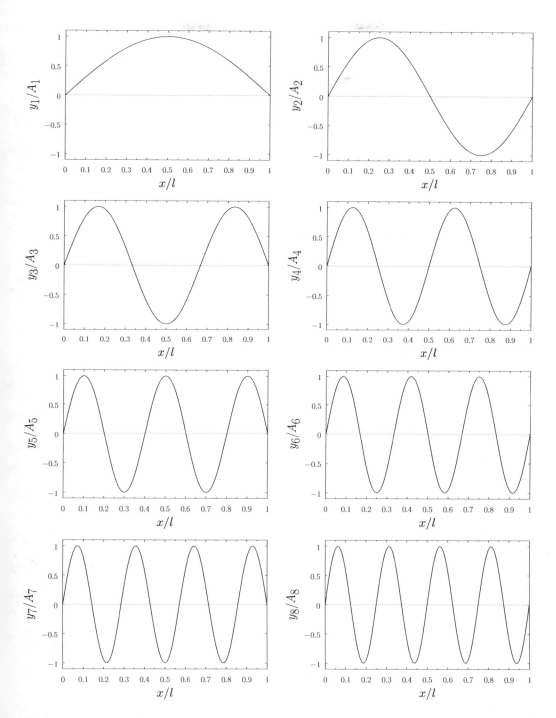

FIGURE 4.6 First eight normal modes of a uniform string. The modes are all shown at the instances in time when they attain their maximum amplitudes; namely, at $\omega_n t - \phi_n = 0$.

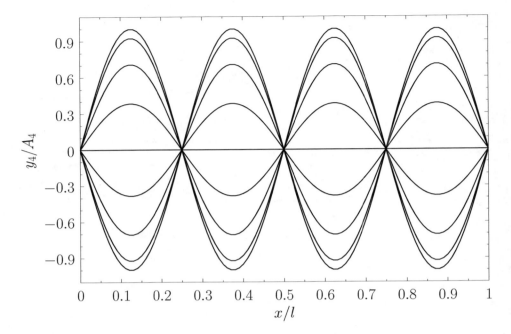

FIGURE 4.7 Time evolution of the $n = 4$ normal mode of a uniform string. The mode is shown at $\omega_4 t - \phi_4 = 0, \pi/8, \pi/4, 3\pi/8, \pi/2, 5\pi/8, 3\pi/2, 7\pi/8$ and π.

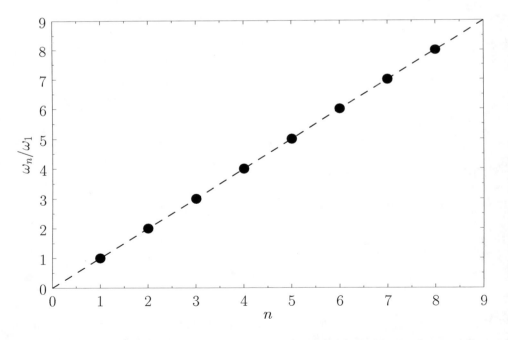

FIGURE 4.8 Normal frequencies of the first eight normal modes of a uniform string.

(4.34) and (4.35). Because the wave equation is linear [i.e., if $y(x, t)$ is a solution then so is $\alpha\, y(x, t)$, where α is an arbitrary constant], it follows that its most general solution is a linear combination of all of the normal modes. In other words,

$$y(x, t) = \sum_{n'=1,\infty} y_{n'}(x, t) = \sum_{n'=1,\infty} A_{n'} \sin\left(n'\,\pi\,\frac{x}{l}\right) \cos\left(n'\,\pi\,\frac{v\,t}{l} - \phi_{n'}\right), \tag{4.44}$$

where use has been made of Equations (4.37) and (4.38). The preceding expression is a solution of Equation (4.30), and also automatically satisfies the boundary conditions (4.34) and (4.35). As we have already mentioned, the constants A_n and ϕ_n are determined by the initial conditions. Let us see how this comes about in more detail.

Suppose that the initial displacement of the string at $t = 0$ is

$$y_0(x) \equiv y(x, 0). \tag{4.45}$$

Likewise, let the initial velocity of the string be

$$v_0(x) \equiv \frac{\partial y(x, 0)}{\partial t}. \tag{4.46}$$

For consistency with the boundary conditions, we must have $y_0(0) = y_0(l) = v_0(0) = v_0(l) = 0$. It follows from Equation (4.44) that

$$y_0(x) = \sum_{n'=1,\infty} A_{n'} \cos\phi_{n'}\,\sin\left(n'\,\pi\,\frac{x}{l}\right), \tag{4.47}$$

$$v_0(x) = \frac{v}{l}\sum_{n'=1,\infty} n'\,\pi\,A_{n'}\,\sin\phi_{n'}\,\sin\left(n'\,\pi\,\frac{x}{l}\right). \tag{4.48}$$

It is readily demonstrated that

$$\frac{2}{l}\int_0^l \sin\left(n\,\pi\,\frac{x}{l}\right)\sin\left(n'\,\pi\,\frac{x}{l}\right)dx = \frac{2}{\pi}\int_0^\pi \sin(n\,\theta)\sin(n'\,\theta)\,d\theta$$

$$= \frac{1}{\pi}\int_0^\pi \cos[(n-n')\,\theta]\,d\theta$$

$$- \frac{1}{\pi}\int_0^\pi \cos[(n+n')\,\theta]\,d\theta$$

$$= \frac{1}{\pi}\left[\frac{\sin[(n-n')\,\theta]}{n-n'} - \frac{\sin[(n+n')\,\theta]}{n+n'}\right]_0^\pi$$

$$= \frac{\sin[(n-n')\,\pi]}{(n-n')\,\pi} - \frac{\sin[(n+n')\,\pi]}{(n+n')\,\pi}, \tag{4.49}$$

where n and n' are (possibly different) positive integers, $\theta = \pi\,x/l$, and use has been made of the trigonometric identity $2\sin a\,\sin b \equiv \cos(a - b) - \cos(a + b)$. (See Appendix B.) Furthermore, if k is a non-zero integer then

$$\frac{\sin(k\,\pi)}{k\,\pi} = 0. \tag{4.50}$$

On the other hand, $k = 0$ is a special case, because both the numerator and the denominator in the preceding expression become zero simultaneously. However, application of *l'Hopital's rule* yields

$$\lim_{x\to 0}\frac{\sin x}{x} = \lim_{x\to 0}\frac{d(\sin x)/dx}{dx/dx} = \lim_{x\to 0}\frac{\cos x}{1} = 1. \tag{4.51}$$

It follows that

$$\frac{\sin(k\pi)}{k\pi} = \begin{cases} 1 & k = 0 \\ 0 & k \neq 0 \end{cases},$$ (4.52)

where k is any integer. This result can be combined with Equation (4.49), recalling that n and n' are both positive integers, to give

$$\frac{2}{l} \int_0^l \sin\left(n\pi\frac{x}{l}\right) \sin\left(n'\pi\frac{x}{l}\right) dx = \delta_{n,n'}.$$ (4.53)

Here, the quantity

$$\delta_{n,n'} = \begin{cases} 1 & n = n' \\ 0 & n \neq n' \end{cases},$$ (4.54)

where n and n' are integers, is known as the *Kronecker delta function*.

Let us multiply Equation (4.47) by $(2/l) \sin(n\pi x/l)$, and integrate over x from 0 to l. We obtain

$$\frac{2}{l} \int_0^l y_0(x) \sin\left(n\pi\frac{x}{l}\right) dx = \sum_{n'=1,\infty} A_{n'} \cos\phi_{n'} \frac{2}{l} \int_0^l \sin\left(n'\pi\frac{x}{l}\right) \sin\left(n\pi\frac{x}{l}\right) dx$$

$$= \sum_{n'=1,\infty} A_{n'} \cos\phi_{n'} \delta_{n,n'} = A_n \cos\phi_n,$$ (4.55)

where use has been made of Equations (4.53) and (4.54). Similarly, Equation (4.48) yields

$$\frac{2}{v} \int_0^l v_0(x) \sin\left(n\pi\frac{x}{l}\right) dx = n\pi A_n \sin\phi_n.$$ (4.56)

Thus, defining the integrals

$$C_n = \frac{2}{l} \int_0^l y_0(x) \sin\left(n\pi\frac{x}{l}\right) dx,$$ (4.57)

$$S_n = \frac{2}{n\pi v} \int_0^l v_0(x) \sin\left(n\pi\frac{x}{l}\right) dx,$$ (4.58)

for $n = 1, \infty$, we obtain

$$C_n = A_n \cos\phi_n,$$ (4.59)

$$S_n = A_n \sin\phi_n,$$ (4.60)

and, hence,

$$A_n = (C_n^2 + S_n^2)^{1/2},$$ (4.61)

$$\phi_n = \tan^{-1}(S_n/C_n).$$ (4.62)

Thus, the constants A_n and ϕ_n, appearing in the general expression (4.44) for the time evolution of a uniform string with fixed ends, are ultimately determined by integrals over the string's initial displacement and velocity of the form (4.57) and (4.58).

As an example, suppose that the string is initially at rest, so that

$$v_0(x) = 0,$$ (4.63)

but has the initial displacement

$$y_0(x) = 2A \begin{cases} x/l & 0 \le x < l/2 \\ 1 - x/l & l/2 \le x \le l \end{cases}. \tag{4.64}$$

This triangular pattern is zero at both ends of the string, rising linearly to the peak value A, halfway along the string, and is designed to mimic the initial displacement of a guitar string that is plucked at its mid-point. See Figure 4.10. A comparison of Equations (4.58) and (4.63) reveals that, in this particular example, all of the S_n coefficients are zero. Hence, from Equations (4.61) and (4.62), $A_n = C_n$ and $\phi_n = 0$ for all n. Thus, making use of Equations (4.44), (4.57), and (4.64), the time evolution of the string is governed by

$$y(x, t) = \sum_{n=1,\infty} A_n \sin\left(n \pi \frac{x}{l}\right) \cos\left(n\, 2\pi \frac{t}{\tau}\right), \tag{4.65}$$

where $\tau = 2\, l/v$ is the oscillation period of the $n = 1$ normal mode, and

$$A_n = \frac{2}{l} \int_0^{l/2} 2A\, \frac{x}{l} \sin\left(n \pi \frac{x}{l}\right) dx + \frac{2}{l} \int_{l/2}^l 2A\left(1 - \frac{x}{l}\right) \sin\left(n \pi \frac{x}{l}\right) dx. \tag{4.66}$$

The preceding expression transforms to

$$A_n = A \left(\frac{2}{\pi}\right)^2 \left\{ \int_0^{\pi/2} \theta \sin(n\,\theta)\, d\theta + \int_{\pi/2}^{\pi} (\pi - \theta)\, \sin(n\,\theta)\, d\theta \right\}, \tag{4.67}$$

where $\theta = \pi\, x/l$. Integration by parts (Riley 1974) yields

$$A_n = 2A\, \frac{\sin(n\,\pi/2)}{(n\,\pi/2)^2}. \tag{4.68}$$

It follows that $A_n = 0$ whenever n is even. We conclude that the triangular initial displacement pattern (4.64) only excites normal modes with odd mode numbers.

When a stringed instrument, such as a guitar, is played, a characteristic pattern of normal mode oscillations is excited on the plucked string. These oscillations excite sound waves of the same frequency, which propagate through the air and are audible to a listener. The normal mode (of appreciable amplitude) with the lowest oscillation frequency is called the *fundamental harmonic*, and determines the *pitch* of the musical note that is heard by the listener. For instance, a fundamental harmonic that oscillates at 261.6 Hz corresponds to "middle C". Those normal modes (of appreciable amplitude) with higher oscillation frequencies than the fundamental harmonic are called *overtone harmonics*, because their frequencies are integer multiples of the fundamental frequency. In general, the amplitudes of the overtone harmonics are much smaller than the amplitude of the fundamental. Nevertheless, when a stringed instrument is played, the particular mix of overtone harmonics that accompanies the fundamental determines the *timbre* of the musical note heard by the listener. For instance, when middle C is played on a piano and a harpsichord, the same frequency fundamental harmonic is excited in both cases. However, the mix of excited overtone harmonics is quite different. This accounts for the fact that middle C played on a piano can be easily distinguished from middle C played on a harpsichord.

Figure 4.9 shows the ratio A_n/A_1 for the first ten overtone harmonics of a uniform guitar string plucked at its midpoint; that is, the ratio A_n/A_1 for odd-n modes, with $n > 1$, calculated from Equation (4.68). It can be seen that the amplitudes of the overtone harmonics are all small compared to the amplitude of the fundamental. Moreover, the amplitudes decrease rapidly in magnitude with increasing mode number, n.

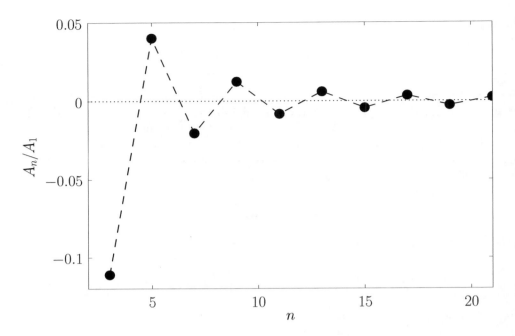

FIGURE 4.9 Relative amplitudes of the overtone harmonics of a uniform guitar string plucked at its mid-point.

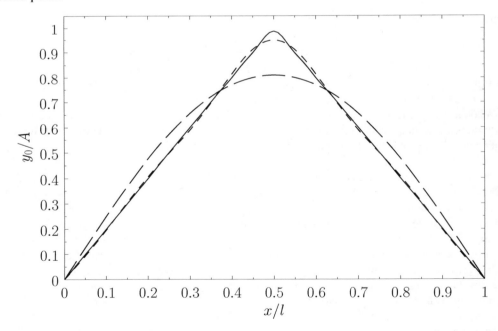

FIGURE 4.10 Reconstruction of the initial displacement of a uniform guitar string plucked at its mid-point. The long-dashed line shows a reconstruction made with only the largest-amplitude normal mode, the short-dashed line shows a reconstruction made with the four largest-amplitude normal modes, and the solid line shows a reconstruction made with the sixteen largest-amplitude normal modes.

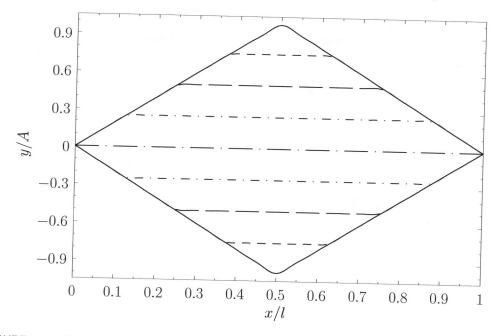

FIGURE 4.11   Time evolution of a uniform guitar string plucked at its mid-point. This evolution is reconstructed from Equation (4.65) using the sixteen largest-amplitude normal modes of the string. The upper solid, upper short-dashed, upper long-dashed, upper dot-short-dashed, dot-long-dashed, lower dot-short-dashed, lower long-dashed, lower short-dashed, and lower solid curves correspond to $t/\tau = 0, 1/16, 1/8, 3/16, 1/4, 5/16, 3/8, 7/16$, and $1/2$, respectively.

In principle, we must include all of the normal modes in the sum on the right-hand side of Equation (4.65). In practice, given that the amplitudes of the normal modes decrease rapidly in magnitude as n increases, we can truncate the sum, by neglecting high-n normal modes, without introducing significant error into our calculation. Figure 4.10 illustrates the effect of such a truncation. In fact, this figure shows the reconstruction of $y_0(x)$, obtained by setting $t = 0$ in Equation (4.65), made with various different numbers of normal modes. It can be seen that sixteen normal modes is sufficient to very accurately reconstruct the triangular initial displacement pattern. Indeed, a reconstruction made with only four normal modes is surprisingly accurate. On the other hand, a reconstruction made with only one normal mode is fairly inaccurate.

Figure 4.11 shows the time evolution of a uniform guitar string plucked at its mid-point. It can be seen that the string oscillates in a rather peculiar fashion. The initial kink in the string at $x = l/2$ splits into two equal kinks that propagate in opposite directions along the string, at the velocity v. The string remains straight and parallel to the x-axis between the kinks, and straight and inclined to the x-axis between each kink and the closest wall. When the two kinks reach the wall, the string is instantaneously found in its undisturbed position. The kinks then reflect off the two walls, with a phase change of π radians. When the two kinks meet again at $x = l/2$ the string is instantaneously found in a state that is an inverted form of its initial state. The kinks subsequently pass through one another, reflect off the walls, with another phase change of π radians, and meet for a second time at $x = l/2$. At this instant, the string is again found in its initial position. The pattern of motion then repeats itself ad infinitum. The period of the oscillation is the time required for a kink to propagate two string lengths, which is $\tau = 2\,l/v$. This is also the oscillation period of the $n = 1$ normal mode.

EXERCISES

4.1 Consider a uniformly-beaded string with N beads that is similar to that pictured in Figure 4.1, except that each end of the string is attached to a massless ring that slides (in the y-direction) on a frictionless rod.

(a) Demonstrate that the normal modes of the system take the form

$$y_{n,i}(t) = A_n \cos\left[\frac{n\,(i-1/2)}{N}\pi\right]\cos(\omega_n t - \phi_n),$$

where

$$\omega_n = 2\,\omega_0 \sin\left(\frac{n}{N}\frac{\pi}{2}\right),$$

ω_0 is as defined in Section 4.2, A_n and ϕ_n are constants, the integer $i = 1, N$ indexes the beads, and the mode number n indexes the modes.

(b) How many unique normal modes does the system possess, and what are their mode numbers?

(c) Show that the lowest frequency mode has an infinite wavelength and zero frequency. Explain this peculiar result.

4.2 Consider a uniformly-beaded string with N beads that is similar to that pictured in Figure 4.1, except that the left end of the string is fixed, and the right end is attached to a massless ring which slides (in the y-direction) on a frictionless rod. Find the normal modes and normal frequencies of the system.

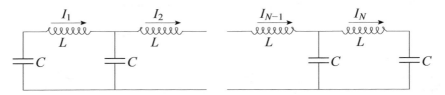

FIGURE 4.12   Figure for Exercise 4.3.

4.3 Figure 4.12 shows the left and right extremities of a linear LC network consisting of N identical inductors of inductance L, and $N + 1$ identical capacitors of capacitance C. Let the instantaneous current flowing through the ith inductor be $I_i(t)$, for $i = 1, N$. Demonstrate from Kirchhoff's circuital laws that the currents evolve in time according to the coupled equations

$$\ddot{I}_i = \omega_0^2\,(I_{i-1} - 2\,I_i + I_{i+1}),$$

for $i = 1, N$, where $\omega_0 = 1/\sqrt{LC}$, and $I_0 = I_{N+1} = 0$. Find the normal frequencies of the system.

4.4 Suppose that the outermost two capacitors in the circuit considered in the previous exercise are short-circuited. Find the new normal frequencies of the system.

4.5 A uniform string of length l, tension T, and mass per unit length ρ, is stretched between two immovable walls. Suppose that the string is initially in its equilibrium state. At $t = 0$ it is struck by a hammer in such a manner as to impart an impulsive velocity u_0 to a small segment of length $a < l$ centered on the mid-point. Find an expression for the subsequent motion of the string. Plot the motion as a function of time in a similar fashion to Figure 4.11, assuming that $a = l/10$.

4.6 A uniform string of length l, tension T, and mass per unit length ρ, is stretched between two massless rings, attached to its ends, that slide (in the y-direction) along frictionless rods.

(a) Demonstrate that the most general solution to the wave equation takes the form

$$y(x,t) = Y_0 + V_0 t + \sum_{n>0} A_n \cos\left(n\pi\frac{x}{l}\right)\cos\left(n\pi\frac{vt}{l} - \phi_n\right),$$

where $v = \sqrt{T/\rho}$, and Y_0, V_0, A_n, and ϕ_n are arbitrary constants.

(b) Show that

$$\frac{2}{l}\int_0^l \cos\left(n\pi\frac{x}{l}\right)\cos\left(n'\pi\frac{x}{l}\right)dx = \delta_{n,n'},$$

where n and n' are integers (that are not both zero).

(c) Use the previous result to demonstrate that the arbitrary constants in the previous solution can be determined from the initial conditions as follows:

$$Y_0 = \frac{1}{l}\int_0^l y_0(x)\,dx,$$

$$V_0 = \frac{1}{l}\int_0^l v_0(x)\,dx,$$

$$A_n = (C_n^2 + S_n^2)^{1/2},$$

$$\phi_n = \tan^{-1}(S_n/C_n),$$

where $y_0(x) \equiv y(x,0)$, $v_0(x) \equiv \partial y(x,0)/\partial t$, and

$$C_n = \frac{2}{l}\int_0^l y_0(x)\cos\left(n\pi\frac{x}{l}\right)dx,$$

$$S_n = \frac{2}{l}\int_0^l v_0(x)\cos\left(n\pi\frac{x}{l}\right)dx.$$

(d) Suppose that the string is initially in its equilibrium state. At $t = 0$ it is struck by a hammer in such a manner as to impart an impulsive velocity u_0 to a small segment of length $a < l$ centered on the mid-point. Find an expression for the subsequent motion of the string.

4.7 The linear LC circuit considered in Exercise 4.3 can be thought of as a discrete model of a uniform lossless transmission line (e.g., a co-axial cable). In this interpretation, $I_i(t)$ represents $I(x_i, t)$, where $x_i = i\,\delta x$. Moreover, $C = \mathcal{C}\,\delta x$, and $L = \mathcal{L}\,\delta x$, where $\mathcal{C}$ and $\mathcal{L}$ are the capacitance per unit length and the inductance per unit length of the line, respectively.

(a) Show that, in the limit $\delta x \to 0$, the evolution equation for the coupled currents given in Exercise 4.3 reduces to the wave equation

$$\frac{\partial^2 I}{\partial t^2} = v^2 \frac{\partial^2 I}{\partial x^2},$$

where $I = I(x,t)$, x measures distance along the line, and $v = 1/\sqrt{\mathcal{L}\mathcal{C}}$.

(b) If $V_i(t)$ is the potential difference (measured from the top to the bottom) across the $i + 1$th capacitor (from the left) in the circuit shown in Exercise 4.3, and $V(x, t)$ is the corresponding voltage in the transmission line, show that the discrete circuit equations relating the $I_i(t)$ and $V_i(t)$ reduce to

$$\frac{\partial V}{\partial t} = -\frac{1}{C}\frac{\partial I}{\partial x},$$

$$\frac{\partial I}{\partial t} = -\frac{1}{\mathcal{L}}\frac{\partial V}{\partial x},$$

in the transmission-line limit.

(c) Demonstrate that the voltage in a transmission line satisfies the wave equation

$$\frac{\partial^2 V}{\partial t^2} = v^2\frac{\partial^2 V}{\partial x^2}.$$

4.8 Consider a uniform string of length l, tension T, and mass per unit length ρ that is stretched between two immovable walls.

(a) Show that the total energy of the string, which is the sum of its kinetic and potential energies, is

$$E = \frac{1}{2}\int_0^l \left[\rho\left(\frac{\partial y}{\partial t}\right)^2 + T\left(\frac{\partial y}{\partial x}\right)^2\right]dx,$$

where $y(x, t)$ is the string's (relatively small) transverse displacement.

(b) The general motion of the string can be represented as a linear superposition of the normal modes; that is,

$$y(x, t) = \sum_{n=1,\infty} A_n \sin\left(n\pi\frac{x}{l}\right)\cos\left(n\pi\frac{vt}{l} - \phi_n\right),$$

where $v = \sqrt{T/\rho}$. Demonstrate that

$$E = \sum_{n=1,\infty} E_n,$$

where

$$E_n = \frac{1}{4}m\omega_n^2 A_n^2$$

is the energy of the nth normal mode. Here, $m = \rho l$ is the mass of the string, and $\omega_n = n\pi v/l$ the angular frequency of the nth normal mode.

CHAPTER 5

Longitudinal Standing Waves

5.1 INTRODUCTION

The aim of this chapter is to generalize the analysis of the previous chapter in order to deal with *longitudinal standing waves*. A standing wave is a disturbance in a physically continuous mechanical system that is periodic in space as well as in time, but which does not propagate. A longitudinal wave is one in which the direction of oscillation is parallel to the direction along which the phase of the waves varies sinusoidally.

5.2 SPRING-COUPLED MASSES

Consider a mechanical system consisting of a linear array of N identical masses m that are free to slide in one dimension over a frictionless horizontal surface. Suppose that the masses are coupled to their immediate neighbors via identical light springs of unstretched length a, and force constant K. (Here, we employ the symbol K to denote the spring force constant, rather than k, because k is already being used to denote wavenumber.) Let x measure distance along the array (from the left to the right). If the array is in its equilibrium state then the x-coordinate of the ith mass is $x_i = i\,a$, for $i = 1, N$. Consider longitudinal oscillations of the masses. Namely, oscillations for which the x-coordinate of the ith mass is

$$x_i = i\,a + \psi_i(t), \tag{5.1}$$

where $\psi_i(t)$ represents longitudinal displacement from equilibrium. It is assumed that all of the displacements are relatively small; that is, $|\psi_i| \ll a$, for $i = 1, N$.

Consider the equation of motion of the ith mass. See Figure 5.1. The extensions of the springs to the immediate left and right of the mass are $\psi_i - \psi_{i-1}$ and $\psi_{i+1} - \psi_i$, respectively. Thus, the x-directed forces that these springs exert on the mass are $-K(\psi_i - \psi_{i-1})$ and $K(\psi_{i+1} - \psi_i)$, respectively. The mass's equation of motion therefore becomes

$$\ddot{\psi}_i = \omega_0^2 \,(\psi_{i-1} - 2\,\psi_i + \psi_{i+1}), \tag{5.2}$$

where $\omega_0 = \sqrt{K/m}$. Because there is nothing special about the ith mass, the preceding equation is assumed to hold for all N masses; that is, for $i = 1, N$. Equation (5.2), which governs the longitudinal oscillations of a linear array of spring-coupled masses, is analogous in form to Equation (4.8), which governs the transverse oscillations of a beaded string. This observation suggests that longitudinal and transverse waves in discrete dynamical systems (i.e., systems with a finite number of degrees of freedom) can be described using the same mathematical equations.

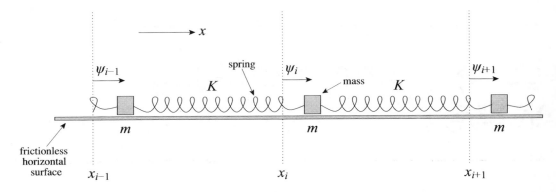

FIGURE 5.1   Detail of a system of spring-coupled masses.

We can interpret the quantities ψ_0 and ψ_{N+1}, that appear in the equations of motion for ψ_1 and ψ_N, respectively, as the longitudinal displacements of the left and right extremities of springs attached to the outermost masses in such a manner as to form the left and right boundaries of the array. The respective equilibrium positions of these extremities are $x_0 = 0$ and $x_{N+1} = (N+1)\,a$. The end displacements, ψ_0 and ψ_{N+1}, must be prescribed, otherwise Equations (5.2) do not constitute a complete set of equations. In other words, there are more unknowns than equations. The particular choice of ψ_0 and ψ_{N+1} depends on the nature of the physical boundary conditions at the two ends of the array. Suppose that the left extremity of the leftmost spring is anchored in an immovable wall. This implies that $\psi_0 = 0$; that is, the left extremity of the spring cannot move. Suppose, on the other hand, that the left extremity of the leftmost spring is not attached to anything. In this case, there is no reason for the spring to become extended, which implies that $\psi_0 = \psi_1$. In other words, if the left end of the array is fixed (i.e., attached to an immovable object) then $\psi_0 = 0$, and if the left end is free (i.e., not attached to anything) then $\psi_0 = \psi_1$. Likewise, if the right end of the array is fixed then $\psi_{N+1} = 0$, and if the right end is free then $\psi_{N+1} = \psi_N$.

Suppose, for the sake of argument, that the left end of the array is free, and the right end is fixed. It follows that $\psi_0 = \psi_1$, and $\psi_{N+1} = 0$. Let us search for normal modes of the general form

$$\psi_i(t) = A\,\cos[k\,(x_i - a/2)]\,\cos(\omega t - \phi), \qquad (5.3)$$

where $A > 0$, $k > 0$, $\omega > 0$, and ϕ are constants. The preceding expression automatically satisfies the boundary condition $\psi_0 = \psi_1$. This follows because $x_0 = 0$ and $x_1 = a$, and, consequently, $\cos[k\,(x_0 - a/2)] = \cos(-k\,a/2) = \cos(k\,a/2) = \cos[k\,(x_1 - a/2)]$. The other boundary condition, $\psi_{N+1} = 0$, is satisfied provided

$$\cos[k\,(x_{N+1} - a/2)] = \cos[(N + 1/2)\,k\,a] = 0, \qquad (5.4)$$

which yields [cf., Equation (4.18)]

$$k\,a = \frac{(n - 1/2)\,\pi}{N + 1/2}, \qquad (5.5)$$

where n is an integer. As before, the imposition of the boundary conditions causes a quantization of the possible mode wavenumbers. (See Section 4.2.) Finally, substitution of Equation (5.3) into Equation (5.2) gives the dispersion relation [cf., Equation (4.16)]

$$\omega = 2\,\omega_0\,\sin(k\,a/2). \qquad (5.6)$$

It follows, from the preceding analysis, that the longitudinal normal modes of a linear array of

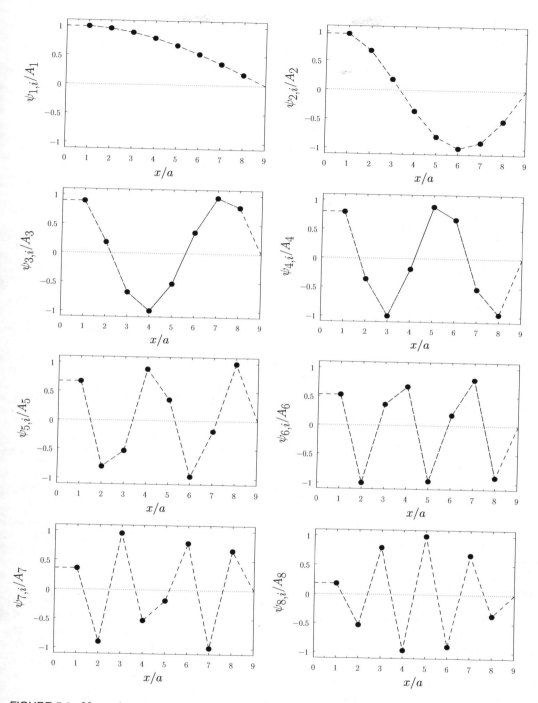

FIGURE 5.2 Normal modes of a system of eight spring-coupled masses with the left end free, and the right end fixed. The modes are all plotted at the instances in time when they attain their maximum amplitudes: namely, when $\cos(\omega_n t - \phi_n) = 1$.

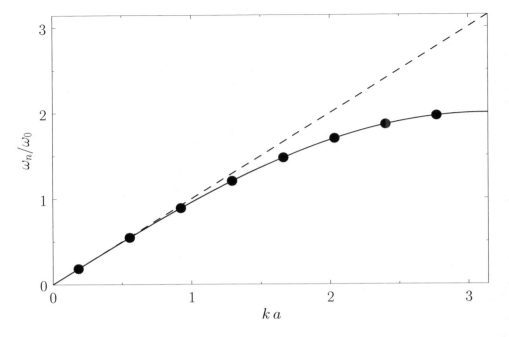

FIGURE 5.3 Normal frequencies of a system of eight spring-coupled masses with the left end free and the right end fixed.

spring-coupled masses, the left end of which is free, and the right end fixed, are associated with the following characteristic displacement patterns,

$$\psi_{n,i}(t) = A_n \cos\left[\frac{(n-1/2)(i-1/2)}{N+1/2}\pi\right]\cos(\omega_n - \phi_n), \tag{5.7}$$

where

$$\omega_n = 2\omega_0 \sin\left(\frac{n-1/2}{N+1/2}\frac{\pi}{2}\right), \tag{5.8}$$

and the A_n and ϕ_n are arbitrary constants determined by the initial conditions. Here, the integer $i = 1, N$ indexes the masses, and the mode number n indexes the normal modes. It can be demonstrated that there are only N unique normal modes, corresponding to mode numbers in the range 1 to N.

Figures 5.2 and 5.3 display the normal modes and normal frequencies of a linear array of eight spring-coupled masses, the left end of which is free, and the right end fixed. The data shown in these figures is obtained from Equations (5.7) and (5.8), respectively, with $N = 8$. It can be seen that normal modes with small wavenumbers—that is, $ka \ll 1$, so that $n \ll N$—have displacements that vary in a fairly smooth sinusoidal manner from mass to mass, and oscillation frequencies that increase approximately linearly with increasing wavenumber. On the other hand, normal modes with large wavenumbers—that is, $ka \sim 1$, so that $n \sim N$—have displacements that exhibit large variations from mass to mass, and oscillation frequencies that do not depend linearly on wavenumber. We conclude that the longitudinal normal modes of an array of spring-coupled masses have analogous properties to the transverse normal modes of a beaded string. (See Section 4.2.)

The dynamical system pictured in Figure 5.1 can be used to model the effect of a planar *sound wave* (i.e., a longitudinal oscillation in position that is periodic in space) on a crystal lattice. In this application, the masses represent parallel planes of atoms, the springs represent the interatomic forces acting between these planes, and the longitudinal oscillations represent the sound wave. Of course, a macroscopic crystal contains a great many atomic planes, so we would expect N to be very large. However, according to Equations (5.5) and (5.8), no matter how large N becomes, $k a$ cannot exceed π (because n cannot exceed N), and ω_n cannot exceed $2\omega_0$. In other words, there is a minimum wavelength that a sound wave in a crystal lattice can have, which turns out to be twice the interplane spacing, and a corresponding maximum oscillation frequency. For waves whose wavelengths are much greater than the interplane spacing (i.e., $k a \ll 1$), the dispersion relation (5.6) reduces to

$$\omega \simeq k v \tag{5.9}$$

where $v = \omega_0\, a = (K/m)^{1/2}\, a$ is a constant that has the dimensions of velocity. It seems plausible that Equation (5.9) is the dispersion relation for sound waves in a continuous elastic medium. Let us investigate such waves.

5.3 LONGITUDINAL WAVES ON THIN ELASTIC ROD

Consider a thin uniform elastic rod of length l and cross-sectional area A. Let us examine the longitudinal oscillations of such a rod. These oscillations are often, somewhat loosely, referred to as sound waves. It is again convenient to let x denote position along the rod. Thus, in equilibrium, the two ends of the rod lie at $x = 0$ and $x = l$. Suppose that a longitudinal wave causes an x-directed displacement $\psi(x, t)$ of the various elements of the rod from their equilibrium positions. Consider a thin section of the rod, of length δx, lying between $x - \delta x/2$ and $x + \delta x/2$. The displacements of the left and right boundaries of the section are $\psi(x - \delta x/2, t)$ and $\psi(x + \delta x/2, t)$, respectively. Thus, the change in length of the section, due to the action of the wave, is $\psi(x + \delta x/2, t) - \psi(x - \delta x/2, t)$. Now, *strain* in an elastic rod is defined as change in length over unperturbed length (Love 1944). Thus, the strain in the section of the rod under consideration is

$$\epsilon(x, t) = \frac{\psi(x + \delta x/2, t) - \psi(x - \delta x/2, t)}{\delta x}. \tag{5.10}$$

In the limit $\delta x \to 0$, this becomes

$$\epsilon(x, t) = \frac{\partial \psi(x, t)}{\partial x}. \tag{5.11}$$

It is assumed that the strain is small compared to unity; that is, $|\epsilon| \ll 1$. *Stress*, $\sigma(x, t)$, in an elastic rod is defined as the elastic force per unit cross-sectional area (ibid.). In a conventional elastic material, the relationship between stress and strain (for relatively small strains) takes the simple form

$$\sigma = Y \epsilon. \tag{5.12}$$

Here, Y is a constant, with the dimensions of pressure, that is known as the *Young's modulus* (ibid.). If the strain in a given element is positive then the stress acts to lengthen the element, and vice versa. (Similarly, in the spring-coupled mass system investigated in the preceding section, the external forces exerted on an individual spring act to lengthen it when its extension is positive, and vice versa.)

Consider the motion of a thin section of the rod lying between $x - \delta x/2$ and $x + \delta x/2$. If ρ is the mass density of the rod then the section's mass is $\rho A\, \delta x$. The stress acting on the left boundary of the section is $\sigma(x - \delta x/2) = Y \epsilon(x - \delta x/2)$. Because stress is force per unit area, the force acting on the left boundary is $A Y \epsilon(x - \delta x/2)$. This force is directed in the minus x-direction, assuming that the strain is positive (i.e., the force acts to lengthen the section). Likewise, the force acting on the right boundary of the section is $A Y \epsilon(x + \delta x/2)$, and is directed in the positive x-direction, assuming that the strain is positive (i.e., the force again acts to lengthen the section). Finally, the mean longitudinal (i.e., x-directed) acceleration of the section is $\partial^2 \psi(x, t)/\partial t^2$. Hence, the section's longitudinal equation of motion becomes

$$\rho\, A\, \delta x\, \frac{\partial^2 \psi(x, t)}{\partial t^2} = A\, Y\, [\epsilon(x + \delta x/2, t) - \epsilon(x - \delta x/2, t)]. \tag{5.13}$$

In the limit $\delta x \to 0$, this expression reduces to

$$\rho\, \frac{\partial^2 \psi(x, t)}{\partial t^2} = Y\, \frac{\partial \epsilon(x, t)}{\partial x}, \tag{5.14}$$

or

$$\frac{\partial^2 \psi}{\partial t^2} = v^2\, \frac{\partial^2 \psi}{\partial x^2}, \tag{5.15}$$

where

$$v = \sqrt{\frac{Y}{\rho}} \tag{5.16}$$

is a constant having the dimensions of velocity, which turns out to be the propagation speed of longitudinal waves along the rod (see Section 6.2), and use has been made of Equation (5.11). Equation (5.15) has the same mathematical form as Equation (4.30), which governs the motion of transverse waves on a uniform string. This implies that longitudinal and transverse waves in continuous dynamical systems (i.e., systems with an infinite number of degrees of freedom) can be described using the same mathematical equation.

In order to solve Equation (5.15), we need to specify boundary conditions at the two ends of the rod. Suppose that the left end of the rod is fixed; that is, it is clamped in place such that it cannot move. This implies that $\psi(0, t) = 0$. Suppose, on the other hand, that the left end of the rod is free; that is, it is not attached to anything. This implies that $\sigma(0, t) = 0$, because there is nothing that the end can exert a force (or a stress) on, and vice versa. It follows from Equations (5.11) and (5.12) that $\partial \psi(0, t)/\partial x = 0$. Likewise, if the right end of the rod is fixed then $\psi(l, t) = 0$, and if the right end is free then $\partial \psi(l, t)/\partial x = 0$.

Suppose, for the sake of argument, that the left end of the rod is free, and the right end is fixed. It follows that $\partial \psi(0, t)/\partial x = 0$, and $\psi(l, t) = 0$. Let us search for normal modes of the form

$$\psi(x, t) = A\, \cos(k\, x)\, \cos(\omega\, t - \phi), \tag{5.17}$$

where $A > 0$, $k > 0$, $\omega > 0$, and ϕ are constants. The preceding expression automatically satisfies the boundary condition $\partial \psi(0, t)/\partial x = 0$. The other boundary condition is satisfied provided

$$\cos(k\, l) = 0, \tag{5.18}$$

which yields

$$k\, l = (n - 1/2)\, \pi, \tag{5.19}$$

where n is an integer. As usual, the imposition of the boundary conditions leads to a quantization of the possible mode wavenumbers. Substitution of Equation (5.17) into the equation of motion, (5.15), yields the normal mode dispersion relation

$$\omega = k\, v. \tag{5.20}$$

The preceding dispersion relation is consistent with the previously derived dispersion relation (5.9), given that $m = \rho\, A\, a$ and $K = A\, Y/a$. Here, a is the interplane spacing, m the mass of a section of the rod containing a single plane of atoms, and K the effective force constant between neighboring atomic planes.

It follows, from the previous analysis, that the nth longitudinal normal mode of an elastic rod, of length l, whose left end is free, and whose right end is fixed, is associated with the characteristic displacement pattern

$$\psi_n(x,t) = A_n \cos\left[(n-1/2)\pi \frac{x}{l}\right] \cos(\omega_n t - \phi_n), \tag{5.21}$$

where

$$\omega_n = (n-1/2)\frac{\pi v}{l}. \tag{5.22}$$

Here, A_n and ϕ_n are constants that are determined by the initial conditions. It can be demonstrated that only those normal modes whose mode numbers are positive integers yield unique displacement patterns. Equation (5.21) describes a standing wave whose nodes (i.e., points where $\psi = 0$ for all t) are evenly spaced a distance $l/(n-1/2)$ apart. The boundary condition $\psi(l,t) = 0$ ensures that the right end of the rod is always coincident with a node. On the other hand, the boundary condition $\partial\psi(0,t)/\partial x = 0$ ensures that the left hand of the rod is always coincident with a point of maximum amplitude oscillation [i.e., a point where $\cos(kx) = \pm 1$]. Such a point is known as an *anti-node*. The anti-nodes associated with a given normal mode lie halfway between the corresponding nodes. According to Equation (5.22), the normal mode oscillation frequencies depend linearly on mode number. Finally, it can be shown that, in the long wavelength limit $ka \ll 1$, the normal modes and normal frequencies of a uniform elastic rod specified in Equations (5.21) and (5.22) are analogous to the normal modes and normal frequencies of a linear array of identical spring-coupled masses specified in Equations (5.7) and (5.8), and pictured in Figures 5.2 and 5.3.

Because Equation (5.15) is linear, its most general solution is a linear combination of all of the normal modes; that is,

$$\psi(x,t) = \sum_{n'=1,\infty} A_{n'} \cos\left[(n'-1/2)\pi \frac{x}{l}\right] \cos\left[(n'-1/2)\pi \frac{vt}{l} - \phi_{n'}\right]. \tag{5.23}$$

The constants A_n and ϕ_n are determined from the initial displacement,

$$\psi(x,0) = \sum_{n'=1,\infty} A_{n'} \cos\phi_{n'} \cos\left[(n'-1/2)\pi \frac{x}{l}\right], \tag{5.24}$$

and the initial velocity,

$$\dot{\psi}(x,0) = \frac{\pi v}{l} \sum_{n'=1,\infty} (n'-1/2) A_{n'} \sin\phi_{n'} \cos\left[(n'-1/2)\pi \frac{x}{l}\right]. \tag{5.25}$$

It can be demonstrated that [cf., Equation (4.53)]

$$\frac{2}{l} \int_0^l \cos\left[(n-1/2)\pi \frac{x}{l}\right] \cos\left[(n'-1/2)\pi \frac{x}{l}\right] dx = \delta_{n,n'}. \tag{5.26}$$

Thus, multiplying Equation (5.24) by $(2/l)\cos[(n-1/2)\pi x/l]$, and integrating in x from 0 to l, we obtain

$$C_n = \frac{2}{l} \int_0^l \psi(x,0) \cos\left[(n-1/2)\pi \frac{x}{l}\right] dx = A_n \cos\phi_n, \tag{5.27}$$

where use has been made of Equations (5.26) and (4.54). Likewise, Equation (5.25) gives

$$S_n = \frac{2}{v(n-1/2)\pi} \int_0^l \dot{\psi}(x,0) \cos\left[(n-1/2)\pi \frac{x}{l}\right] dx = A_n \sin\phi_n. \tag{5.28}$$

Finally, $A_n = (C_n^2 + S_n^2)^{1/2}$ and $\phi_n = \tan^{-1}(S_n/C_n)$.

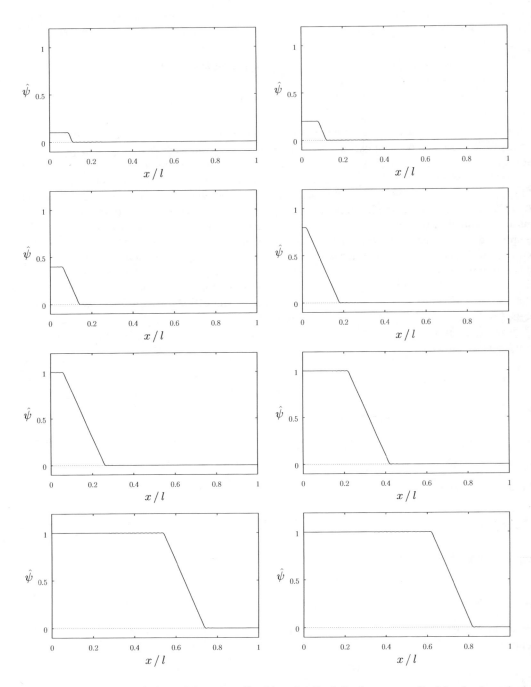

FIGURE 5.4 Time evolution of the normalized longitudinal displacement of a thin elastic rod whose left end is free (and is struck with a hammer at $t = 0$), and whose right end is fixed. The top-left, top-right, middle-left, middle-right, bottom-left, and bottom-right panels correspond to $t/\tau = 0.01$, 0.02, 0.04, 0.08, 0.16, 0.32, 0.64, and 1.28, respectively.

Suppose, for the sake of example, that the rod is initially at rest, and that its left end is hit with a hammer at $t = 0$ in such a manner that a section of the rod lying between $x = 0$ and $x = a$ (where $a < l$) acquires an instantaneous velocity V_0. It follows that $\psi(x,0) = 0$. Furthermore, $\dot\psi(x,0) = V_0$ if $0 \le x \le a$, and $\dot\psi(x,0) = 0$ otherwise. It can be demonstrated that these initial conditions yield $C_n = 0$, $\phi_n = \pi/2$,

$$A_n = S_n = \frac{V_0\,a}{v}\,\frac{2}{\pi}\,\frac{\sin[(n-1/2)\,\pi\,a/l]}{(n-1/2)^2\,\pi\,a/l}, \tag{5.29}$$

and

$$\psi(x,t) = \sum_{n=1,\infty} A_n \cos\left[(n-1/2)\pi\frac{x}{l}\right]\sin\left[(n-1/2)\pi\frac{t}{\tau}\right], \tag{5.30}$$

where $\tau = l/v$. Figure 5.4 shows the time evolution of the normalized rod displacement, $\hat\psi(x,t) = (v/V_0\,a)\,\psi(x,t)$, calculated from the preceding equations using the first 100 normal modes (i.e., $n = 1, 100$), and choosing $a/l = 0.1$. It can be seen that the hammer blow generates a displacement wave that initially develops at the free end of the rod ($x/l = 0$), which is the end that is struck, propagates along the rod at the velocity v, and reflects off the fixed end ($x/l = 1$) at time $t/\tau = 1$ with no phase shift. (The wave front is traveling from the left to the right in all panels except the final one, where it is traveling from right to left.)

5.4 SOUND WAVES IN IDEAL GAS

Consider a uniform ideal gas of equilibrium mass density ρ and equilibrium pressure p. Let us investigate the longitudinal oscillations of such a gas. Such oscillations are known as sound waves. Generally speaking, a sound wave in an ideal gas oscillates sufficiently rapidly that heat is unable to flow fast enough to smooth out any temperature perturbations generated by the wave. Under these circumstances, the gas obeys the *adiabatic gas law* (Reif 2008),

$$p\,V^\gamma = \text{constant}, \tag{5.31}$$

where p is the pressure, V the volume, and γ the *ratio of specific heats* (i.e., the ratio of the gas's specific heat at constant pressure to its specific heat at constant volume) (ibid.). This ratio is 1.40 for ordinary air (Haynes and Lide 2011b).

Consider a sound wave in a column of gas of cross-sectional area A. Let x measure distance along the column. Suppose that the wave generates an x-directed displacement of the column, $\psi(x,t)$. Consider a small section of the column lying between $x - \delta x/2$ and $x + \delta x/2$. The change in volume of the section is $\delta V = A\,[\psi(x + \delta x/2, t) - \psi(x - \delta x/2, t)]$. Hence, the relative change in volume, which is assumed to be small compared to unity, is

$$\frac{\delta V}{V} = \frac{A\,[\psi(x + \delta x/2, t) - \psi(x - \delta x/2, t)]}{A\,\delta x}. \tag{5.32}$$

In the limit $\delta x \to 0$, this becomes

$$\frac{\delta V(x,t)}{V} = \frac{\partial \psi(x,t)}{\partial x}. \tag{5.33}$$

The pressure perturbation $\delta p(x,t)$ associated with the volume perturbation $\delta V(x,t)$ follows from Equation (5.31), which yields

$$(p + \delta p)(V + \delta V)^\gamma = p\,V^\gamma, \tag{5.34}$$

or

$$(1 + \delta p/p)(1 + \delta V/V)^\gamma \simeq 1 + \delta p/p + \gamma\,\delta V/V = 1, \tag{5.35}$$

giving

$$\delta p = -\gamma p \frac{\delta V}{V} = -\gamma p \frac{\partial \psi}{\partial x}, \tag{5.36}$$

where use has been made of Equation (5.33). Here, we have neglected terms that are second order, or higher, in the small quantities $\delta V/V$ and $\delta p/p$.

Consider a section of the gas column lying between $x - \delta x/2$ and $x + \delta x/2$. The mass of this section is $\rho A \, \delta x$. The x-directed force acting on its left boundary is $A \, [p + \delta p(x - \delta x/2, t)]$, whereas the x-directed force acting on its right boundary is $-A \, [p + \delta p(x + \delta x/2, t)]$. Finally, the average longitudinal (i.e., x-directed) acceleration of the section is $\partial^2 \psi(x, t)/\partial t^2$. Thus, the section's longitudinal equation of motion is written

$$\rho A \, \delta x \frac{\partial^2 \psi(x, t)}{\partial t^2} = -A \, [\delta p(x + \delta x/2, t) - \delta p(x - \delta x/2, t)]. \tag{5.37}$$

In the limit $\delta x \to 0$, this equation reduces to

$$\rho \frac{\partial^2 \psi(x, t)}{\partial t^2} = -\frac{\partial \, \delta p(x, t)}{\partial x}. \tag{5.38}$$

Finally, Equation (5.36) yields

$$\frac{\partial^2 \psi}{\partial t^2} = v^2 \frac{\partial^2 \psi}{\partial x^2}, \tag{5.39}$$

where

$$v = \sqrt{\frac{\gamma p}{\rho}} \tag{5.40}$$

is a constant with the dimensions of velocity, which turns out to be the sound speed in the gas. (See Section 6.2.)

As an example, suppose that a standing wave is excited in a uniform organ pipe of length l. Let the closed end of the pipe lie at $x = 0$, and the open end at $x = l$. The standing wave satisfies the wave equation (5.39), where v represents the speed of sound in air. The boundary conditions are that $\psi(0, t) = 0$—that is, there is zero longitudinal displacement of the air at the closed end of the pipe—and $\partial \psi(l, t)/\partial x = 0$—that is, there is zero pressure perturbation at the open end of the pipe (because the small pressure perturbation in the pipe is not intense enough to modify the pressure of the atmosphere external to the pipe). Let us write the displacement pattern associated with the standing wave in the form

$$\psi(x, t) = A \, \sin(k \, x) \, \cos(\omega t - \phi), \tag{5.41}$$

where $A > 0$, $k > 0$, $\omega > 0$, and ϕ are constants. This expression automatically satisfies the boundary condition $\psi(0, t) = 0$. The other boundary condition is satisfied provided

$$\cos(k \, l) = 0, \tag{5.42}$$

which yields

$$k \, l = (n - 1/2) \, \pi, \tag{5.43}$$

where the mode number n is a positive integer. Equations (5.39) and (5.41) give the dispersion relation

$$\omega = k \, v. \tag{5.44}$$

Hence, the nth normal mode has a wavelength

$$\lambda_n = \frac{4 \, l}{2 \, n - 1}, \tag{5.45}$$

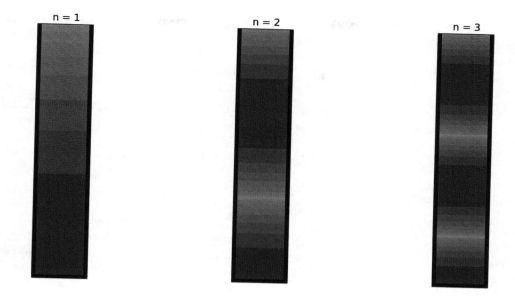

FIGURE 5.5 First three normal modes of an organ pipe with one open end. The plots show contours of the magnitude of the pressure perturbation associated with each mode. Dark/light contours correspond to high/low magnitude.

and an oscillation frequency (in hertz)

$$f_n = (2n - 1) f_1,$$ (5.46)

where $f_1 = v/(4l)$ is the frequency of the fundamental harmonic (i.e., the normal mode with the lowest oscillation frequency). Figure 5.5 illustrates the characteristic displacement patterns and oscillation frequencies of the pipe's first three normal modes (i.e., $n = 1, 2,$ and 3). In fact, the figure plots the magnitude of the pressure perturbation associated with each mode. It can be seen that the modes all have an anti-node in the pressure (which corresponds to a node in the displacement, and vice versa) at the closed end of the pipe, and a node at the open end. The fundamental harmonic has a wavelength that is four times the length of the pipe. The first overtone harmonic has a wavelength that is 4/3rds the length of the pipe, and a frequency that is three times that of the fundamental. Finally, the second overtone has a wavelength that is 4/5ths the length of the pipe, and a frequency that is five times that of the fundamental. By contrast, the normal modes of a guitar string have nodes at either end of the string. (See Figure 4.6.) Thus, the fundamental harmonic has a wavelength that is twice the length of the string. The first overtone harmonic has a wavelength that is the length of the string, and a frequency that is twice that of the fundamental. Finally, the second overtone harmonic has a wavelength that is 2/3rds the length of the string, and a frequency that is three times that of the fundamental.

5.5 FOURIER ANALYSIS

Playing a musical instrument, such as a guitar or an organ, generates a set of standing waves that cause a sympathetic oscillation in the surrounding air. Such an oscillation consists of a fundamental harmonic, whose frequency determines the pitch of the musical note heard by the listener, accompanied by a set of overtone harmonics that determine the timbre of the note. By definition, the oscillation frequencies of the overtone harmonics are integer multiples of that of the fundamental. Thus, we expect the pressure perturbation generated in a listener's ear to have the general form

$$\delta p(t) = \sum_{n=1,\infty} A_n \cos(n\omega t - \phi_n), \tag{5.47}$$

where ω is the angular frequency of the fundamental (i.e., $n = 1$) harmonic, and the A_n and ϕ_n are the amplitudes and phases of the various harmonics. The preceding expression can also be written

$$\delta p(t) = \sum_{n=1,\infty} [C_n \cos(n\omega t) + S_n \sin(n\omega t)], \tag{5.48}$$

where $C_n = A_n \cos\phi_n$ and $S_n = A_n \sin\phi_n$. The function $\delta p(t)$ is periodic in time with period $\tau = 2\pi/\omega$. In other words, $\delta p(t + \tau) = \delta p(t)$ for all t. This follows because of the mathematical identities $\cos(\theta + n\,2\pi) \equiv \cos\theta$ and $\sin(\theta + n\,2\pi) \equiv \sin\theta$, where n is an integer. [Moreover, there is no $\tau' < \tau$ for which $\delta p(t + \tau') = \delta p(t)$ for all t.] Can any periodic waveform be represented as a linear superposition of sine and cosine waveforms, whose periods are integer subdivisions of that of the waveform, such as that specified in Equation (5.48)? To put it another way, given an arbitrary periodic waveform $\delta p(t)$, can we uniquely determine the constants C_n and S_n appearing in Equation (5.48)? It turns out that we can. Incidentally, the decomposition of a periodic waveform into a linear superposition of sinusoidal waveforms is commonly known as *Fourier analysis*.

The problem under investigation is as follows. Given a periodic waveform $y(t)$, where $y(t + \tau) = y(t)$ for all t, we need to determine the constants C_n and S_n in the expansion

$$y(t) = \sum_{n'=1,\infty} [C_{n'} \cos(n'\omega t) + S_{n'} \sin(n'\omega t)], \tag{5.49}$$

where

$$\omega = \frac{2\pi}{\tau}. \tag{5.50}$$

It can be demonstrated that [cf., Equation (4.53)]

$$\frac{2}{\tau} \int_0^\tau \cos(n\omega t) \cos(n'\omega t)\,dt = \delta_{n,n'}, \tag{5.51}$$

$$\frac{2}{\tau} \int_0^\tau \sin(n\omega t) \sin(n'\omega t)\,dt = \delta_{n,n'}, \tag{5.52}$$

$$\frac{2}{\tau} \int_0^\tau \cos(n\omega t) \sin(n'\omega t)\,dt = 0, \tag{5.53}$$

where n and n' are positive integers. Thus, multiplying Equation (5.49) by $(2/\tau)\cos(n\omega t)$, and then integrating over t from 0 to τ, we obtain

$$C_n = \frac{2}{\tau} \int_0^\tau y(t) \cos(n\omega t)\,dt, \tag{5.54}$$

where use has been made of Equation (5.51)–(5.53), as well as Equation (4.54). Likewise, multiplying Equation (5.49) by $(2/\tau)\sin(n\omega t)$, and then integrating over t from 0 to τ, we obtain

$$S_n = \frac{2}{\tau} \int_0^\tau y(t) \sin(n\omega t)\,dt. \tag{5.55}$$

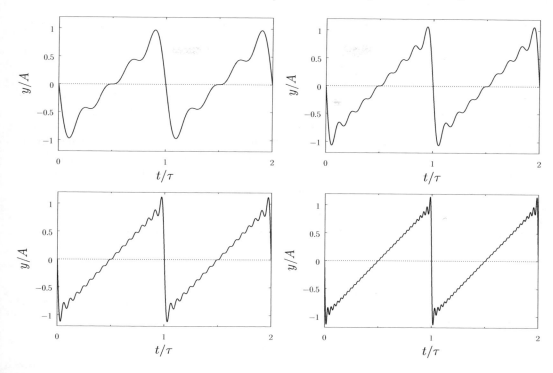

FIGURE 5.6 Fourier reconstruction of a periodic sawtooth waveform. The top-left, top-right, bottom-left, and bottom-right panels correspond to reconstructions using 4, 8, 32, and 64 terms, respectively, in the Fourier series.

Hence, we have uniquely determined the constants C_n and S_n in the expansion (5.49). These constants are generally known as *Fourier coefficients*, whereas the expansion itself is known as either a *Fourier expansion* or a *Fourier series*.

In principle, there is no restriction on the waveform $y(t)$ in the previous analysis, other than the requirement that it be periodic in time. In other words, we ought to be able to Fourier analyze any periodic waveform. Let us see how this works. Consider the periodic sawtooth waveform

$$y(t) = A\,(2\,t/\tau - 1) \quad 0 \le t/\tau \le 1, \tag{5.56}$$

with $y(t + \tau) = y(t)$ for all t. See Figure 5.6. This waveform rises linearly from an initial value $-A$ at $t = 0$ to a final value $+A$ at $t = \tau$, discontinuously jumps back to its initial value, and then repeats ad infinitum. According to Equations (5.54) and (5.55), the Fourier harmonics of the waveform are

$$C_n = \frac{2}{\tau} \int_0^\tau A\,(2\,t/\tau - 1)\,\cos(n\,\omega\,t)\,dt = \frac{A}{\pi^2} \int_0^{2\pi} (\theta - \pi)\,\cos(n\,\theta)\,d\theta, \tag{5.57}$$

$$S_n = \frac{2}{\tau} \int_0^\tau A\,(2\,t/\tau - 1)\,\sin(n\,\omega\,t)\,dt = \frac{A}{\pi^2} \int_0^{2\pi} (\theta - \pi)\,\sin(n\,\theta)\,d\theta, \tag{5.58}$$

where $\theta = \omega\,t$. Integration by parts (Riley 1974) yields

$$C_n = 0, \tag{5.59}$$

$$S_n = -\frac{2\,A}{n\,\pi}. \tag{5.60}$$

Hence, the Fourier reconstruction of the waveform is written

$$y(t) = -\frac{2A}{\pi} \sum_{n=1,\infty} \frac{\sin(n\,2\pi\,t/\tau)}{n}. \tag{5.61}$$

Given that the Fourier coefficients fall off like $1/n$, as n increases, it seems plausible that the preceding series can be truncated after a finite number of terms without unduly affecting the reconstructed waveform. Figure 5.6 shows the result of truncating the series after 4, 8, 16, and 32 terms. It can be seen that the reconstruction becomes increasingly accurate as the number of terms retained in the series increases. The annoying oscillations in the reconstructed waveform at $t = 0$, τ, and $2\,\tau$ are known as *Gibbs' phenomena*, and are the inevitable consequence of trying to represent a discontinuous waveform as a Fourier series (Riley 1974). In fact, it can be demonstrated mathematically that, no matter how many terms are retained in the series, the Gibbs' phenomena never entirely go away (Zygmund 1955).

We can slightly generalize the Fourier series (5.49) by including an $n = 0$ term. In other words,

$$y(t) = C_0 + \sum_{n'=1,\infty} [C_{n'}\cos(n'\,\omega\,t) + S_{n'}\sin(n'\,\omega\,t)], \tag{5.62}$$

which allows the waveform to have a non-zero average. There is no term involving S_0, because $\sin(n\,\omega\,t) = 0$ when $n = 0$. It can be demonstrated that

$$\frac{2}{\tau}\int_0^\tau \cos(n\,\omega\,t)\,dt = 0, \tag{5.63}$$

$$\frac{2}{\tau}\int_0^\tau \sin(n\,\omega\,t)\,dt = 0, \tag{5.64}$$

where $\omega = 2\pi/\tau$, and n is a positive integer. Making use of the preceding expressions, as well as Equations (5.51)–(5.53), we can show that

$$C_0 = \frac{1}{\tau}\int_0^\tau y(t)\,dt, \tag{5.65}$$

and also that Equations (5.54) and (5.55) still hold for $n > 0$.

As an example, consider the periodic "tent" waveform

$$y(t) = 2A \begin{cases} t/\tau & 0 \le t/\tau \le 1/2 \\ 1 - t/\tau & 1/2 < t/\tau \le 1 \end{cases}, \tag{5.66}$$

where $y(t+\tau) = y(t)$ for all t. See Figure 5.7. This waveform rises linearly from zero at $t = 0$, reaches a peak value A at $t = \tau/2$, falls linearly, becomes zero again at $t = \tau$, and repeats ad infinitum. Moreover, the waveform has a non-zero average. It can be demonstrated, from Equations (5.54), (5.55), (5.65), and (5.66), that

$$C_0 = \frac{A}{2}, \tag{5.67}$$

and

$$C_n = -A\,\frac{\sin^2(n\,\pi/2)}{(n\,\pi/2)^2} \tag{5.68}$$

for $n > 1$, with $S_n = 0$ for $n > 1$. In fact, only the odd-n Fourier harmonics are non-zero. Thus,

$$y(t) = \frac{A}{2} - \frac{4A}{\pi^2} \sum_{n=1,3,5,\cdots} \frac{\cos(n\,2\pi\,t/\tau)}{n^2}. \tag{5.69}$$

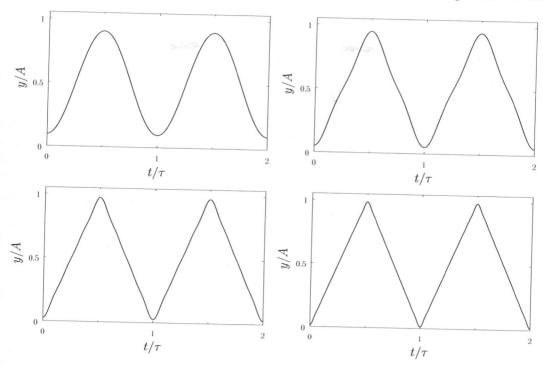

FIGURE 5.7 Fourier reconstruction of a periodic "tent" waveform. The top-left, top-right, bottom-left, and bottom-right panels correspond to reconstruction using the first 1, 2, 4, and 8 terms, respectively, in the Fourier series (in addition to the C_0 term).

Figure 5.7 shows a Fourier reconstruction of the "tent" waveform using the first 1, 2, 4, and 8 terms (in addition to the C_0 term) in the Fourier series The reconstruction becomes increasingly accurate as the number of terms in the series increases. Moreover, in this example, there is no sign of Gibbs' phenomena, because the tent waveform is completely continuous.

In our first example—that is, the sawtooth waveform—all of the C_n Fourier coefficients are zero, whereas in our second example—that is, the tent waveform—all of the S_n coefficients are zero. This occurs because the sawtooth waveform is odd in t—that is, $y(-t) = -y(t)$ for all t—whereas the tent waveform is even—that is, $y(-t) = y(t)$ for all t. It is a general rule that waveforms that are even in t only have cosines in their Fourier series, whereas waveforms that are odd only have sines (Riley 1974). Waveforms that are neither even nor odd in t have both cosines and sines in their Fourier series.

Fourier series arise quite naturally in the theory of standing waves, because the normal modes of oscillation of any uniform continuous system possessing linear equations of motion (e.g., a uniform string, an elastic rod, an ideal gas) take the form of spatial cosine and sine waves whose wavelengths are rational fractions of one another. Thus, the instantaneous spatial waveform of such a system can always be represented as a linear superposition of cosine and sine waves; that is, a Fourier series in space, rather than in time. In fact, the process of determining the amplitudes and phases of the normal modes of oscillation from the initial conditions is essentially equivalent to Fourier analyzing the initial conditions in space. (See Sections 4.4 and 5.3.)

EXERCISES

5.1 Estimate the highest possible frequency (in hertz), and the smallest possible wavelength, of a longitudinal wave in a thin aluminum rod, due to the discrete atomic structure of this material. The mass density, Young's modulus, and atomic weight of aluminum are $2.7 \times 10^3 \, \text{kg m}^{-3}$, $7 \times 10^{10} \, \text{N m}^{-2}$, and 27, respectively. (Assume, for the sake of simplicity, that aluminum has a simple cubic lattice.)

5.2 A simple model of an ionic crystal consists of a linear array of a great many equally-spaced atoms of alternating masses M and m, where $m < M$. The masses are connected by identical chemical bonds that are modeled as springs of spring constant K.

   (a) Show that the frequencies of the system's longitudinal modes of vibration either lie in the band 0 to $(2 \, K/M)^{1/2}$ or in the band $(2 \, K/m)^{1/2}$ to $[2 \, K \, (1/M + 1/m)]^{1/2}$.

   (b) Show that, in the long-wavelength limit, modes whose frequencies lie in the lower band are such that neighboring atomics move in the same direction, whereas modes whose frequencies lie in the upper band are such that neighboring atoms move in opposite directions. The lower band is known as the *acoustic branch*, whereas the upper band is known as the *optical branch*.

5.3 Consider a linear array of N identical simple pendula of mass m and length l that are suspended from equal-height points, evenly-spaced a distance a apart. Suppose that each pendulum bob is attached to its two immediate neighbors by means of light springs of unstretched length a and spring constant K. Figure 5.8 shows a small part of such an array. Let $x_i = i \, a$ be the equilibrium position of the ith bob, for $i = 1, N$, and let $\psi_i(t)$ be its horizontal displacement. It is assumed that $|\psi_i|/a \ll 1$ for all i.

   (a) Demonstrate that the equation of motion of the ith pendulum bob is

   $$\ddot{\psi}_i = -\frac{g}{l} \psi_i + \frac{K}{m} (\psi_{i-1} - 2 \psi_i + \psi_{i+1}).$$

   (b) Consider a general normal mode of the form

   $$\psi_i(t) = [A \, \sin(k \, x_i) + B \, \cos(k \, x_i)] \, \cos(\omega t - \phi).$$

   Show that the associated dispersion relation is

   $$\omega^2 = \frac{g}{l} + \frac{4 \, K}{m} \sin^2(k \, a/2).$$

   (c) Suppose that the first and last pendulums in the array are attached to immovable walls, located a horizontal distance a away, by means of light springs of unstretched length a and spring constant K. Find the normal modes of the system.

   (d) Suppose that the first and last pendulums are not attached to anything on their outer sides. Find the normal modes of the system.

5.4 Consider a periodic waveform $y(t)$ of period τ, where $y(t + \tau) = y(t)$ for all t, which is represented as a Fourier series,

$$y(t) = C_0 + \sum_{n>0} [C_n \, \cos(n \, \omega t) + S_n \, \sin(n \, \omega t)],$$

where $\omega = 2\pi/\tau$.

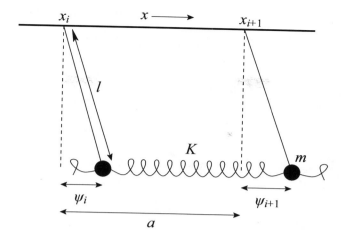

FIGURE 5.8 Figure for Exercise 5.3.

(a) Demonstrate that

$$y(-t) = C_0 + \sum_{n>0} [C'_n \cos(n\,\omega\,t) + S'_n \sin(n\,\omega\,t)],$$

where $C'_n = C_n$ and $S'_n = -S_n$.

(b) Show that

$$y(t+T) = C_0 + \sum_{n>0} [C''_n \cos(n\,\omega\,t) + S''_n \sin(n\,\omega\,t)],$$

where

$$C''_n = C_n \cos(n\,\omega\,T) + S_n \sin(n\,\omega\,T),$$
$$S''_n = S_n \cos(n\,\omega\,T) - C_n \sin(n\,\omega\,T).$$

5.5 Demonstrate that the periodic square-wave

$$y(t) = A \begin{cases} -1 & 0 \le t/\tau \le 1/2 \\ +1 & 1/2 < t/\tau \le 1 \end{cases},$$

where $y(t+\tau) = y(t)$ for all t, has the Fourier representation

$$y(t) = -\frac{4A}{\pi} \left[\frac{\sin(\omega t)}{1} + \frac{\sin(3\,\omega t)}{3} + \frac{\sin(5\,\omega t)}{5} + \cdots \right].$$

Here, $\omega = 2\pi/\tau$. Plot the reconstructed waveform, retaining the first 4, 8, 16, and 32 terms in the Fourier series.

5.6 Show that the periodically repeated pulse waveform

$$y(t) = A \begin{cases} 1 & |t - T/2| \le \tau/2 \\ 0 & \text{otherwise} \end{cases},$$

where $y(t + T) = y(t)$ for all t, and $\tau < T$, has the Fourier representation

$$y(t) = A \frac{\tau}{T} + \frac{2A}{\pi} \sum_{n=1,\infty} (-1)^n \frac{\sin(n\,\pi\,\tau/T)}{n} \cos(n\,2\pi\,t/T).$$

Demonstrate that if $\tau \ll T$ then the most significant terms in the preceding series have frequencies (in hertz) that range from the fundamental frequency $1/T$ to a frequency of order $1/\tau$.

CHAPTER **6**

Traveling Waves

6.1 INTRODUCTION

In this chapter, we shall generalize the analysis of the previous two chapters in order to deal with longitudinal and transverse waves (in continuous media) that *propagate*; in other words, waves that carry a net energy flux in a particular direction.

6.2 STANDING WAVES IN FINITE CONTINUOUS MEDIUM

We saw earlier, in Sections 4.3 and 5.2, that a small-amplitude transverse wave on a uniform string, and a small-amplitude longitudinal wave on a thin elastic rod, are both governed by the wave equation, which (in one dimension) takes the general form

$$\frac{\partial^2 \psi}{\partial t^2} = v^2 \frac{\partial^2 \psi}{\partial x^2},\tag{6.1}$$

where $\psi(x,t)$ represents the wave disturbance, and $v > 0$ is a constant, with the dimensions of velocity, that is a property of the particular medium that supports the wave. Up to now, we have only considered media of finite length. That is, media that extend from (say) $x = 0$ to $x = (l > 0)$. Generally speaking, we have encountered two distinct types of physical constraint that hold at the boundaries of such media. First, if a given boundary is *fixed* then the wave disturbance is constrained to be zero there. For instance, if the left boundary is fixed then $\psi(0,t) = 0$. Second, if a given boundary is *free* then the spatial derivative of the disturbance (which usually corresponds to some sort of force) is constrained to be zero there. For instance, if the right boundary is free then $\partial \psi(l,t)/\partial x = 0$. It follows that a fixed boundary corresponds to a node; that is, a point where the amplitude of the wave disturbance is always zero. On the other hand, a free boundary corresponds to an anti-node; that is, a point where the amplitude of the wave disturbance is always locally maximal. Consequently, the nodes and the anti-nodes of a wave, of definite wavelength, supported in a medium of finite length that has stationary boundaries (which can be either fixed or free) are constrained to be stationary. The only simple solution of the wave equation, (6.1), that has stationary nodes and anti-nodes is a *standing wave* of the general form

$$\psi(x,t) = [A \cos(k x) + B \sin(k x)] \cos(\omega t - \phi).\tag{6.2}$$

The associated nodes are located at the values of x that satisfy

$$A \cos(k x) + B \sin(k x) = 0,\tag{6.3}$$

which implies that they are stationary, and also evenly spaced a distance $\lambda/2$ apart, where $\lambda = 2\pi/k$ is the wavelength. Moreover, the anti-nodes are situated halfway between the nodes. For example, suppose that both boundaries of the medium are fixed. It follows that the points $x = 0$ and $x = l$ must each correspond to a node. This is only possible if the length of the medium, l, is a half-integer number of wavelengths; that is, $l = n\lambda/2$, where n is a positive integer. We conclude that, in this case, the possible wavenumbers of standing wave solutions to the wave equation are quantized such that

$$k l = n\pi. \tag{6.4}$$

Moreover, the same is true if both boundaries are free. Finally, if one boundary is free, and the other fixed, then the quantization of wavenumbers takes the slightly different form

$$k l = (n - 1/2)\pi. \tag{6.5}$$

Those standing wave solutions that satisfy the appropriate quantization criterion are known as the *normal modes* of the system. (See Chapter 3.) Moreover, substitution of Equation (6.2) into the wave equation, (6.1), yields the standing wave dispersion relation

$$\omega = k v. \tag{6.6}$$

The fact that normal mode wavenumbers are quantized immediately implies that the associated oscillation frequencies are also quantized. Finally, because the wave equation is linear, the most general solution that satisfies the boundary conditions is a linear superposition of all of the normal modes. Such a solution has the appropriate node or anti-node at each of the boundaries, but does not necessarily have any stationary nodes or anti-nodes in the interior of the medium.

6.3 TRAVELING WAVES IN INFINITE CONTINUOUS MEDIUM

Consider solutions of the wave equation, (6.1), in an *infinite* medium. Such a medium does not possess any spatial boundaries, and so is not subject to boundary constraints. Hence, there is no particular reason why a wave of definite wavelength should have stationary nodes or anti-nodes. In other words, Equation (6.2) may not be the only permissible type of wave solution in an infinite medium. What other kind of solution could we have? Suppose that

$$\psi(x, t) = A \cos(\omega t - k x - \phi), \tag{6.7}$$

where $A > 0$, $k > 0$, $\omega > 0$, and ϕ are constants. This solution is interpreted as a wave of amplitude A, wavenumber k, wavelength $\lambda = 2\pi/k$, angular frequency ω, frequency (in hertz) $f = \omega/2\pi$, period $T = 1/f$, and phase angle ϕ. It can be seen that $\psi(x + \lambda, t) = \psi(x, t)$, and $\psi(x, t + T) = \psi(x, t)$, for all x and t. In other words, the wave is periodic in space with period λ, and periodic in time with period T. A wave maximum corresponds to a point at which $\cos(\omega t - k x - \phi) = 1$. It follows, from the well-known properties of the cosine function, that the various wave maxima are located at

$$\omega t - k x - \phi = n\,2\pi, \tag{6.8}$$

where n is an integer. Differentiating the previous expression with respect to t, and rearranging, the equation of motion of a particular maximum becomes

$$\frac{dx}{dt} = \frac{\omega}{k}. \tag{6.9}$$

We conclude that the wave maximum in question propagates along the x-axis at the velocity

$$v_p = \frac{\omega}{k}. \tag{6.10}$$

Metal	$Y\,(\mathrm{N\,m^{-2}})$	$\rho\,(\mathrm{kg\,m^{-3}})$	$\sqrt{Y/\rho}\,(\mathrm{m\,s^{-1}})$	$v\,(\mathrm{m\,s^{-1}})$
Aluminum	7.0×10^{10}	2.7×10^{3}	5100	5000
Nickel	2.0×10^{11}	8.9×10^{3}	4700	4900
Zinc	1.1×10^{11}	7.1×10^{3}	3900	3900
Copper	1.2×10^{11}	8.9×10^{3}	3600	3800
Silver	8.3×10^{10}	1.1×10^{4}	2800	2700
Tin	5.0×10^{10}	7.4×10^{3}	2600	2700
Lead	1.6×10^{10}	1.1×10^{4}	1100	1100

TABLE 6.1 Calculated and measured longitudinal wave speeds in thin rods made up of common metals. Sources: Haynes and Lide 2011c, Wikipedia contributors 2018.

It can be shown that the other wave maxima (as well as the wave minima and the wave zeros) also propagate along the x-axis at the same velocity. In fact, the whole wave pattern propagates in the positive x-direction without changing shape. The characteristic propagation velocity v_p is known as the *phase velocity*, because it is the velocity of constant phase points on the wave disturbance (i.e., points that satisfy $\omega t - k x - \phi = $ constant). For obvious reasons, the type of wave solution specified in Equation (6.7) is called a *traveling wave*.

Substitution of Equation (6.7) into the wave equation, (6.1), yields the familiar dispersion relation

$$\omega = k v. \tag{6.11}$$

We conclude that the traveling wave solution (6.7) satisfies the wave equation provided

$$v_p = \frac{\omega}{k} = v; \tag{6.12}$$

that is, provided the phase velocity of the wave takes the fixed value v. It follows that the constant v^2, that appears in the wave equation, (6.1), can be interpreted as the square of the velocity with which traveling waves propagate through the medium in question. Hence, from the discussions in Sections 4.3, 5.2, and 5.3, transverse waves propagate along strings of tension T and mass per unit length ρ at the phase velocity $\sqrt{T/\rho}$, longitudinal waves propagate along thin elastic rods of Young's modulus Y and mass density ρ at the phase velocity $\sqrt{Y/\rho}$, and sound waves propagate through ideal gases of pressure p, mass density ρ, and ratio of specific heats γ, at the phase velocity $\sqrt{\gamma p/\rho}$.

Table 6.1 displays calculated and measured longitudinal wave speeds in thin rods made up of various common metals. It can be seen that the agreement between the two is excellent.

An ideal gas of mass m and molecular weight M satisfies the *ideal gas equation of state*,

$$p V = \frac{m}{M} R T, \tag{6.13}$$

where p is the pressure, V the volume, $R = 8.3145\,\mathrm{J\,mol^{-1}\,K^{-1}}$ the molar ideal gas constant, and T the absolute temperature (Reif 2008). Because the ratio m/V is equal to the density, ρ, the expression for the sound speed, $v = \sqrt{\gamma p/\rho}$, yields

$$v = \left(\frac{\gamma R T}{M}\right)^{1/2}. \tag{6.14}$$

We conclude that the speed of sound in an ideal gas is independent of the pressure or the density, proportional to the square root of the absolute temperature, and inversely proportional to the square root of the molecular mass. Incidentally, the root-mean-square molecular speed in an ideal gas in thermal equilibrium is $v_{rms} = \sqrt{3RT/M}$ (Reif 2008). Hence, the speed of sound in an ideal gas is of order, but slightly less than (because the maximum possible value of γ is 5/3), the mean molecular speed.

A comparison between Equations (6.6) and (6.11) reveals that standing waves and traveling waves in a given medium satisfy the same dispersion relation. However, because traveling waves in infinite media are not subject to boundary constraints, there is no restriction on the possible wavenumbers, or wavelengths, of such waves. Hence, any traveling wave solution whose wavenumber, k, and angular frequency, ω, are related according to the dispersion relation (6.11) is a valid solution of the wave equation. In other words, any traveling wave solution whose wavelength, $\lambda = 2\pi/k$, and frequency, $f = \omega/2\pi$, are related according to

$$v = f\lambda \tag{6.15}$$

is a valid solution. We conclude that relatively high-frequency traveling waves propagating through a given medium possess relatively short wavelengths, and vice versa.

Consider the alternative wave solution

$$\psi(x,t) = A\,\cos(\omega t + kx - \phi), \tag{6.16}$$

where $A > 0$, $k > 0$, $\omega > 0$, and ϕ are constants. As before, this solution is interpreted as a wave of amplitude A, wavenumber k, angular frequency ω, and phase angle ϕ. However, the wave maxima are now located at

$$\omega t + kx - \phi = n\,2\pi, \tag{6.17}$$

where n is an integer, and have equations of motion of the form

$$\frac{dx}{dt} = -\frac{\omega}{k}. \tag{6.18}$$

Equation (6.16), thus, represents a traveling wave that propagates in the minus x-direction at the phase velocity $v_p = \omega/k$. Moreover, substitution of Equation (6.16) into the wave equation, (6.1), again yields the dispersion relation (6.11), which implies that $v_p = v$. It follows that traveling wave solutions to the wave equation, (6.1), can propagate in either the positive or the negative x-direction, as long as they always travel at the fixed speed v.

6.4 WAVE INTERFERENCE

What is the relationship between traveling wave and standing wave solutions to the wave equation, (6.1), in an infinite medium? To help answer this question, let us form a superposition of two traveling wave solutions of equal amplitude A, and zero phase angle ϕ, that have the same wavenumber k, but are moving in opposite directions. In other words,

$$\psi(x,t) = A\,\cos(\omega t - kx) + A\,\cos(\omega t + kx). \tag{6.19}$$

Because the wave equation, (6.1), is linear, the previous superposition is a valid solution provided the two component waves are also valid solutions; that is, provided $\omega = kv$, which we shall assume to be the case. Making use of the trigonometric identity $\cos a + \cos b \equiv 2\,\cos[(a+b)/2]\,\cos[(a-b)/2]$ (see Appendix B), the previous expression can also be written

$$\psi(x,t) = 2A\,\cos(kx)\,\cos(\omega t), \tag{6.20}$$

which is a standing wave [cf., Equation (6.2)]. Evidently, a standing wave is a linear superposition of two, otherwise identical, traveling waves that propagate in opposite directions. The two waves completely cancel one another out at the nodes, which are situated at $kx = (n - 1/2)\pi$, where n is an integer. This process is known as total *destructive interference*. On the other hand, the waves reinforce one another at the anti-nodes, which are situated at $kx = n\pi$, generating a wave whose amplitude is twice that of the component waves. This process is known as *constructive interference*.

As a more general example of wave interference, consider a superposition of two traveling waves of unequal amplitudes which again have the same wavenumber and zero phase angle, and are moving in opposite directions; that is,

$$\psi(x, t) = A_1 \cos(\omega t - kx) + A_2 \cos(\omega t + kx), \tag{6.21}$$

where $A_1, A_2 > 0$. In this case, the trigonometric identities $\cos(a - b) \equiv \cos a \cos b + \sin a \sin b$ and $\cos(a + b) \equiv \cos a \cos b - \sin a \sin b$ (see Appendix B) yield

$$\psi(x, t) = (A_1 + A_2) \cos(kx) \cos(\omega t) + (A_1 - A_2) \sin(kx) \sin(\omega t). \tag{6.22}$$

Thus, the two waves interfere destructively at $kx = (n - 1/2)\pi$ [i.e., at points where $\cos(kx) = 0$ and $|\sin(kx)| = 1$] to produce a minimum wave amplitude $|A_1 - A_2|$, and interfere constructively at $kx = n\pi$ [i.e., at points where $|\cos(kx)| = 1$ and $\sin(kx) = 0$] to produce a maximum wave amplitude $A_1 + A_2$. It can be seen that the destructive interference is incomplete unless $A_1 = A_2$. Incidentally, it is a general result that if two waves of amplitude $A_1 > 0$ and $A_2 > 0$ interfere then the maximum and minimum possible values of the resulting wave amplitude are $A_1 + A_2$ and $|A_1 - A_2|$, respectively.

6.5 ENERGY CONSERVATION

Consider a small-amplitude transverse wave propagating along a uniform string of infinite length, tension T, and mass per unit length ρ. (See Section 4.3.) Let x measure distance along the string, and let $y(x, t)$ be the transverse wave displacement. As we have seen, $y(x, t)$ satisfies the wave equation

$$\frac{\partial^2 y}{\partial t^2} = v^2 \frac{\partial^2 y}{\partial x^2}, \tag{6.23}$$

where $v = \sqrt{T/\rho}$ is the phase velocity of traveling waves on the string.

Consider a section of the string lying between $x = x_1$ and $x = x_2$. The kinetic energy of this section is

$$K = \int_{x_1}^{x_2} \frac{1}{2} \rho \left(\frac{\partial y}{\partial t}\right)^2 dx, \tag{6.24}$$

because $\partial y/\partial t$ is the string's transverse velocity (and the longitudinal velocity is assumed to be negligibly small). The potential energy is the work done in stretching the section, which is $T \, \Delta s$, where Δs is the difference between the section's stretched and unstretched lengths. Here, it is assumed that the tension remains approximately constant as the section is stretched. An element of length of the string is written

$$ds = (dx^2 + dy^2)^{1/2} = \left[1 + \left(\frac{\partial y}{\partial x}\right)^2\right]^{1/2} dx. \tag{6.25}$$

Hence,

$$\Delta s = \int_{x_1}^{x_2} \left\{\left[1 + \left(\frac{\partial y}{\partial x}\right)^2\right]^{1/2} - 1\right\} dx \tag{6.26}$$

$$= \int_{x_1}^{x_2} \left\{\left[1 + \frac{1}{2}\left(\frac{\partial y}{\partial x}\right)^2 + O\left(\frac{\partial y}{\partial x}\right)^4\right] - 1\right\} dx \simeq \int_{x_1}^{x_2} \frac{1}{2}\left(\frac{\partial y}{\partial x}\right)^2 dx, \tag{6.27}$$

because it is assumed that $|\partial y/\partial x| \ll 1$ (i.e., the transverse displacement is sufficiently small that the string remains almost parallel to the x-axis). Thus, the potential energy of the section is $U = T \, \Delta s$, or

$$U = \int_{x_1}^{x_2} \frac{1}{2} T \left(\frac{\partial y}{\partial x}\right)^2 dx. \tag{6.28}$$

It follows that the total energy of the section is

$$E = \int_{x_1}^{x_2} \frac{1}{2} \left[\rho \left(\frac{\partial y}{\partial t}\right)^2 + T \left(\frac{\partial y}{\partial x}\right)^2\right] dx. \tag{6.29}$$

Multiplying the wave equation, (6.23), by $\rho \, (\partial y/\partial t)$, we obtain

$$\rho \frac{\partial y}{\partial t} \frac{\partial^2 y}{\partial t^2} = T \frac{\partial y}{\partial t} \frac{\partial^2 y}{\partial x^2}, \tag{6.30}$$

because $v^2 = T/\rho$. This expression yields

$$\rho \frac{\partial y}{\partial t} \frac{\partial^2 y}{\partial t^2} + T \frac{\partial y}{\partial x} \frac{\partial^2 y}{\partial t \, \partial x} = T \frac{\partial y}{\partial t} \frac{\partial^2 y}{\partial x^2} + T \frac{\partial y}{\partial x} \frac{\partial^2 y}{\partial t \, \partial x}, \tag{6.31}$$

which can be written in the form

$$\frac{1}{2} \frac{\partial}{\partial t} \left[\rho \left(\frac{\partial y}{\partial t}\right)^2 + T \left(\frac{\partial y}{\partial x}\right)^2\right] = \frac{\partial}{\partial x} \left(T \frac{\partial y}{\partial t} \frac{\partial y}{\partial x}\right), \tag{6.32}$$

or

$$\frac{\partial \mathcal{E}}{\partial t} + \frac{\partial \mathcal{I}}{\partial x} = 0, \tag{6.33}$$

where

$$\mathcal{E}(x, t) = \frac{1}{2} \left[\rho \left(\frac{\partial y}{\partial t}\right)^2 + T \left(\frac{\partial y}{\partial x}\right)^2\right] \tag{6.34}$$

is the *energy density* (i.e., the energy per unit length) of the string, and

$$\mathcal{I}(x, t) = -T \frac{\partial y}{\partial t} \frac{\partial y}{\partial x}. \tag{6.35}$$

Finally, integrating Equation (6.33) in x from x_1 to x_2, we obtain

$$\frac{d}{dt} \int_{x_1}^{x_2} \mathcal{E} \, dx + \mathcal{I}(x_2, t) - \mathcal{I}(x_1, t) = 0, \tag{6.36}$$

or

$$\frac{dE}{dt} = \mathcal{I}(x_1, t) - \mathcal{I}(x_2, t). \tag{6.37}$$

Here, $E(t)$ is the energy stored in the section of the string lying between $x = x_1$ and $x = x_2$. [See Equation (6.29).] If we interpret $\mathcal{I}(x, t)$ as the instantaneous *energy flux* (i.e., rate of energy flow) in the positive-x direction, at position x and time t, then the previous equation can be recognized as a declaration of *energy conservation*. Basically, the equation states that the rate of increase in the energy stored in the section of the string lying between $x = x_1$ and $x = x_2$, which is dE/dt, is equal to the difference between the rate at which energy flows into the left end of the section, which is $\mathcal{I}(x_1, t)$, and the rate at which it flows out of the right end, which is $\mathcal{I}(x_2, t)$. Incidentally, the string conserves energy because it lacks any mechanism for energy dissipation. The same is true of the other wave media discussed in this chapter.

Consider a wave propagating in the positive x-direction of the form

$$y(x, t) = A \, \cos(\omega t - k \, x - \phi). \tag{6.38}$$

According to Equation (6.35), the energy flux associated with this wave is

$$I(x, t) = T k \omega A^2 \, \sin^2(\omega t - k x - \phi). \tag{6.39}$$

Thus, the mean energy flux is written

$$\langle I \rangle = \frac{1}{2} \omega^2 Z A^2, \tag{6.40}$$

where $\langle A \rangle(x) \equiv (\omega/2\pi) \int_t^{t+2\pi/\omega} A(x, t') \, dt'$ represents an average over a period of the wave oscillation. Here, use has been made of $\omega/k = \sqrt{T/\rho}$, as well as the result that $\langle \sin^2(\omega t + \theta) \rangle = 1/2$ for all θ. Moreover, the quantity

$$Z = \sqrt{\rho T} \tag{6.41}$$

is known as the characteristic *impedance* of the string. The units of Z are force over velocity. Thus, the string impedance measures the typical tension required to produce a unit transverse velocity. Finally, according to Equation (6.40), a traveling wave propagating in the positive x-direction is associated with a positive energy flux. In other words, the wave transports energy in the positive x-direction.

Consider a wave propagating in the negative x-direction of the general form

$$y(x, t) = A \, \cos(\omega t + k \, x - \phi). \tag{6.42}$$

It can be demonstrated, from Equation (6.35), that the mean energy flux associated with this wave is

$$\langle I \rangle = -\frac{1}{2} \omega^2 Z A^2. \tag{6.43}$$

The fact that the energy flux is negative means that the wave transports energy in the negative x-direction.

Suppose that we have a superposition of a wave of amplitude A_+ propagating in the positive x-direction, and a wave of amplitude A_- propagating in the negative x-direction, so that

$$y(x, t) = A_+ \cos(\omega t - k x - \phi_+) + A_- \cos(\omega t + k x - \phi_-). \tag{6.44}$$

According to Equation (6.35), the instantaneous energy flux is written

$$\begin{aligned}
I(x, t) &= \omega^2 Z \left[A_+ \sin(\omega t - k x - \phi_+) + A_- \sin(\omega t + k x - \phi_-) \right] \\
&\quad \left[A_+ \sin(\omega t - k x - \phi_+) - A_- \sin(\omega t + k x - \phi_-) \right] \\
&= \omega^2 Z \left[A_+^2 \sin^2(\omega t - k x - \phi_+) - A_-^2 \sin^2(\omega t + k x - \phi_-) \right].
\end{aligned} \tag{6.45}$$

Hence, the mean energy flux,

$$\langle I \rangle = \frac{1}{2} \omega^2 Z A_+^2 - \frac{1}{2} \omega^2 Z A_-^2, \tag{6.46}$$

is the difference between the independently calculated mean fluxes associated with the waves traveling to the right (i.e., in the positive x-direction) and to the left. Recall, from the previous section, that a standing wave is a superposition of two traveling waves, of equal amplitude and frequency, propagating in opposite directions. It immediately follows, from the previous expression, that a standing wave has zero associated net energy flux. In other words, a standing wave does not give rise to net energy transport.

We saw earlier, in Section 5.3, that a small-amplitude longitudinal wave in a thin elastic rod satisfies the wave equation,

$$\frac{\partial^2 \psi}{\partial t^2} = v^2 \frac{\partial^2 \psi}{\partial x^2}, \tag{6.47}$$

where $\psi(x,t)$ is the longitudinal wave displacement, $v = \sqrt{Y/\rho}$ the phase velocity of traveling waves along the rod, Y the Young's modulus, and ρ the mass density. Using similar analysis to that just employed, we can derive an energy conservation equation of the form (6.33) from this wave equation, where

$$\mathcal{E} = \frac{1}{2} \left[\rho \left(\frac{\partial \psi}{\partial t} \right)^2 + Y \left(\frac{\partial \psi}{\partial x} \right)^2 \right] \tag{6.48}$$

is the wave energy density (i.e., the energy per unit volume), and

$$\mathcal{I} = -Y \frac{\partial \psi}{\partial t} \frac{\partial \psi}{\partial x} \tag{6.49}$$

the wave energy flux (i.e., the rate of energy flow per unit area) in the positive x-direction. For a traveling wave of the form $\psi(x,t) = A \cos(\omega t - k x - \phi)$, the previous expression yields

$$\langle \mathcal{I} \rangle = \frac{1}{2} \omega^2 Z A^2, \tag{6.50}$$

where

$$Z = \sqrt{\rho Y} \tag{6.51}$$

is the impedance of the rod. The units of Z are pressure over velocity. Hence, in this case, the impedance measures the typical pressure in the rod required to produce a unit longitudinal velocity. Analogous arguments reveal that the impedance of an ideal gas of density ρ, pressure p, and ratio of specific heats γ, is

$$Z = \sqrt{\rho \gamma p}. \tag{6.52}$$

(See Section 5.4.)

6.6 TRANSMISSION LINES

In Sections 1.4, 2.4, and 2.6, we examined simple alternating current (ac) circuits consisting of a single loop. In analyzing these circuits, we made the assumption that the current oscillations were in phase with one another at all points on the loop. In other words, the current passes through zero, and attains maximum/minimum values, *simultaneously* at all points on the loop. Let us now consider under which circumstances this single-phase assumption is justified. Let the current oscillate at the frequency f (in hertz), and let L be the typical spatial extent of the circuit. Information is presumably carried around the circuit via electromagnetic fields, which cannot transmit information faster than the velocity of light in vacuum, c. Hence, it is only possible for the current oscillations around the circuit to be synchronized with one another if the time required for information to propagate around the circuit, L/c, is much less than the period of the oscillation, $1/f$. Thus, for a given oscillation frequency, there is a maximum circuit size, $L_c = c/f$, above which synchronization is not possible, and the phase of the current oscillations presumably starts to vary around the circuit. Of course, $c/f = \lambda$, where λ is the free-space wavelength of an electromagnetic wave of frequency f. Hence, we deduce that the single-phase assumption breaks down when the size of an ac circuit exceeds the free-space wavelength of an electromagnetic wave that oscillates at the alternation frequency.

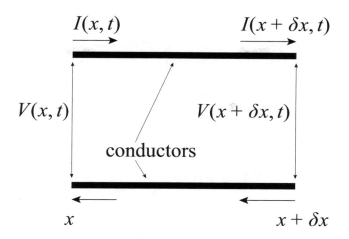

FIGURE 6.1   A section of a transmission line.

For the case of a standard "mains" circuit, which oscillates at 60 Hz (in the US), we find that $L_c \sim 5000$ km. Hence, we deduce that the single-phase assumption is reasonable for such circuits. For the case of a telephone circuit, which typically oscillates at 10 kHz (i.e., the typical frequency of human speech), we find that $L_c \sim 30$ km. Thus, the single-phase assumption definitely breaks down for long-distance telephone lines. For the case of internet cables, which typically oscillate at 10 MHz, we find that $L_c \sim 30$ m. Hence, the single-phase assumption is not valid in most internet networks. Finally, for the case of TV circuits, which typically oscillate at 10 GHz, we find that $L_c \sim 3$ cm. Thus, the single-phase approximation breaks down completely in TV circuits. Roughly speaking, the single-phase approximation is unlikely to hold in the type of electrical circuits involved in *communication*, because these invariably require high-frequency signals to be transmitted over large distances.

A so-called *transmission line* is typically used to carry high-frequency electromagnetic signals over long distances; that is, distances sufficiently large that the phase of the signal varies significantly along the line (which implies that the line is much longer than the free-space wavelength of the signal). In its simplest form, a transmission line consists of *two* parallel conductors that carry *equal and opposite electrical currents*, $I(x,t)$, where x measures distance along the line. See Figure 6.1. (This combination of two conductors carrying equal and opposite currents is necessary to prevent intolerable losses due to electromagnetic radiation.) Let $V(x,t)$ be the instantaneous voltage difference between the two conductors at position x. Consider a small section of the line lying between x and $x + \delta x$. If $Q(t)$ is the electric charge on one of the conducting sections, and $-Q(t)$ the charge on the other, then charge conservation implies that $dQ/dt = I(x,t) - I(x+\delta x, t)$. However, according to standard electrical circuit theory (Fitzpatrick 2008), $Q(t) = C\,\delta x\,V(x,t)$, where C is the *capacitance per unit length* of the line. Standard circuit theory also yields $V(x+\delta x, t) - V(x,t) = -\mathcal{L}\,\delta x\,\partial I(x,t)/\partial t$ (ibid.), where $\mathcal{L}$ is the *inductance per unit length* of the line. Taking the limit $\delta x \to 0$, we obtain the so-called *Telegrapher's equations* (ibid.),

$$\frac{\partial V}{\partial t} = -\frac{1}{C}\frac{\partial I}{\partial x}, \tag{6.53}$$

$$\frac{\partial I}{\partial t} = -\frac{1}{\mathcal{L}}\frac{\partial V}{\partial x}. \tag{6.54}$$

(See Exercise 4.7.) These two equations can be combined to give

$$\frac{\partial^2 V}{\partial t^2} = \frac{1}{\mathcal{L}C} \frac{\partial^2 V}{\partial x^2},$$ (6.55)

together with an analogous equation for I. In other words, $V(x,t)$ and $I(x,t)$ both obey a wave equation of the form (6.23) in which the associated phase velocity is

$$v = \frac{1}{\sqrt{\mathcal{L}C}}.$$ (6.56)

Multiplying (6.53) by CV, (6.54) by $\mathcal{L}I$, and then adding the two resulting expressions, we obtain the energy conservation equation

$$\frac{\partial \mathcal{E}}{\partial t} + \frac{\partial \mathcal{I}}{\partial x} = 0,$$ (6.57)

where

$$\mathcal{E} = \frac{1}{2}\mathcal{L}I^2 + \frac{1}{2}CV^2$$ (6.58)

is the electromagnetic energy density (i.e., energy per unit length) of the line, and

$$\mathcal{I} = IV$$ (6.59)

is the electromagnetic energy flux along the line (i.e., the energy per unit time that passes a given point) in the positive x-direction (ibid.). Consider a signal propagating along the line, in the positive x-direction, whose associated current takes the form

$$I(x,t) = I_0 \cos(\omega t - kx - \phi),$$ (6.60)

where ω and k are related according to the dispersion relation

$$\omega = kv.$$ (6.61)

It can be demonstrated, from Equation (6.53), that the corresponding voltage is

$$V(x,t) = V_0 \cos(\omega t - kx - \phi),$$ (6.62)

where

$$V_0 = I_0 Z.$$ (6.63)

Here,

$$Z = \sqrt{\frac{\mathcal{L}}{C}}$$ (6.64)

is the characteristic impedance of the line, and has units of ohms. It follows that the mean energy flux associated with the signal is written

$$\langle \mathcal{I} \rangle = \langle IV \rangle = \frac{1}{2}I_0 V_0 = \frac{1}{2}Z I_0^2 = \frac{1}{2}\frac{V_0^2}{Z}.$$ (6.65)

Likewise, for a signal propagating along the line in the negative x-direction,

$$I(x,t) = I_0 \cos(\omega t + kx - \phi),$$ (6.66)

$$V(x,t) = -V_0 \cos(\omega t + kx - \phi),$$ (6.67)

and the mean energy flux is

$$\langle I \rangle = -\frac{1}{2} I_0 V_0 = -\frac{1}{2} Z I_0^2 = -\frac{1}{2} \frac{V_0^2}{Z}. \tag{6.68}$$

As a specific example, consider a transmission line consisting of two uniform parallel conducting strips of width w and perpendicular distance apart d, where $d \ll w$. It can be demonstrated, using standard electrostatic theory (Grant and Philips 1975), that the capacitance per unit length of the line is

$$C = \epsilon_0 \frac{w}{d}, \tag{6.69}$$

where $\epsilon_0 = 8.8542 \times 10^{-12} \, C^2 \, N^{-1} \, m^{-2}$ is the *electric permittivity of free space*. Likewise, according to standard magnetostatic theory (ibid.), the line's inductance per unit length takes the form

$$\mathcal{L} = \mu_0 \frac{d}{w}, \tag{6.70}$$

where $\mu_0 = 4\pi \times 10^{-7} \, N \, A^{-2}$ is the *magnetic permeability of free space*. Thus, the phase velocity of a signal propagating down the line is

$$v = \frac{1}{\sqrt{\mathcal{L} C}} = \frac{1}{\sqrt{\epsilon_0 \mu_0}}, \tag{6.71}$$

which, of course, is the *velocity of light in vacuum*. [See Equation (6.120).] This is not a coincidence. In fact, it can be demonstrated that the inductance per unit length and the capacitance per unit length of any (vacuum-filled) transmission line satisfy (Jackson 1975)

$$\frac{\mathcal{L}}{\mu_0} = \frac{\epsilon_0}{C}. \tag{6.72}$$

Hence, signals always propagate down such lines at the velocity of light in vacuum. The impedance of a parallel strip transmission line is

$$Z = \sqrt{\frac{\mathcal{L}}{C}} = \frac{Z_0}{C/\epsilon_0} = \frac{d}{w} Z_0, \tag{6.73}$$

where the quantity

$$Z_0 = \sqrt{\frac{\mu_0}{\epsilon_0}} = 376.73 \, \Omega \tag{6.74}$$

is known as the *impedance of free space*. However, because we have assumed that $d \ll w$, it follows that the impedance of the line is impractically low (i.e., $Z \ll Z_0$). In fact, it is clear from Exercise 6.11 that a signal sent down a transmission line will attenuate significantly unless the impedance of the line greatly exceeds its resistance. Given that it is impossible to construct a practical transmission line of usable length whose resistance is less than a few ohms, it follows that practical transmission lines must have impedances that are much greater than an ohm ($50 \, \Omega$ is a typical value).

Practical transmission lines generally consist of two parallel wires twisted about one another (for example, twisted-pair ethernet cables), or two concentric cylindrical conductors (for example, co-axial TV cables). For the case of two parallel wires of radius a and distance apart d (where $d > 2a$), the capacitance per unit length is (Wikipedia contributors 2018)

$$C = \frac{\pi \epsilon_0}{\cosh^{-1}(d/2a)}. \tag{6.75}$$

Thus, the impedance of a parallel wire transmission line becomes

$$Z = \frac{Z_0}{C/\epsilon_0} = \pi^{-1} \cosh^{-1}\left(\frac{d}{2a}\right) Z_0. \tag{6.76}$$

For the case of a co-axial cable in which the radii of the inner and outer conductors are a and b, respectively, the capacitance per unit length is (Fitzpatrick 2008)

$$C = \frac{2\pi \epsilon_0}{\ln(b/a)}. \tag{6.77}$$

Thus, the impedance of a co-axial transmission line becomes

$$Z = (2\pi)^{-1} \ln\left(\frac{b}{a}\right) Z_0. \tag{6.78}$$

It follows that a parallel wire transmission line of given impedance can be fabricated by choosing the appropriate ratio of the wire spacing to the wire radius. Likewise, a co-axial transmission line of given impedance can be fabricated by choosing the appropriate ratio of the radii of the outer and inner conductors.

6.7 NORMAL REFLECTION AND TRANSMISSION AT INTERFACES

Consider two uniform semi-infinite strings that run along the x-axis, and are tied together at $x = 0$. Let the first string be of density per unit length ρ_1, and occupy the region $x < 0$, and let the second string be of density per unit length ρ_2, and occupy the region $x > 0$. The tensions in the two strings must be equal, otherwise the string interface would not be in force balance in the x-direction. Let T be the common tension. Suppose that a transverse wave of angular frequency ω is launched from a wave source at $x = -\infty$, and propagates towards the interface. Assuming that $\rho_1 \neq \rho_2$, we would expect the wave incident on the interface to be partially reflected, and partially transmitted. The frequencies of the incident, reflected, and transmitted waves are all the same, because this property of the waves is ultimately determined by the oscillation frequency of the wave source. Hence, in the region $x < 0$, the wave displacement takes the form

$$y(x, t) = A_i \cos(\omega t - k_1 x - \phi_i) + A_r \cos(\omega t + k_1 x - \phi_r). \tag{6.79}$$

In other words, the displacement is a linear superposition of an *incident wave* and a *reflected wave*. The incident wave propagates in the positive x-direction, and is of amplitude A_i, wavenumber $k_1 = \omega/v_1$, and phase angle ϕ_i. The reflected wave propagates in the negative x-direction, and is of amplitude A_r, wavenumber $k_1 = \omega/v_1$, and phase angle ϕ_r. Here, $v_1 = \sqrt{T/\rho_1}$ is the phase velocity of traveling waves on the first string. In the region $x > 0$, the wave displacement takes the form

$$y(x, t) = A_t \cos(\omega t - k_2 x - \phi_t). \tag{6.80}$$

In other words, the displacement is solely due to a *transmitted wave* that propagates in the positive x-direction, and is of amplitude A_t, wavenumber $k_2 = \omega/v_2$, and phase angle ϕ_t. Here, $v_2 = \sqrt{T/\rho_2}$ is the phase velocity of traveling waves on the second string. Incidentally, there is no backward traveling wave in the region $x > 0$ because such a wave could only originate from a source at $x = \infty$, and there is no such source in the problem under consideration.

Let us consider the matching conditions at the interface between the two strings; that is, at $x = 0$. First, because the two strings are tied together at $x = 0$, their transverse displacements at this point must equal one another. In other words,

$$y(0_-, t) = y(0_+, t), \tag{6.81}$$

or

$$A_i \cos(\omega t - \phi_i) + A_r \cos(\omega t - \phi_r) = A_t \cos(\omega t - \phi_t). \tag{6.82}$$

The only way that the previous equation can be satisfied for all values of t is if $\phi_i = \phi_r = \phi_t$. This being the case, the common $\cos(\omega t - \phi_i)$ factor cancels out, and we are left with

$$A_i + A_r = A_t. \tag{6.83}$$

Second, because the two strings lack an energy dissipation mechanism, the energy flux into the interface must match that out of the interface. In other words,

$$\frac{1}{2} \omega^2 Z_1 (A_i^2 - A_r^2) = \frac{1}{2} \omega^2 Z_2 A_t^2, \tag{6.84}$$

where $Z_1 = \sqrt{\rho_1 T}$ and $Z_2 = \sqrt{\rho_2 T}$ are the impedances of the first and second strings, respectively. The previous expression reduces to

$$Z_1 (A_i + A_r)(A_i - A_r) = Z_2 A_t^2, \tag{6.85}$$

which, when combined with Equation (6.83), yields

$$Z_1 (A_i - A_r) = Z_2 A_t. \tag{6.86}$$

Equations (6.83) and (6.86) can be solved to give

$$A_r = \left(\frac{Z_1 - Z_2}{Z_1 + Z_2}\right) A_i, \tag{6.87}$$

$$A_t = \left(\frac{2 Z_1}{Z_1 + Z_2}\right) A_i. \tag{6.88}$$

The *coefficient of reflection*, R, is defined as the ratio of the reflected to the incident energy flux; that is,

$$R = \left(\frac{A_r}{A_i}\right)^2 = \left(\frac{Z_1 - Z_2}{Z_1 + Z_2}\right)^2. \tag{6.89}$$

The *coefficient of transmission*, T, is defined as the ratio of the transmitted to the incident energy flux; that is,

$$T = \frac{Z_2}{Z_1} \left(\frac{A_t}{A_i}\right)^2 = \frac{4 Z_1 Z_2}{(Z_1 + Z_2)^2}. \tag{6.90}$$

It can be seen that

$$R + T = 1. \tag{6.91}$$

In other words, any incident wave energy that is not reflected is transmitted.

Suppose that the density per unit length of the second string, ρ_2, tends to infinity, so that $Z_2 = \sqrt{\rho_2 T} \to \infty$. It follows from Equations (6.87) and (6.88) that $A_r = -A_i$ and $A_t = 0$. Likewise, Equations (6.89) and (6.90) yield $R = 1$ and $T = 0$. Hence, the interface between the two strings is stationary (because it oscillates with amplitude A_t), and there is no transmitted energy. In other words, the second string acts exactly like a fixed boundary. It follows that when a transverse wave on a string is incident at a fixed boundary then it is perfectly reflected with a phase shift of π radians. In other words, $A_r = -A_i$. Thus, the resultant wave displacement on the string becomes

$$y(x, t) = A_i \cos(\omega t - k_1 x - \phi_i) - A_i \cos(\omega t + k_1 x + \phi_i)$$

$$= 2 A_i \sin(k_1 x) \sin(\omega t - \phi_i), \tag{6.92}$$

where use has been made of the trigonometric identity $\cos a - \cos b \equiv 2\,\sin[(a+b)/2]\,\sin[(b-a)/2]$. (See Appendix B.) We conclude that the incident and reflected waves interfere in such a manner as to produce a standing wave with a node at the fixed boundary.

Suppose that the density per unit length of the second string, ρ_2, tends to zero, so that $Z_2 = \sqrt{\rho_2 T} \to 0$. It follows from Equations (6.87) and (6.88) that $A_r = A_i$ and $A_t = 2\,A_i$. Likewise, Equations (6.89) and (6.90) yield $R = 1$ and $T = 0$. Hence, the interface between the two strings oscillates at twice the amplitude of the incident wave (that is, the interface is a point of maximal amplitude oscillation), and there is no transmitted energy. In other words, the second string acts exactly like a free boundary. It follows that when a transverse wave on a string is incident at a free boundary then it is perfectly reflected with zero phase shift. In other words, $A_r = A_i$. Thus, the resultant wave displacement on the string becomes

$$y(x,t) = A_i\,\cos(\omega t - k_1\,x - \phi_i) + A_i\,\cos(\omega t + k_1\,x + \phi_i)$$
$$= 2\,A_i\,\cos(k_1\,x)\,\cos(\omega t - \phi_i), \tag{6.93}$$

where use has been made of the trigonometric identity $\cos a + \cos b \equiv 2\,\cos[(a+b)/2]\,\cos[(a-b)/2]$. (See Appendix B.) We conclude that the incident and reflected waves interfere in such a manner as to produce a standing wave with an anti-node at the free boundary.

Suppose that two strings of mass per unit length ρ_1 and ρ_2 are separated by a short section of string of mass per unit length ρ_3. Let all three strings have the common tension T. Suppose that the first and second strings occupy the regions $x < 0$ and $x > a$, respectively. Thus, the middle string occupies the region $0 \leq x \leq a$. Moreover, the interface between the first and middle strings is at $x = 0$, and the interface between the middle and second strings is at $x = a$. Suppose that a wave of angular frequency ω is launched from a wave source at $x = -\infty$, and propagates towards the two interfaces. We would expect this wave to be partially reflected and partially transmitted at the first interface ($x = 0$), and the resulting transmitted wave to then be partially reflected and partially transmitted at the second interface ($x = a$). Thus, we can write the wave displacement in the region $x < 0$ as

$$y(x,t) = A_i\,\cos(\omega t - k_1\,x) + A_r\,\cos(\omega t + k_1\,x), \tag{6.94}$$

where A_i is the amplitude of the incident wave, A_r is the amplitude of the reflected wave, and $k_1 = \omega/(T/\rho_1)^{1/2}$. Here, the phase angles of the two waves have been chosen so as to facilitate the matching process at $x = 0$. The wave displacement in the region $x > a$ takes the form

$$y(x,t) = A_t\,\cos(\omega t - k_2\,x - \phi_t), \tag{6.95}$$

where A_t is the amplitude of the final transmitted wave, and $k_2 = \omega/(T/\rho_2)^{1/2}$. Finally, the wave displacement in the region $0 \leq x \leq a$ is written

$$y(x,t) = A_+\,\cos(\omega t - k_3\,x) + A_-\,\cos(\omega t + k_3\,x), \tag{6.96}$$

where A_+ and A_- are the amplitudes of the forward- and backward-propagating waves on the middle string, respectively, and $k_3 = \omega/(T/\rho_3)^{1/2}$. Continuity of the transverse displacement at $x = 0$ yields

$$A_i + A_r = A_+ + A_-, \tag{6.97}$$

where a common factor $\cos(\omega t)$ has cancelled out. Continuity of the energy flux at $x = 0$ gives

$$Z_1\,(A_i^2 - A_r^2) = Z_3\,(A_+^2 - A_-^2), \tag{6.98}$$

so the previous two expressions can be combined to produce

$$Z_1\,(A_i - A_r) = Z_3\,(A_+ - A_-). \tag{6.99}$$

Continuity of the transverse displacement at $x = a$ yields

$$A_+ \cos(\omega t - k_3\, a) + A_- \cos(\omega t + k_3\, a) = A_t \cos(\omega t - k_2\, a - \phi_t). \tag{6.100}$$

Suppose that the length of the middle string is one quarter of a wavelength; that is, $k_3\, a = \pi/2$. Furthermore, let $\phi_t = k_3\, a - k_2\, a$. It follows that $\cos(\omega t - k_3\, a) = \sin(\omega t)$, $\cos(\omega t + k_3\, a) = -\sin(\omega t)$, and $\cos(\omega t - k_2\, a - \phi_t) = \sin(\omega t)$. Thus, canceling out a common factor $\sin(\omega t)$, the previous expression yields

$$A_+ - A_- = A_t. \tag{6.101}$$

Continuity of the energy flux at $x = a$ gives

$$Z_3\,(A_+^2 - A_-^2) = Z_2\, A_t^2. \tag{6.102}$$

so the previous two equations can be combined to generate

$$Z_3\,(A_+ + A_-) = Z_2\, A_t. \tag{6.103}$$

Equations (6.97) and (6.103) yield

$$A_i + A_r = \frac{Z_2}{Z_3}\, A_t, \tag{6.104}$$

whereas Equations (6.99) and (6.101) give

$$A_i - A_r = \frac{Z_3}{Z_1}\, A_t. \tag{6.105}$$

Hence, combining the previous two expression, we obtain

$$A_r = \left(\frac{Z_1\, Z_2 - Z_3^2}{Z_1\, Z_2 + Z_3^2}\right) A_i, \tag{6.106}$$

$$A_t = \left(\frac{2\, Z_1\, Z_3}{Z_1\, Z_2 + Z_3^2}\right) A_i. \tag{6.107}$$

Finally, the overall coefficient of reflection is

$$R = \left(\frac{A_r}{A_i}\right)^2 = \left(\frac{Z_1\, Z_2 - Z_3^2}{Z_1\, Z_2 + Z_3^2}\right)^2, \tag{6.108}$$

whereas the overall coefficient of transmission becomes

$$T = \frac{Z_2}{Z_1}\left(\frac{A_t}{A_i}\right)^2 = \frac{4\, Z_1\, Z_2\, Z_3^2}{(Z_1\, Z_2 + Z_3^2)^2} = 1 - R. \tag{6.109}$$

Suppose that the impedance of the middle string is the geometric mean of the impedances of the two outer strings; that is, $Z_3 = \sqrt{Z_1\, Z_2}$. In this case, it follows, from the previous two equations, that $R = 0$ and $T = 1$. In other words, there is no reflection of the incident wave, and all of the incident energy ends up being transmitted across the middle string from the leftmost to the rightmost string. Thus, if we want to transmit transverse wave energy from a string of impedance Z_1 to a string of impedance Z_2 (where $Z_2 \neq Z_1$) in the most efficient manner possible—that is, with no reflection of the incident energy flux—then we can achieve this by connecting the two strings via a short section of string whose length is one quarter of a wavelength, and whose impedance is $\sqrt{Z_1\, Z_2}$. This procedure is known as *impedance matching*. Incidentally, impedance matching works because the one quarter of a wavelength (in length) middle string introduces a phase difference of π radians (i.e., the phase increase of a wave that travels from $x = 0$ to $x = a$, and back again) between the backward traveling wave reflected from the first interface (at $x = 0$) and the backward traveling wave reflected from the second interface (at $x = a$). Furthermore, if $Z_3 = \sqrt{Z_1 Z_2}$ then the two backward traveling waves have the same amplitude in the region $x < 0$. It follows that the two waves undergo total destructive interference in this region, which implies that there is no reflected wave on the first string.

The previous analysis of the reflection and transmission of transverse waves at an interface between two strings is also applicable to the reflection and transmission of other types of wave incident on an interface between two media of differing impedances. For example, consider a transmission line, such as a co-axial cable. Suppose that the line occupies the region $x < 0$, and is terminated (at $x = 0$) by a load resistor of resistance R_L. Such a resistor might represent a radio antenna [which acts just like a resistor in an electrical circuit, except that the dissipated energy is radiated, rather than being converted into heat energy (Fitzpatrick 2088)]. Suppose that a signal of angular frequency ω is sent down the line from a wave source at $x = -\infty$. The current and voltage on the line can be written

$$I(x, t) = I_i \cos(\omega t - k x) + I_r \cos(\omega t + k x), \tag{6.110}$$

$$V(x, t) = I_i Z \cos(\omega t - k x) - I_r Z \cos(\omega t + k x), \tag{6.111}$$

where I_i is the amplitude of the incident signal, I_r the amplitude of the signal reflected by the load, Z the characteristic impedance of the line, and $k = \omega/v$. Here, v is the characteristic phase velocity at which signals propagate down the line. (See Section 6.6.) The resistor obeys Ohm's law, which yields

$$V(0, t) = I(0, t) R_L. \tag{6.112}$$

It follows, from the previous three equations, that

$$I_r = \left(\frac{Z - R_L}{Z + R_L}\right) I_i. \tag{6.113}$$

Hence, the coefficient of reflection, which is the ratio of the power reflected by the load to the power sent down the line, is

$$R = \left(\frac{I_r}{I_i}\right)^2 = \left(\frac{Z - R_L}{Z + R_L}\right)^2. \tag{6.114}$$

Furthermore, the coefficient of transmission, which is the ratio of the power absorbed by the load to the power sent down the line, takes the form

$$T = 1 - R = \frac{4 Z R_L}{(Z + R_L)^2}. \tag{6.115}$$

It can be seen, by comparison with Equations (6.89) and (6.90), that the load terminating the line acts just like another transmission line of impedance R_L. It follows that power can only be efficiently sent down a transmission line, and transferred to a terminating load, when the impedance of the line matches the effective impedance of the load (which, in this case, is the same as the resistance of the load). In other words, when $Z = R_L$ there is no reflection of the signal sent down the line (i.e., $R = 0$), and all of the signal energy is therefore absorbed by the load (i.e., $T = 1$). As an example a *half-wave antenna* (i.e., an antenna whose length is half the wavelength of the emitted radiation) has a characteristic impedance of $73\,\Omega$ (Held and Marion 1995). Hence, a transmission line used to feed energy into such an antenna should also have a characteristic impedance of $73\,\Omega$. Suppose however, that we encounter a situation in which the impedance of a transmission line, Z_1, does no match that of its terminating load, Z_2. Can anything be done to avoid reflection of the signal sent down the line? It turns out, by analogy with the previous analysis, that if the line is connected to the load via a short section of transmission line whose length is one quarter of the wavelength of the signal, and whose characteristic impedance is $Z_3 = \sqrt{Z_1 Z_2}$, then there is no reflection of the signal; that is, all of the signal power is absorbed by the load. A short section of transmission line used in this manner is known as a *quarter-wave transformer*.

6.8   ELECTROMAGNETIC WAVES

Consider a plane *electromagnetic wave* propagating through a vacuum in the z-direction. Incidentally, electromagnetic waves are the only commonly-occurring waves that do not require a medium through which to propagate. Suppose that the wave is linearly polarized in the x-direction; that is, its electric component oscillates in the x-direction. It follows that the magnetic component of the wave oscillates in the y-direction (Fitzpatrick 2008). According to standard electromagnetic theory (see Appendix C), the wave is described by the following pair of coupled partial differential equations:

$$\frac{\partial E_x}{\partial t} = -\frac{1}{\epsilon_0}\frac{\partial H_y}{\partial z}, \tag{6.116}$$

$$\frac{\partial H_y}{\partial t} = -\frac{1}{\mu_0}\frac{\partial E_x}{\partial z}, \tag{6.117}$$

where $E_x(z, t)$ is the *electric field-strength*, and $H_y(z, t)$ is the *magnetic intensity* (i.e., the magnetic field-strength divided by μ_0). Observe that Equations (6.116) and (6.117), which govern the propagation of electromagnetic waves through a vacuum, are analogous to Equations (6.53) and (6.54), which govern the propagation of electromagnetic signals down a transmission line. In particular, E_x has units of voltage over length, H_y has units of current over length, ϵ_0 has units of capacitance per unit length, and μ_0 has units of inductance per unit length.

Equations (6.116) and (6.117) can be combined to give

$$\frac{\partial^2 E_x}{\partial t^2} = \frac{1}{\epsilon_0 \mu_0}\frac{\partial^2 E_x}{\partial z^2}, \tag{6.118}$$

$$\frac{\partial^2 H_y}{\partial t^2} = \frac{1}{\epsilon_0 \mu_0}\frac{\partial^2 H_y}{\partial z^2}. \tag{6.119}$$

It follows that the electric and the magnetic components of an electromagnetic wave propagating through a vacuum both separately satisfy a wave equation of the form (6.1). Furthermore, the phase velocity of the wave is the velocity of light in vacuum,

$$c = \frac{1}{\sqrt{\epsilon_0 \mu_0}} = 2.998 \times 10^8 \text{ m s}^{-1}. \tag{6.120}$$

Let us search for a traveling wave solution of Equations (6.116) and (6.117), propagating in the positive z-direction, whose electric component has the form

$$E_x(z, t) = E_0 \cos(\omega t - k z - \phi). \tag{6.121}$$

This is a valid solution provided that

$$\omega = k c. \tag{6.122}$$

According to Equation (6.117), the magnetic component of the wave is written

$$H_y(z, t) = Z^{-1} E_0 \cos(\omega t - k z - \phi), \tag{6.123}$$

where

$$Z = Z_0 \equiv \sqrt{\frac{\mu_0}{\epsilon_0}}, \tag{6.124}$$

and Z_0 is the impedance of free space. [See Equation (6.74).] Thus, the electric and magnetic components of an electromagnetic wave propagating through a vacuum are mutually perpendicular, and also perpendicular to the direction of propagation. Moreover, the two components oscillate in phase (i.e., they have simultaneous maxima and zeros), and the amplitude of the magnetic component is that of the electric component divided by the impedance of free space.

Multiplying Equation (6.116) by $\epsilon_0 E_x$, Equation (6.117) by $\mu_0 H_y$, and adding the two resulting expressions, we obtain the energy conservation equation

$$\frac{\partial \mathcal{E}}{\partial t} + \frac{\partial \mathcal{I}_z}{\partial z} = 0, \tag{6.125}$$

where

$$\mathcal{E} = \frac{1}{2}\left(\epsilon_0 E_x^2 + \mu_0 H_y^2\right) \tag{6.126}$$

is the energy density (i.e., energy per unit volume) of the wave (Fitzpatrick 2008), whereas

$$\mathcal{I}_z = E_x H_y \tag{6.127}$$

is the energy flux (i.e., power per unit area) in the positive z-direction. (See Appendix C.) The mean energy flux associated with the z-directed electromagnetic wave specified in Equations (6.121) and (6.123) is thus

$$\langle \mathcal{I}_z \rangle = \frac{1}{2}\frac{E_0^2}{Z}. \tag{6.128}$$

For a similar wave propagating in the negative z-direction, it can be demonstrated that

$$E_x(z, t) = E_0 \cos(\omega t + k z - \phi), \tag{6.129}$$

$$H_y(z, t) = -Z^{-1} E_0 \cos(\omega t + k z - \phi), \tag{6.130}$$

and

$$\langle \mathcal{I}_z \rangle = -\frac{1}{2}\frac{E_0^2}{Z}. \tag{6.131}$$

Consider a plane electromagnetic wave, linearly polarized in the x-direction, that propagates in the z-direction through a transparent *dielectric medium*, such as glass or water. As is well-known (Fitzpatrick 2008), the electric component of the wave causes the neutral molecules making up the medium to polarize; that is, it causes a small separation to develop between the mean positions of the positively and negatively charged constituents of the molecules (i.e., the atomic nuclei and the orbiting electrons). [Incidentally, it can be shown that the magnetic component of the wave has a negligible influence on the molecules, provided the wave amplitude is sufficiently small that the wave electric field does not cause the electrons and nuclei to move with relativistic velocities (ibid.).] If the mean position of the positively charged constituents of a given molecule, of net charge $+q$, develops a vector displacement $\mathbf{d}$ with respect to the mean position of the negatively charged constituents, of net charge $-q$, in response to a wave electric field $\mathbf{E}$, then the associated *electric dipole moment* is $\mathbf{p} = q\,\mathbf{d}$, where $\mathbf{d}$ is generally parallel to $\mathbf{E}$ (ibid.). Furthermore, if there are N such molecules per unit volume then the *electric dipole moment per unit volume* is written $\mathbf{P} = N q\,\mathbf{d}$. In a linear, isotropic, dielectric medium (ibid.),

$$\mathbf{P} = \epsilon_0\,(\epsilon - 1)\,\mathbf{E}, \tag{6.132}$$

where $\epsilon > 1$ is a dimensionless quantity, known as the *relative dielectric constant,* that is a property of the medium in question. In the presence of a dielectric medium, Equations (6.116) and (6.117) generalize to give

$$\frac{\partial E_x}{\partial t} = -\frac{1}{\epsilon_0}\left(\frac{\partial P_x}{\partial t} + \frac{\partial H_y}{\partial z}\right),$$

(6.133)

$$\frac{\partial H_y}{\partial t} = -\frac{1}{\mu_0}\frac{\partial E_x}{\partial z}.$$

(6.134)

(See Appendix C.) When combined with Equation (6.132), these expressions yield

$$\frac{\partial E_x}{\partial t} = -\frac{1}{\epsilon \epsilon_0}\frac{\partial H_y}{\partial z},$$

(6.135)

$$\frac{\partial H_y}{\partial t} = -\frac{1}{\mu_0}\frac{\partial E_x}{\partial z}.$$

(6.136)

It can be seen that the previous equations are just like the corresponding vacuum equations, (6.116) and (6.117), except that ϵ_0 has been replaced by $\epsilon \epsilon_0$. It immediately follows that the phase velocity of an electromagnetic wave propagating through a dielectric medium is

$$v = \frac{1}{\sqrt{\epsilon \epsilon_0 \mu_0}} = \frac{c}{n},$$

(6.137)

where $c = 1/(\epsilon_0 \mu_0)^{1/2}$ is the velocity of light in vacuum, and the dimensionless quantity

$$n = \sqrt{\epsilon}$$

(6.138)

is known as the *refractive index* of the medium. Thus, an electromagnetic wave propagating through a transparent dielectric medium does so at a phase velocity that is less than the velocity of light in vacuum by a factor n (where $n > 1$). The dispersion relation of the wave is thus

$$\omega = k v = \frac{k c}{n}.$$

(6.139)

Furthermore, the impedance of a transparent dielectric medium becomes

$$Z = \sqrt{\frac{\mu_0}{\epsilon \epsilon_0}} = \frac{Z_0}{n},$$

(6.140)

where Z_0 is the impedance of free space.

Incidentally, the signal that travels down a transmission line is a form of guided electromagnetic wave. It follows that if the space between the two conductors that constitute the line is filled with dielectric material of relative dielectric constant ϵ then the signal propagates down the line at the reduced phase velocity

$$v = \frac{c}{\sqrt{\epsilon}}.$$

(6.141)

This occurs because the dielectric material increases the capacitance per unit length of the line by a factor ϵ, but leaves the inductance per unit length unchanged. (See Section 6.6.) For the same reason, the presence of the dielectric material decreases the impedance of the line by a factor $\sqrt{\epsilon}$. Hence, the impedance of a dielectric filled co-axial cable is [cf., Equation (6.78)]

$$Z = \frac{1}{2\pi \sqrt{\epsilon}} \ln\left(\frac{b}{a}\right) Z_0.$$

(6.142)

Here, a and b are the radii of the inner and outer conductors, respectively.

Suppose that the plane $z = 0$ forms the interface between two transparent dielectric media of refractive indices n_1 and n_2. Let the first medium occupy the region $z < 0$, and the second the region $z > 0$. Suppose that a plane electromagnetic wave, linearly polarized in the x-direction, and propagating in the positive z-direction, is launched toward the interface from a wave source of angular frequency ω situated at $z = -\infty$. We expect the wave incident on the interface to be partly reflected, and partly transmitted. The wave electric and magnetic fields in the region $z < 0$ are written

$$E_x(z, t) = E_i \cos(\omega t - k_1 z) + E_r \cos(\omega t + k_1 z), \tag{6.143}$$

$$H_y(z, t) = Z_1^{-1} E_i \cos(\omega t - k_1 z) - Z_1^{-1} E_r \cos(\omega t + k_1 z), \tag{6.144}$$

where E_i is the amplitude of (the electric component of) the incident wave, E_r the amplitude of the reflected wave, $k_1 = n_1 \omega/c$, and $Z_1 = Z_0/n_1$. The wave electric and magnetic fields in the region $z > 0$ take the form

$$E_x(z, t) = E_t \cos(\omega t - k_2 z), \tag{6.145}$$

$$H_y(z, t) = Z_2^{-1} E_t \cos(\omega t - k_2 z), \tag{6.146}$$

where E_t is the amplitude of the transmitted wave, $k_2 = n_2 \omega/c$, and $Z_2 = Z_0/n_2$. According to standard electromagnetic theory (see Appendix C), the appropriate matching conditions at the interface ($z = 0$) are that E_x and H_y are both continuous. Thus, continuity of E_x yields

$$E_i + E_r = E_t, \tag{6.147}$$

whereas continuity of H_y gives

$$n_1 (E_i - E_r) = n_2 E_t, \tag{6.148}$$

because $Z^{-1} \propto n$. It follows that

$$E_r = \left(\frac{n_1 - n_2}{n_1 + n_2}\right) E_i, \tag{6.149}$$

$$E_t = \left(\frac{2 n_1}{n_1 + n_2}\right) E_i. \tag{6.150}$$

The coefficient of reflection, R, is defined as the ratio of the reflected to the incident energy flux, so that

$$R = \left(\frac{E_r}{E_i}\right)^2 = \left(\frac{n_1 - n_2}{n_1 + n_2}\right)^2. \tag{6.151}$$

Likewise, the coefficient of transmission, T, is the ratio of the transmitted to the incident energy flux, so that

$$T = \frac{Z_2^{-1}}{Z_1^{-1}} \left(\frac{E_t}{E_i}\right)^2 = \frac{n_2}{n_1} \left(\frac{E_t}{E_i}\right)^2 = \frac{4 n_1 n_2}{(n_1 + n_2)^2} = 1 - R. \tag{6.152}$$

It can be seen, first of all, that if $n_1 = n_2$ then $E_r = 0$ and $E_t = E_i$. In other words, if the two media have the same indices of refraction then there is no reflection at the interface between them, and the transmitted wave is consequently equal in amplitude to the incident wave. On the other hand, if $n_1 \neq n_2$ then there is always some reflection at the interface. Indeed, the amplitude of the reflected wave is roughly proportional to the difference between n_1 and n_2. This has important practical consequences. We can only see a clean pane of glass in a window because some of the light incident on an air/glass interface is reflected, as a consequence of the different refractive indices of air and glass. As is well-known, it is a lot more difficult to see glass when it is submerged in water. This is because the refractive indices of glass and water are quite similar, and so there is very little reflection of light incident on a water/glass interface.

According to Equation (6.149), $E_r/E_i < 0$ when $n_2 > n_1$. The negative sign indicates a π radian phase shift of the (electric component of the) reflected wave, with respect to the incident wave. We conclude that there is a π radian phase shift of the reflected wave, relative to the incident wave, on reflection from an interface with a medium of greater refractive index. Conversely, there is zero phase shift on reflection from an interface with a medium of lesser refractive index. (This effect is important in thin-film interference. See Section 10.5.)

Equations (6.149)–(6.152) are analogous to Equations (6.87)–(6.90), with the inverse of the refractive index playing the role of impedance. This suggests, by analogy with earlier analysis, that we can prevent reflection of an electromagnetic wave normally incident at an interface between two transparent dielectric media of different refractive indices by separating the media in question by a thin transparent layer whose thickness is one quarter of a wavelength, and whose refractive index is the geometric mean of the refractive indices of the two media. This is the physical principle behind the *non-reflective lens coatings* used in high-quality optical instruments. (See Exercise 6.15.)

6.9 DOPPLER EFFECT

Consider a sinusoidal wave of angular frequency ω and wavenumber k that is propagating in the $+x$-direction. We can represent the wave in terms of a *wavefunction* of the form

$$\psi(x, t) = \psi_0 \cos(\omega t - k x). \tag{6.153}$$

The wavelength and frequency of the wave, as seen by a stationary observer, are $\lambda = 2\pi/k$ and $f = \omega/2\pi$, respectively. Consider a second observer moving with uniform speed u_o in the $+x$-direction. What are the wavelength and frequency of the wave seen by the latter observer? Assuming non-relativistic motion, the x-coordinate in the moving observer's frame of reference is given by the standard Gallilean transformation formula $x' = x - u_0 t$ (Rindler 1997). Both observers measure the same time. Hence, in the second observer's frame of reference, the wavefunction is written

$$\psi(x', t) = \psi_0 \cos(\omega' t - k x'), \tag{6.154}$$

where

$$\omega' = \omega - k u_o. \tag{6.155}$$

Here, we have replaced x by $x' + u_o t$ in Equation (6.153). Thus, the moving observer sees a wave possessing the same wavelength (i.e., the same k) but a different frequency (i.e., a different ω) to that seen by the stationary observer. This phenomenon is known as the *Doppler effect*. If f is the wave frequency (in hertz) seen by the stationary observer then the wave frequency seen by the moving observer is

$$f' = \left(1 - \frac{u_o}{v}\right) f, \tag{6.156}$$

where $v = \omega/k$ is the characteristic wave speed. Thus, an observer moving in the same direction as a wave sees a lower frequency than a stationary observer. On the other hand, an observer moving in the opposite direction to a wave sees a higher frequency than a stationary observer. Hence, the general Doppler shift formula (for a moving observer and a stationary wave source) is

$$f' = \left(1 \mp \frac{u_o}{v}\right) f, \tag{6.157}$$

where the upper/lower signs correspond to the observer moving in the same/opposite direction to the wave.

Consider a stationary observer measuring a wave emitted by a source that is moving towards the observer with speed u_s. Let v be the characteristic propagation speed of the wave. Consider two neighboring wave crests emitted by the source. Suppose that the first is emitted at time $t = 0$, and the second at time $t = T$, where $T = 1/f$ is the wave period in the frame of reference of the source. At time t, the first wave crest has traveled a distance $d_1 = vt$ towards the observer, whereas the second wave crest has traveled a distance $d_2 = v(t - T) + u_s T$ (measured from the position of the source at $t = 0$). Here, we have taken into account the fact that the source is a distance $u_s T$ closer to the observer when the second wave crest is emitted. The effective wavelength, λ', seen by the observer is the distance between neighboring wave crests. Hence,

$$\lambda' = d_1 - d_2 = (v - u_s) T. \tag{6.158}$$

Because $v = f' \lambda'$, the effective frequency f' seen by the observer is

$$f' = \frac{f}{1 - u_s/v}, \tag{6.159}$$

where f is the wave frequency in the frame of reference of the source. We conclude that if the source is moving towards the observer then the wave frequency is shifted upwards. Likewise, if the source is moving away from the observer then the frequency is shifted downwards. This manifestation of the Doppler effect is familiar from everyday experience. When an ambulance passes us on the street, its siren has a higher pitch (i.e., a high frequency) when it is coming towards us than when it is moving away from us. In fact, the oscillation frequency of the siren never changes. It is the Doppler shift induced by the motion of the siren with respect to a stationary listener that causes the frequency change.

The general formula for the shift in wave frequency induced by relative motion of an observer and a source is

$$f' = \left(\frac{1 \mp u_o/v}{1 \pm u_s/v} \right) f, \tag{6.160}$$

where u_o is the speed of the observer, and u_s is the speed of the source (both measured relative to the wave medium). The upper/lower signs correspond to relative motion by which the observer and the source move apart/together. If the observer and source are not moving directly toward or directly away from one another then the quantities u_o and u_s in the above formulae correspond to the components of the observer and source velocities, respectively, along the straight line that instantaneously joins them.

An important proviso to the previous formula is that it is strictly classical, and only holds for non-relativistic motion (i.e., u_0, $u_s \ll c$, where c is the velocity of light in vacuum.) In fact, when applied to light propagation in a vacuum, the formula is only accurate up to first-order in u_0/c and u_s/c (Rindler 1997). In other words, for light propagation the previous equation reduces to

$$f' = \left[1 - \frac{u}{c} + O\left(\frac{u^2}{c^2} \right) \right] f, \tag{6.161}$$

where u is the relative radial velocity of the source with respect to the observer (being positive when the source and observer are moving apart, and vice versa).

Probably the most well-known use of the Doppler effect in everyday life is in police speed traps. In such a trap, a police officer fires radar waves (i.e., electromagnetic waves of centimeter wavelength) of fixed frequency at an oncoming car. These waves reflect off the car, which effectively becomes a moving source. Hence, by measuring the frequency increase of the reflected waves, the officer can determine the car's speed.

6.10 WAVE PROPAGATION IN INHOMOGENEOUS MEDIA

We saw in Section 6.7 that if a wave is normally incident at an interface between two media of substantially different impedances then a significant fraction of the wave energy is reflected. Moreover, this reflection takes place in such a manner that the net energy flux on either side of the interface is the same. On the other hand, if the two media have similar impedances then the reflected energy fraction is small. In this case, the incident and transmitted waves have the same frequency (because this property of the waves is determined by the source) but slightly different wavenumbers (assuming that the characteristic wave propagation speeds on either side of the interface are slightly different). Suppose that a wave travels though a medium whose impedance and characteristic wave propagation speed are slowly-varying functions of position. We would expect such a wave to propagate through the medium without significant reflection. Moreover, the frequency of the wave should remain fixed, while its wavenumber slowly varies with position, in accordance with the dispersion relation. Finally, the amplitude of the wave should vary slowly with position in such a manner that the energy flux remains constant. Let us investigate further.

Suppose that the medium has a characteristic wave propagation speed $v(x)$, and a wave impedance $Z(x)$. The local wave dispersion relation is assumed to take the form

$$\omega = k(x)\, v(x). \tag{6.162}$$

Hence, at fixed wave frequency, the wavenumber varies with position as

$$k(x) = \frac{\omega}{v(x)}. \tag{6.163}$$

If $\phi(x, t)$ is the phase of the wave then the most general definitions of angular frequency and wavenumber are

$$\omega = \frac{\partial \phi}{\partial t}, \tag{6.164}$$

$$k = -\frac{\partial \phi}{\partial x}, \tag{6.165}$$

respectively. (See Section 9.11.) It follows that, in the present case,

$$\phi(x, t) = \omega \left(t - \int^x [v(x')]^{-1}\, dx' \right). \tag{6.166}$$

Hence, the wavefunction can be written

$$\psi(x, t) = A(x)\, \cos[\phi(x, t)] = A(x)\, \cos \left[\omega \left(t - \int^x [v(x')]^{-1}\, dx' \right) \right], \tag{6.167}$$

where $A(x)$ is the wave amplitude. The mean energy flux associated with the wave is

$$\langle I \rangle = \frac{1}{2}\, \omega^2\, Z(x)\, [A(x)]^2. \tag{6.168}$$

(See Section 6.5.) We would expect this flux to be position independent (otherwise, energy would not be conserved), which implies that $A(x) \propto [Z(x)]^{-1/2}$. Thus, we obtain

$$\psi(x, t) = \frac{A_0}{[Z(x)]^{1/2}}\, \cos \left[\omega \left(t - \int^x [v(x')]^{-1}\, dx' \right) \right], \tag{6.169}$$

where A_0 is a constant. The previous expression, which is known as the *WKB solution* (WKB stands for Wentzel–Kramers–Brillouin), is not exact. However, it can be demonstrated that the WKB so-lution is a good approximation to the true solution of the inhomogeneous wave equation in the limit that the variations in $v(x)$ and $Z(x)$ take place on lengthscales that are much longer than the wave-length of the wave (Schiff 1955). On the other hand, the WKB solution breaks down completely if the variations in $v(x)$ and $Z(x)$ take place on lengthscales smaller than, or similar to, the wavelength.

EXERCISES

6.1 (a) Write the traveling wave $\psi(x,t) = A \cos(\omega t - k x)$ as a superposition of two standing waves.

(b) Write the standing wave $\psi(x,t) = A \cos(k x) \cos(\omega t)$ as a superposition of two travel-ing waves propagating in opposite directions.

(c) Show that the following superposition of traveling waves,

$$\psi(x,t) = A \cos(\omega t - k x) + A R \cos(\omega t + k x),$$

can be written as the following superposition of standing waves,

$$\psi(x,t) = A (1 + R) \cos(k x) \cos(\omega t) + A (1 - R) \sin(k x) \sin(\omega t).$$

6.2 Show that the solution of the wave equation,

$$\frac{\partial^2 y}{\partial t^2} = v^2 \frac{\partial^2 y}{\partial x^2},$$

subject to the initial conditions

$$y(x, 0) = F(x),$$
$$\dot{y}(x, 0) = G(x),$$

for $-\infty \le x \le \infty$, can be written

$$y(x, t) = \frac{1}{2} \left[F(x - v t) + F(x + v t) + \frac{1}{v} \int_{x-vt}^{x+vt} G(x') \, dx' \right].$$

This is known as the *d'Alembert solution*.

6.3 Demonstrate that for a transverse traveling wave propagating on a stretched string,

$$\langle I \rangle = v \langle \mathcal{E} \rangle,$$

where $\langle I \rangle$ is the mean energy flux along the string due to the wave, $\langle \mathcal{E} \rangle$ is the mean wave energy per unit length, and v is the phase velocity of the wave.

6.4 A transmission line of characteristic impedance Z occupies the region $x < 0$, and is termi-nated at $x = 0$. Suppose that the current carried by the line takes the form

$$I(x, t) = I_i \cos(\omega t - k x) + I_r \cos(\omega t + k x)$$

for $x \le 0$, where I_i is the amplitude of the incident signal, and I_r the amplitude of the signal reflected at the end of the line. Let the end of the line be open circuited, such that the line is effectively terminated by an infinite resistance. Find the relationship between I_r and I_i. Show that the current and voltage oscillate $\pi/2$ radians out of phase everywhere along the line. Demonstrate that there is zero net flux of electromagnetic energy along the line.

6.5 Suppose that the transmission line in the previous exercise is short circuited, such that the line is effectively terminated by a negligible resistance. Find the relationship between I_r and I_i. Show that the current and voltage oscillate $\pi/2$ radians out of phase everywhere along the line. Demonstrate that there is zero net flux of electromagnetic energy along the line.

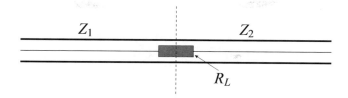

FIGURE 6.2 Figure for Exercise 6.6.

6.6 Two co-axial transmission lines of impedances Z_1 and Z_2 are connected as indicated in Figure 6.2. That is, the outer conductors are continuous, whereas the inner wires are connected to either side of a resistor of resistance R_L. The length of the resistor is negligible compared to the wavelengths of the signals propagating down the line. Suppose that $Z_1 > Z_2$. Suppose, further, that a signal is incident on the junction along the line whose impedance is Z_1.

(a) Show that the coefficients of reflection and transmission are

$$R = \left(\frac{R_L - Z_1 + Z_2}{R_L + Z_1 + Z_2}\right)^2,$$

$$T = \frac{4 Z_1 Z_1}{(R_L + Z_1 + Z_2)^2},$$

respectively.

(b) Hence, deduce that the choice

$$R_L = Z_1 - Z_2$$

suppresses reflection at the junction. Demonstrate that, in this case, the fraction of the incident power absorbed by the resistor is

$$A = 1 - \frac{Z_2}{Z_1}.$$

(c) Show, finally, that if the signal is, instead, incident along the line with impedance Z_2 (and $R_L = Z_1 - Z_2$) then the coefficient of reflection is

$$R = \left(1 - \frac{Z_2}{Z_1}\right)^2,$$

and the fraction of the incident power absorbed by the resistor is

$$A = \left(1 - \frac{Z_2}{Z_1}\right)\frac{Z_2}{Z_1}.$$

This analysis suggests that a resistor can be used to suppress reflection at a junction between two transmission lines, but that some of the incident power is absorbed by the resistor, and the suppression of the reflected signal only works if the signal is incident on the junction from one particular direction.

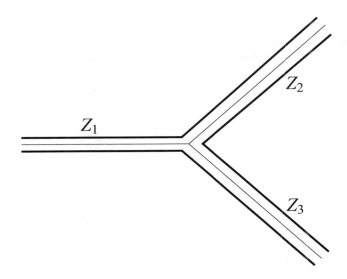

FIGURE 6.3   Figure for Exercise 6.7.

6.7 Consider a junction of three co-axial transmission lines of impedances Z_1, Z_2, and Z_3 whose inner and outer conductors are connected as shown in Figure 6.3. Suppose that a signal is incident on the junction along the line whose impedance is Z_1.

(a) Show that the coefficient of reflection is

$$R = \left[\frac{Z_2 Z_3 - Z_1 (Z_2 + Z_3)}{Z_2 Z_3 + Z_1 (Z_2 + Z_3)} \right]^2 ,$$

and that the fractions of the incident power that are transmitted down the lines with impedances Z_2 and Z_3 are

$$T_2 = \frac{4 Z_1 Z_2 Z_3^2}{[Z_2 Z_3 + Z_1 (Z_2 + Z_3)]^2},$$

$$T_3 = \frac{4 Z_1 Z_2^2 Z_3}{[Z_2 Z_3 + Z_1 (Z_2 + Z_3)]^2},$$

respectively.

(b) Hence, deduce that if the lines all have the same impedance then $R = 1/9$ and $T_2 = T_3 = 4/9$.

(c) Demonstrate, further, that if

$$\frac{1}{Z_1} = \frac{1}{Z_2} + \frac{1}{Z_3}$$

then there is no reflection, and

$$\frac{T_2}{T_3} = \frac{Z_3}{Z_2}.$$

This analysis suggests how one might construct a non-reflecting junction between three transmission lines that diverts a given fraction of the incident power into one of the outgoing lines, and the remainder of the power into the other outgoing line.

6.8 Consider the problem investigated in the previous question. Suppose that

$$\frac{1}{Z_1} = \frac{1}{Z_2} + \frac{1}{Z_3},$$

which implies that there is no reflection when a signal is incident on the junction along the transmission line whose impedance is Z_1. Suppose, however, that the signal is incident along the transmission line whose impedance is Z_2.

(a) Show that, in this case, the coefficient of reflection is

$$R = \left(\frac{Z_2}{Z_2 + Z_3}\right)^2.$$

(b) Likewise, show that the coefficient of reflection is

$$R = \left(\frac{Z_3}{Z_2 + Z_3}\right)^2$$

if the signal is incident along the transmission line whose coefficient of reflection is Z_3.

This analysis indicates that the lossless junction considered in the previous question can only be made non-reflecting when the signal is incident along one particular transmission line.

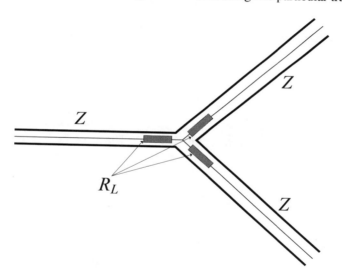

FIGURE 6.4   Figure for Exercise 6.9.

6.9 Consider a junction of three identical co-axial transmission lines of impedance Z that are connected in the manner shown in Figure 6.4. That is, the outer conductors are continuous, whereas the inner wires are connected via three identical resistors of resistance R_L. The lengths of the resistors are negligible compared to the wavelengths of the signals propagating down the lines.

(a) Show that if a signal is incident on the junction along a particular line then a fraction

$$R = \left(\frac{R_L - Z/3}{R_L + Z}\right)^2$$

of the incident power is reflected, and a fraction

$$T = 4\left(\frac{Z/3}{R_L + Z}\right)^2$$

is transmitted along each outgoing line.

(b) Hence, deduce that the choice

$$R_L = \frac{Z}{3}$$

suppresses reflection at the junction. Show that, in this case, half of the incident power is absorbed by the resistors.

This analysis indicates how one might construct a lossy junction between three identical transmission lines which is such that there is zero reflection no matter along which line the signal is incident.

6.10 Consider a generalization of the problem considered in the previous question in which $N > 1$ identical transmission lines of impedance Z meet at a common junction, and are connected via identical resistors of resistance R_L. Show that the choice

$$R_L = \left(1 - \frac{2}{N}\right)Z$$

suppresses reflection at the junction. Show, further, that, in this case, a fraction

$$A = \frac{N-2}{N-1}$$

of the incident power is absorbed by the resistors.

6.11 A lossy transmission line has a resistance per unit length $\mathcal{R}$, in addition to an inductance per unit length $\mathcal{L}$, and a capacitance per unit length C. The resistance can be considered to be in series with the inductance.

(a) Demonstrate that the Telegrapher's equations generalize to

$$\frac{\partial V}{\partial t} = -\frac{1}{C}\frac{\partial I}{\partial x},$$

$$\frac{\partial I}{\partial t} = -\frac{\mathcal{R}}{\mathcal{L}}I - \frac{1}{\mathcal{L}}\frac{\partial V}{\partial x},$$

where $I(x,t)$ and $V(x,t)$ are the voltage and current along the line.

(b) Derive an energy conservation equation of the form

$$\frac{\partial \mathcal{E}}{\partial t} + \frac{\partial \mathcal{I}}{\partial x} = -\mathcal{R}I^2,$$

where $\mathcal{E}$ is the energy per unit length along the line, and $\mathcal{I}$ the energy flux. Give expressions for $\mathcal{E}$ and $\mathcal{I}$. What does the right-hand side of the previous equation represent?

(c) Show that the current obeys the wave-diffusion equation

$$\frac{\partial^2 I}{\partial t^2} + \frac{\mathcal{R}}{\mathcal{L}}\frac{\partial I}{\partial t} = \frac{1}{\mathcal{L}C}\frac{\partial^2 I}{\partial x^2}.$$

(d) Consider the low-resistance, high-frequency, limit $\omega \gg R/\mathcal{L}$. Demonstrate that a signal propagating down the line varies as

$$I(x,t) \simeq I_0 \cos[k(vt-x)]e^{-x/\delta},$$

$$V(x,t) \simeq Z I_0 \cos[k(vt-x)-1/(k\delta)]e^{-x/\delta},$$

where $k = \omega/v$, $v = 1/\sqrt{\mathcal{L}C}$, $\delta = 2Z/\mathcal{R}$, and $Z = \sqrt{\mathcal{L}/C}$. Show that $k\delta \gg 1$; that is, the decay length of the signal is much longer than its wavelength. Estimate the maximum useful length of a low-resistance, high-frequency, lossy transmission line.

6.12 Suppose that a transmission line consisting of two uniform parallel conducting strips of width w and perpendicular distance apart d, where $d \ll w$, is terminated by a strip of material of uniform resistance per square meter $\sqrt{\mu_0/\epsilon_0} = 376.73\,\Omega$. Such material is known as *spacecloth*. Demonstrate that a signal sent down the line is completely absorbed, with no reflection, by the spacecloth. Incidentally, the resistance of a uniform strip of material is proportional to its length, and inversely proportional to its cross-sectional area.

6.13 At normal incidence, the mean radiant power from the Sun illuminating one square meter of the Earth's surface is $1.35\,\mathrm{kW}$. Show that the peak amplitude of the electric component of solar electromagnetic radiation at the Earth's surface is $1010\,\mathrm{V\,m^{-1}}$. Demonstrate that the corresponding peak amplitude of the magnetic component is $2.7\,\mathrm{A\,m^{-1}}$. [From Pain 1999.]

6.14 According to Einstein's famous formula, $E = mc^2$, where E is energy, m is mass, and c is the velocity of light in vacuum. This formula implies that anything that possesses energy also has an effective mass. Use this idea to show that an electromagnetic wave of mean intensity (energy per unit time per unit area) $\langle I \rangle$ has an associated mean pressure (momentum per unit time per unit area) $\langle \mathcal{P} \rangle = \langle I \rangle/c$. Hence, estimate the mean pressure due to sunlight at the Earth's surface (assuming that the sunlight is completely absorbed).

6.15 A glass lens is coated with a non-reflecting coating of thickness one quarter of a wavelength (in the coating) of light whose wavelength in air is λ_0. The index of refraction of the glass is n, and that of the coating is $\sqrt{n}$. The refractive index of air can be taken to be unity.

(a) Show that the coefficient of reflection for light normally incident on the lens from air is

$$R = \frac{(n-1)^2 \cos^2[(\pi/2)(\lambda_0/\lambda)]}{4n + (n-1)^2 \cos^2[(\pi/2)(\lambda_0/\lambda)]},$$

where λ is the wavelength of the incident light in air.

(b) Assume that $n = 1.5$, and that this value remains approximately constant for light whose wavelengths lie in the visible band. Suppose that $\lambda_0 = 550\,\mathrm{nm}$, which corresponds to green light. It follows that $R = 0$ for green light. What is R for blue light of wavelength $\lambda = 450\,\mathrm{nm}$, and for red light of wavelength $650\,\mathrm{nm}$? Comment on how effective the coating is at suppressing unwanted reflection of visible light incident on the lens.

[From Crawford 1968.]

6.16 A glass lens is coated with a non-reflective coating whose thickness is one quarter of a wavelength (in the coating) of light whose frequency is f_0. Demonstrate that the coating also suppresses reflection from light whose frequency is $3 f_0$, $5 f_0$, et cetera, assuming that the refractive index of the coating and the glass is frequency independent.

6.17 A plane electromagnetic wave, linearly polarized in the x-direction, and propagating in the z-direction through an electrical conducting medium of conductivity σ, is governed by

$$\frac{\partial H_y}{\partial t} = -\frac{1}{\mu_0}\frac{\partial E_x}{\partial z},$$

$$\frac{\partial E_x}{\partial t} = -\frac{\sigma}{\epsilon_0}E_x - \frac{1}{\epsilon_0}\frac{\partial H_y}{\partial z},$$

where $E_x(z,t)$ and $H_y(z,t)$ are the electric and magnetic components of the wave. (See Appendix C.)

(a) Derive an energy conservation equation of the form

$$\frac{\partial \mathcal{E}}{\partial t} + \frac{\partial I}{\partial z} = -\sigma E_x^2,$$

where $\mathcal{E}$ is the electromagnetic energy per unit volume, and I the electromagnetic energy flux. Give expressions for $\mathcal{E}$ and I. What does the right-hand side of the previous equation represent?

(b) Demonstrate that E_x obeys the wave-diffusion equation

$$\frac{\partial^2 E_x}{\partial t^2} + \frac{\sigma}{\epsilon_0}\frac{\partial E_x}{\partial t} = c^2\frac{\partial^2 E_x}{\partial z^2},$$

where $c = 1/\sqrt{\epsilon_0\mu_0}$.

(c) Consider the high-frequency, low-conductivity, limit $\omega \gg \sigma/\epsilon_0$. Show that a wave propagating into the medium varies as

$$E_x(z,t) \simeq E_0 \cos[k(v\,t - z)]\,e^{-z/\delta},$$

$$H_y(z,t) \simeq Z_0^{-1} E_0 \cos[k(v\,t - z) - 1/(k\,\delta)]\,e^{-z/\delta},$$

where $k = \omega/c$, $\delta = 2/(Z_0\,\sigma)$, and $Z_0 = \sqrt{\mu_0/\epsilon_0}$. Demonstrate that $k\,\delta \ll 1$; that is, the wave penetrates many wavelengths into the medium. Estimate how far a high-frequency electromagnetic wave penetrates into a low-conductivity conducting medium.

6.18 Sound waves travel horizontally from a source to a receiver. Let the source have the speed u and the receiver the speed v (in the same direction). In addition, suppose that a wind of speed w (in the same direction) is blowing from the source to the receiver. Show that if the source emits sound whose frequency is f_0 in still air then the frequency recorded by the receiver is

$$f = \left(\frac{V - v + w}{V - u + w}\right) f_0,$$

where V is the speed of sound in still air. Note that if the velocities of the source and receiver are the same then the wind makes no difference to the frequency of the recorded signal. [Modified from French 1971.]

CHAPTER 7

Multi-Dimensional Waves

7.1 INTRODUCTION

In this chapter, we shall generalize the analysis of the previous chapter in order to deal with plane waves that propagate in arbitrary directions, as well as propagating waves whose wave fronts are not necessarily planar.

7.2 PLANE WAVES

As we saw in the previous chapter, a sinusoidal wave of amplitude $\psi_0 > 0$, wavenumber $k > 0$, and angular frequency $\omega > 0$, propagating in the positive x-direction, can be represented in terms of a wavefunction of the form

$$\psi(x, t) = \psi_0 \cos(\omega t - k x). \tag{7.1}$$

This type of wave is conventionally termed a *one-dimensional plane wave*. It is one-dimensional because its associated wavefunction only depends on a single Cartesian coordinate. Furthermore, it is a plane wave because the wave maxima, which are located at

$$\omega t - k x = j 2\pi, \tag{7.2}$$

where j is an integer, consist of a series of parallel planes, normal to the x-axis, that are equally spaced a distance $\lambda = 2\pi/k$ apart, and propagate along the x-axis at the fixed phase velocity $\omega/k = v$, where v is the characteristic wave speed. These conclusions follow because Equation (7.2) can be rewritten in the form

$$x = d, \tag{7.3}$$

where $d = vt - j\lambda$. Moreover, Equation (7.3) is the equation of a plane, normal to the x-axis, whose distance of closest approach to the origin is d.

The previous equation can also be written in the coordinate-free form

$$\mathbf{n} \cdot \mathbf{r} = d, \tag{7.4}$$

where $\mathbf{n} = (1, 0, 0)$ is a unit vector directed along the x-axis, and $\mathbf{r} = (x, y, z)$ represents the vector displacement of a general point from the origin. Because there is nothing special about the x-direction, it follows that if $\mathbf{n}$ is re-interpreted as a unit vector pointing in an arbitrary direction then Equation (7.4) can be re-interpreted as the general equation of a plane (Fitzpatrick 2008). As before, the plane is normal to $\mathbf{n}$, and its distance of closest approach to the origin is d. See Figure 7.1. This observation allows us to write the three-dimensional equivalent to the wavefunction (7.1) as

$$\psi(x, y, z, t) = \psi_0 \cos(\omega t - \mathbf{k} \cdot \mathbf{r}), \tag{7.5}$$

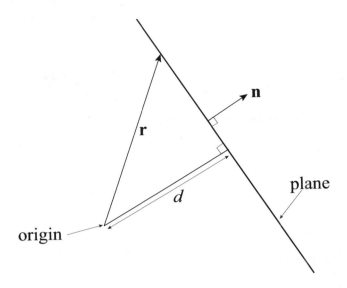

FIGURE 7.1 The solution of $\mathbf{n} \cdot \mathbf{r} = d$ is a plane.

where the constant vector $\mathbf{k} = (k_x, k_y, k_z) = k\mathbf{n}$ is known as the *wavevector*. The wave represented in the previous expression is conventionally termed a *three-dimensional plane wave*. It is
three-dimensional because its wavefunction, $\psi(x, y, z, t)$, depends on all three Cartesian coordinates.
Moreover, it is a plane wave because the wave maxima are located at

$$\omega t - \mathbf{k} \cdot \mathbf{r} = j\, 2\pi, \tag{7.6}$$

or

$$\mathbf{n} \cdot \mathbf{r} = v\, t - j\, \lambda, \tag{7.7}$$

where $\lambda = 2\pi/k$, and $\omega/k = v$. The wavenumber, k, is the magnitude of the wavevector, $\mathbf{k}$; that is,
$k = |\mathbf{k}|$. It follows, by comparison with Equation (7.4), that the wave maxima consist of a series of
parallel planes, normal to the wavevector, that are equally spaced a distance λ apart, and propagate
in the $\mathbf{k}$-direction at the fixed phase velocity $\omega/k = v$. See Figure 7.2. Hence, the direction of
the wavevector corresponds to the direction of wave propagation. The most general expression for
the wavefunction of a three-dimensional plane wave is $\psi = \psi_0 \cos(\omega t - \mathbf{k} \cdot \mathbf{r} - \phi)$, where ϕ is a
constant phase angle. As is readily demonstrated, the inclusion of a non-zero phase angle in the
wavefunction shifts all the wave maxima a distance $-(\phi/2\pi)\, \lambda$ in the $\mathbf{k}$-direction. In the following,
whenever possible, ϕ is set to zero, for the sake of simplicity.

7.3 THREE-DIMENSIONAL WAVE EQUATION

We have already seen that the one-dimensional plane wave solution, (7.1), satisfies the *one-dimensional wave equation*,

$$\frac{\partial^2 \psi}{\partial t^2} = v^2 \frac{\partial^2 \psi}{\partial x^2}, \tag{7.8}$$

where v is the characteristic wave speed of the medium through which the wave propagates.
(See Section 6.3.) Likewise, the three-dimensional plane wave solution, (7.5), satisfies the *three-*

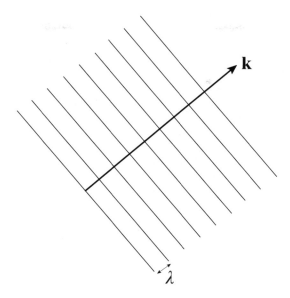

FIGURE 7.2   Wave maxima associated with a plane wave.

dimensional wave equation (see Exercise 7.1),

$$\frac{\partial^2 \psi}{\partial t^2} = v^2 \left(\frac{\partial^2}{\partial x^2} + \frac{\partial^2}{\partial y^2} + \frac{\partial^2}{\partial z^2} \right) \psi. \tag{7.9}$$

7.4   CYLINDRICAL WAVES

Consider a cylindrically-symmetric (about the z-axis) wavefunction $\psi(\rho, t)$, where $\rho = (x^2 + y^2)^{1/2}$ is a standard radial cylindrical coordinate (Fitzpatrick 2008). Assuming that this function satisfies the three-dimensional wave equation, (7.9), which can be rewritten (see Exercise 7.2)

$$\frac{\partial^2 \psi}{\partial t^2} = v^2 \left(\frac{\partial^2 \psi}{\partial \rho^2} + \frac{1}{\rho} \frac{\partial \psi}{\partial \rho} \right), \tag{7.10}$$

it can be shown (see Exercise 7.2) that a sinusoidal cylindrical wave of phase angle ϕ, wavenumber k, and angular frequency $\omega = k v$, has the approximate wavefunction

$$\psi(\rho, t) \simeq \frac{\psi_0}{\rho^{1/2}} \cos(\omega t - k\rho - \phi) \tag{7.11}$$

in the limit $k\rho \gg 1$. Here, $\psi_0/\rho^{1/2}$ is the amplitude of the wave. The associated wavefronts (i.e., the surfaces of constant phase) are a set of concentric cylinders that propagate radially outward, from their common axis ($\rho = 0$), at the phase velocity $\omega/k = v$. See Figure 7.3. The wave amplitude attenuates as $\rho^{-1/2}$. Such behavior can be understood as a consequence of energy conservation, according to which the power flowing across the various $\rho = $ const. surfaces must be constant. (The areas of such surfaces scale as $A \propto \rho$. Moreover, the power flowing across them is proportional to $\psi^2 A$, because the energy flux associated with a wave is generally proportional to ψ^2, and is directed normal to the wavefronts.) The cylindrical wave specified in expression (7.11) is such as would be generated by a uniform line source located at $\rho = 0$. See Figure 7.3.

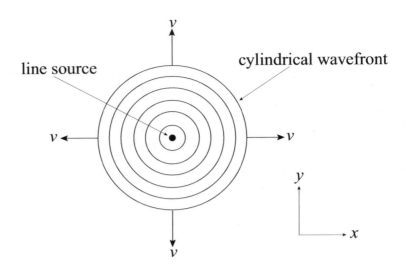

FIGURE 7.3   A cylindrical wave.

7.5   SPHERICAL WAVES

Consider a spherically-symmetric (about the origin) wavefunction $\psi(r, t)$, where $r = (x^2+y^2+z^2)^{1/2}$ is a standard radial spherical coordinate (Fitzpatrick 2008). Assuming that this function satisfies the three-dimensional wave equation, (7.9), which can be rewritten (see Exercise 7.3)

$$\frac{\partial^2 \psi}{\partial t^2} = v^2 \left(\frac{\partial^2 \psi}{\partial r^2} + \frac{2}{r} \frac{\partial \psi}{\partial r} \right), \tag{7.12}$$

it can be shown (see Exercise 7.3) that a sinusoidal spherical wave of phase angle ϕ, wavenumber k, and angular frequency $\omega = kv$, has the wavefunction

$$\psi(r, t) = \frac{\psi_0}{r} \cos(\omega t - k r - \phi). \tag{7.13}$$

Here, ψ_0/r is the amplitude of the wave. The associated wavefronts are a set of concentric spheres that propagate radially outward, from their common center ($r = 0$), at the phase velocity $\omega/k = v$. The wave amplitude attenuates as r^{-1}. Such behavior can again be understood as a consequence of energy conservation, according to which the power flowing across the various $r = $ const. surfaces must be constant. (The area of a constant-r surface scales as $A \propto r^2$, and the power flowing across such a surface is proportional to $\psi^2 A$.) The spherical wave specified in expression (7.13) is such as would be generated by a point source located at $r = 0$.

7.6   OSCILLATION OF AN ELASTIC SHEET

A straightforward generalization of the analysis of Section 4.3 reveals that the transverse oscillation of a uniform elastic sheet, stretched over a rigid frame, is governed by the two-dimensional wave equation (Pain 1999)

$$\frac{\partial^2 \psi}{\partial t^2} = v^2 \left(\frac{\partial^2 \psi}{\partial x^2} + \frac{\partial^2 \psi}{\partial y^2} \right). \tag{7.14}$$

Here, $\psi(x, y, t)$ is the sheet's transverse (i.e., in the z-direction) displacement, $v = \sqrt{T/\rho}$ the characteristic speed of elastic waves on the sheet, T the tension, and ρ the mass per unit area. In equilib-

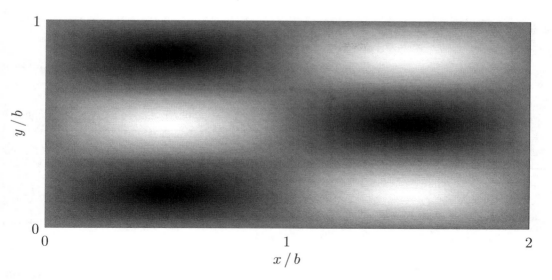

FIGURE 7.4 Density plot illustrating the spatial variation of the $m = 2$, $n = 3$ normal mode of a rectangular elastic sheet with $a = 2b$. Dark/light regions indicate positive/negative displacements.

rium, the sheet is assumed to lie in the x-y plane. The boundary condition is that $\psi = 0$ at the rigid frame.

Suppose that the frame is rectangular, extending from $x = 0$ to $x = a$, and from $y = 0$ to $y = b$. Let us search for a normal mode of the form

$$\psi(x, y, t) = \hat{\psi}(x, y) \cos(\omega t - \phi). \tag{7.15}$$

Substitution into Equation (7.14) yields

$$\frac{\partial^2 \hat{\psi}}{\partial x^2} + \frac{\partial^2 \hat{\psi}}{\partial y^2} + \frac{\omega^2}{v^2} \hat{\psi} = 0, \tag{7.16}$$

subject to the boundary conditions $\hat{\psi}(0, y) = \hat{\psi}(a, y) = \hat{\psi}(x, 0) = \hat{\psi}(x, b) = 0$. Let us search for a separable solution of the form

$$\hat{\psi}(x, y) = X(x)\, Y(y). \tag{7.17}$$

Such a solution satisfies the boundary conditions provided $X(0) = X(a) = Y(0) = Y(b) = 0$. It follows that

$$\frac{1}{X}\frac{d^2 X}{dx^2} + \frac{1}{Y}\frac{d^2 Y}{dy^2} + \frac{\omega^2}{v^2} = 0. \tag{7.18}$$

The only way that the preceding equation can be satisfied at all x and y is if

$$\frac{1}{X}\frac{d^2 X}{dx^2} = -\alpha^2, \tag{7.19}$$

$$\frac{1}{Y}\frac{d^2 Y}{dy^2} = -\beta^2, \tag{7.20}$$

$$\omega^2 = (\alpha^2 + \beta^2)\, v^2, \tag{7.21}$$

where α^2 and β^2 are positive constants. (The constants have to be positive, rather than negative, to give oscillatory solutions that are capable of satisfying the boundary conditions.) Appropriate solutions of Equations (7.19) and (7.20) are

$$X(x) = X_0 \sin(\alpha x), \qquad (7.22)$$

$$Y(y) = Y_0 \sin(\beta y), \qquad (7.23)$$

where X_0 and Y_0 are arbitrary constants. These solutions automatically satisfy the boundary conditions $X(0) = Y(0) = 0$. The boundary conditions $X(a) = Y(b) = 0$ are satisfied provided

$$\alpha = \frac{m\pi}{a}, \qquad (7.24)$$

$$\beta = \frac{n\pi}{b}, \qquad (7.25)$$

where m and n are positive integers. Thus, the normal modes of a rectangular elastic sheet, which are indexed by the mode numbers m and n, take the form

$$\psi_{m,n}(x, y, t) = A_{m,n} \sin\left(m\pi \frac{x}{a}\right) \sin\left(n\pi \frac{y}{b}\right) \cos(\omega_{m,n} t - \phi_{m,n}), \qquad (7.26)$$

where

$$\omega_{m,n} = \pi \left(\frac{m^2}{a^2} + \frac{n^2}{b^2}\right)^{1/2} v. \qquad (7.27)$$

Here, $A_{m,n}$ and $\phi_{m,n}$ are arbitrary constants. Because Equation (7.14) is linear, its solutions are superposable. Hence, the most general solution is a superposition of all of the normal modes; that is,

$$\psi(x, y, t) = \sum_{m=1,\infty} \sum_{n=1,\infty} A_{m,n} \sin\left(m\pi \frac{x}{a}\right) \sin\left(n\pi \frac{y}{b}\right) \cos(\omega_{m,n} t - \phi_{m,n}). \qquad (7.28)$$

The amplitudes $A_{m,n}$, and the phase angles $\phi_{n,m}$, are determined by the initial conditions. (See Exercise 7.4.) Figure 7.4 illustrates the spatial variation of the $m = 2, n = 3$ normal mode of a rectangular elastic sheet with $a = 2b$.

Suppose that an elastic sheet is stretched over a circular frame of radius a. Defining the radial cylindrical coordinate $\rho = (x^2 + y^2)^{1/2}$, where $\rho = a$ corresponds to the location of the frame, the axisymmetric oscillations of the sheet are governed by the cylindrical wave equation (see Section 7.4)

$$\frac{\partial^2 \psi}{\partial t^2} = v^2 \left(\frac{\partial^2 \psi}{\partial \rho^2} + \frac{1}{\rho} \frac{\partial \psi}{\partial \rho}\right), \qquad (7.29)$$

subject to the boundary condition $\psi = 0$ at $\rho = a$. Let us search for a normal mode of the form

$$\psi(\rho, t) = \hat{\psi}(\rho) \cos(\omega t - \phi). \qquad (7.30)$$

Substitution into Equation (7.29) yields

$$\frac{d^2 \hat{\psi}}{d\rho^2} + \frac{1}{\rho} \frac{d\hat{\psi}}{d\rho} + \frac{\omega^2}{v^2} \hat{\psi} = 0, \qquad (7.31)$$

subject to the boundary condition $\hat{\psi}(a) = 0$. Let us define the scaled radial coordinate $z = (\omega/v)\rho$. When written in terms of this new coordinate, the previous equation transforms to

$$\frac{d^2 \hat{\psi}}{dz^2} + \frac{1}{z} \frac{d\hat{\psi}}{dz} + \hat{\psi} = 0. \qquad (7.32)$$

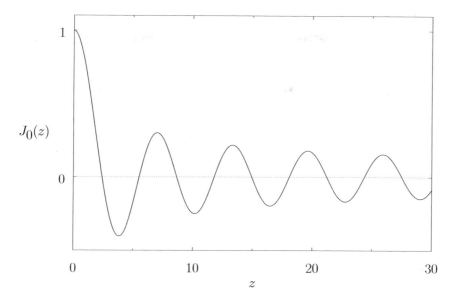

FIGURE 7.5   The Bessel function $J_0(z)$.

This well-known equation is called *Bessel's equation* (of degree zero), and has the standard solution (that is well-behaved at $z = 0$) (Abramowitz and Stegun 1965)

$$J_0(z) = \frac{1}{\pi} \int_0^\pi \cos(z \sin v)\, dv. \tag{7.33}$$

Here, $J_0(z)$ is termed a *Bessel function* (of degree zero), and is plotted in Figure 7.5. It can be seen that the function is broadly similar to a cosine function, except that its zeros are not quite evenly spaced, and its amplitude gradually decreases as z increases. The first few values of z at which $J_0(z) = 0$ are listed in Table 7.1. Let the jth zero be located at $z = z_j$. In order to satisfy the boundary condition, we require that $(\omega/v)\,a = z_j$. Hence, the axisymmetric normal modes of an elastic sheet, stretched over a circular frame of radius a, are indexed by the mode number j (which is a positive integer), and take the form

$$\psi_j(\rho, t) = A_j J_0\left(\frac{z_j \rho}{a}\right) \cos(\omega_j t - \phi_j), \tag{7.34}$$

where

$$\omega_j = z_j \frac{v}{a}, \tag{7.35}$$

and A_j, ϕ_j are arbitrary constants. Figure 7.6 illustrates the spatial variation of the $j = 5$ normal mode.

7.7   POLARIZATION OF ELECTROMAGNETIC WAVES

The electric component of an electromagnetic plane wave can oscillate in any direction normal to the direction of wave propagation (which is parallel to the **k**-vector) (Fitzpatrick 2008). Suppose that the wave is propagating in the z-direction. It follows that the electric field can oscillate in any direction that lies in the x-y plane. The actual direction of oscillation determines the *polarization* of the wave. For instance, a vacuum electromagnetic wave of angular frequency ω that is polarized in

j	z_j	j	z_j
1	2.40482	6	18.07106
2	5.52007	7	21.21163
3	8.65372	8	24.35247
4	11.79153	9	27.49347
5	14.93091	10	30.63460

TABLE 7.1   First few zeros of the Bessel function $J_0(z)$. Source: Abramowitz and Stegun 1965.

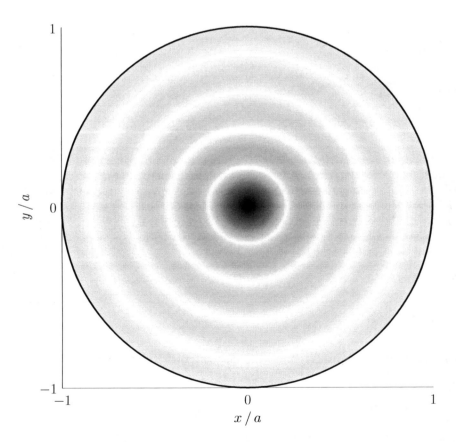

FIGURE 7.6   Density plot illustrating the spatial variation of the $j = 5$ normal mode of a circular elastic sheet of radius a. Dark/light regions indicate large/small displacement amplitudes.

the x-direction has the associated electric field

$$\mathbf{E} = E_0 \cos(\omega t - kz)\,\mathbf{e}_x, \tag{7.36}$$

where $\omega = kc$. Likewise, a wave polarized in the y-direction has the electric field

$$\mathbf{E} = E_0 \cos(\omega t - kz)\,\mathbf{e}_y. \tag{7.37}$$

These two waves are termed *linearly polarized*, because the electric field vector oscillates in a straight line. However, other types of polarization are possible. For instance, if we combine two linearly polarized waves of equal amplitude, one polarized in the x-direction, and one in the y-direction, that oscillate $\pi/2$ radians out of phase, then we obtain a *circularly polarized* wave:

$$\mathbf{E} = E_0 \cos(\omega t - kz)\,\mathbf{e}_x + E_0 \sin(\omega t - kz)\,\mathbf{e}_y. \tag{7.38}$$

This nomenclature arises from the fact that the tip of the electric field vector traces out a circle in the plane normal to the direction of wave propagation. To be more exact, the previous wave is a *right-hand* circularly polarized wave, because if the thumb of the right hand points in the direction of wave propagation then the electric field vector rotates in the same sense as the fingers of this hand. Conversely, a left-hand circularly polarized wave takes the form

$$\mathbf{E} = E_0 \cos(\omega t - kz)\,\mathbf{e}_x - E_0 \sin(\omega t - kz)\,\mathbf{e}_y. \tag{7.39}$$

Finally, if the x- and y-components of the electric field in the previous two expressions have different (non-zero) amplitudes then we obtain right-hand and left-hand *elliptically polarized* waves, respectively. This nomenclature arises from the fact that the tip of the electric field vector traces out an ellipse in the plane normal to the direction of wave propagation.

7.8 LAWS OF GEOMETRIC OPTICS

Suppose that the region $z < 0$ is occupied by a transparent dielectric medium of refractive index n_1, whereas the region $z > 0$ is occupied by a second transparent dielectric medium of refractive index n_2. See Figure 7.7. Let a plane light wave be launched toward positive z from a light source of angular frequency ω located at large negative z. Furthermore, suppose that this wave, which has the wavevector $\mathbf{k}_i$, is obliquely incident on the interface between the two media. We would expect the incident plane wave to be partially reflected and partially refracted (i.e., transmitted) by the interface. (See Section 6.8.) Let the reflected and refracted plane waves have the wavevectors $\mathbf{k}_r$ and $\mathbf{k}_t$, respectively. See Figure 7.7. Hence, we can write

$$\psi(x, y, z, t) = \psi_i \cos(\omega t - \mathbf{k}_i \cdot \mathbf{r}) + \psi_r \cos(\omega t - \mathbf{k}_r \cdot \mathbf{r}) \tag{7.40}$$

in the region $z < 0$, and

$$\psi(x, y, z, t) = \psi_t \cos(\omega t - \mathbf{k}_t \cdot \mathbf{r}) \tag{7.41}$$

in the region $z > 0$. Here, $\psi(x, y, z, t)$ represents the magnetic component of the resultant light wave, ψ_i the amplitude of the incident wave, ψ_r the amplitude of the reflected wave, and ψ_t the amplitude of the refracted wave. All of the component waves have the same angular frequency, ω, because this property is ultimately determined by the wave source. Furthermore, according to standard electromagnetic theory (Fitzpatrick 2008), if the magnetic component of an electromagnetic wave is specified then the electric component of the wave is fully determined, and vice versa.

In general, the wavefunction, ψ, must be continuous at $z = 0$, because, according to standard electromagnetic theory (see Appendix C), there cannot be a discontinuity in either the normal or the

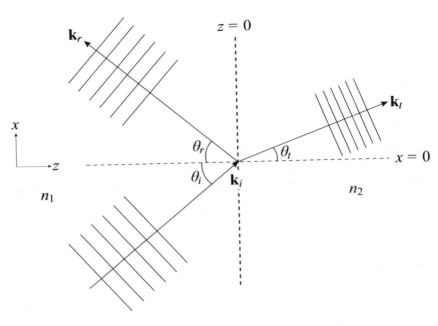

FIGURE 7.7   Reflection and refraction of a plane wave at a plane boundary.

tangential component of a magnetic field across an interface between two (non-magnetic) dielectric media. [The same is not true of an electric field, which can have a normal discontinuity across an interface between two dielectric media (ibid.). This explains why we have chosen ψ to represent the magnetic, rather than the electric, component of the wave.] Thus, the matching condition at $z = 0$ takes the form

$$\psi_i \cos(\omega t - k_{ix}\, x - k_{iy}\, y)$$
$$+ \psi_r \cos(\omega t - k_{rx}\, x - k_{ry}\, y) = \psi_t \cos(\omega t - k_{tx}\, x - k_{ty}\, y). \tag{7.42}$$

This condition must be satisfied at all values of x, y, and t. This is only possible if

$$k_{ix} = k_{rx} = k_{tx}, \tag{7.43}$$

and

$$k_{iy} = k_{ry} = k_{ty}. \tag{7.44}$$

Suppose that the direction of propagation of the incident wave lies in the x-z plane, so that $k_{iy} = 0$. It immediately follows, from Equation (7.44), that $k_{ry} = k_{ty} = 0$. In other words, the directions of propagation of the reflected and the refracted waves also lie in the x-z plane, which implies that $\mathbf{k}_i$, $\mathbf{k}_r$ and $\mathbf{k}_t$ are co-planar vectors. This constraint is implicit in the well-known laws of geometric optics (Hecht 1974).

Assuming that the previously mentioned constraint is satisfied, let the incident, reflected, and refracted wave normals subtend angles θ_i, θ_r, and θ_t with the z-axis, respectively. See Figure 7.7. It follows that

$$\mathbf{k}_i = n_1\, k_0\, (\sin\theta_i,\, 0,\, \cos\theta_i), \tag{7.45}$$
$$\mathbf{k}_r = n_1\, k_0\, (\sin\theta_r,\, 0,\, -\cos\theta_r), \tag{7.46}$$
$$\mathbf{k}_t = n_2\, k_0\, (\sin\theta_t,\, 0,\, \cos\theta_t), \tag{7.47}$$

where $k_0 = \omega/c$ is the vacuum wavenumber, and c the velocity of light in vacuum. Here, we have made use of the fact that wavenumber (i.e., the magnitude of the wavevector) of a light wave propagating through a dielectric medium of refractive index n is $n\,k_0$. (See Section 6.8.)

According to Equation (7.43), $k_{ix} = k_{rx}$, which yields

$$\sin\theta_i = \sin\theta_r, \tag{7.48}$$

and $k_{ix} = k_{tx}$, which reduces to

$$n_1\,\sin\theta_i = n_2\,\sin\theta_t. \tag{7.49}$$

The first of these relations states that the angle of incidence, θ_i, is equal to the angle of reflection, θ_r. This is the familiar *law of reflection* (Hecht 1974). Furthermore, the second relation corresponds to the equally familiar *law of refraction*, otherwise known as *Snell's law* (ibid.).

Incidentally, the fact that a plane wave propagates through a uniform dielectric medium with a constant wavevector, and, therefore, a constant direction of motion, is equivalent to the well-known *law of rectilinear propagation*, which states that a *light ray* (i.e., the normal to a constant phase surface) propagates through a uniform medium in a straight line (Hecht 1974).

It follows, from the previous discussion, that the laws of geometric optics (i.e., the law of rectilinear propagation, the law of reflection, and the law of refraction) are fully consistent with the wave properties of light, despite the fact that they do not seem to explicitly depend on these properties.

7.9 FRESNEL RELATIONS

The theory described in the previous section is sufficient to determine the directions of the reflected and refracted waves, when a light wave is obliquely incident on a plane interface between two dielectric media. However, it cannot determine the fractions of the incident energy that are reflected and refracted, respectively. In order to calculate the coefficients of reflection and transmission, it is necessary to take into account both the electric and the magnetic components of the various waves. It turns out that there are two independent wave polarizations that behave slightly differently. The first of these is such that the magnetic components of the incident, reflected, and refracted waves are all parallel to the interface. The second is such that the electric components of these waves are all parallel to the interface.

Consider the first polarization. Let the interface correspond to the plane $z = 0$, let the region $z < 0$ be occupied by material of refractive index n_1, and let the region $z > 0$ be occupied by material of refractive index n_2. Suppose that the incident, reflected, and refracted waves are plane waves, of angular frequency ω, whose wavevectors lie in the x-z plane. See Figure 7.7. The equations governing oblique electromagnetic wave propagation through a uniform dielectric medium are a generalization of Equations (6.135)–(6.136), and take the form (see Appendix C)

$$\frac{\partial D_x}{\partial t} = -\frac{\partial H_y}{\partial z}, \tag{7.50}$$

$$\frac{\partial D_z}{\partial t} = \frac{\partial H_y}{\partial x}, \tag{7.51}$$

$$\frac{\partial H_y}{\partial t} = v^2\left(\frac{\partial D_z}{\partial x} - \frac{\partial D_x}{\partial z}\right), \tag{7.52}$$

where

$$\mathbf{D} = \epsilon_0\,\epsilon\,\mathbf{E} \tag{7.53}$$

is the *electric displacement*, $v = c/n$ the characteristic wave speed, and $n = \sqrt{\epsilon}$ the refractive index.

Suppose that, as described in the previous section,

$$H_y(x, z, t) = \psi_i \cos(\omega t - n_1 k_0 \sin \theta_i \, x - n_1 k_0 \cos \theta_i \, z)$$
$$+ \psi_r \cos(\omega t - n_1 k_0 \sin \theta_i \, x + n_1 k_0 \cos \theta_i \, z) \quad (7.54)$$

in the region $z < 0$, and

$$H_y(x, z, t) = \psi_t \cos(\omega t - n_1 k_0 \sin \theta_i \, x - n_2 k_0 \cos \theta_t \, z). \quad (7.55)$$

in the region $z > 0$. Here, $k_0 = \omega/c$ is the vacuum wavenumber, θ_i the angle of incidence, and θ_t the angle of refraction. See Figure 7.7. In writing the previous expressions, we have made use of the law of reflection (i.e., $\theta_r = \theta_i$), as well as the law of refraction (i.e., $n_1 \sin \theta_i = n_2 \sin \theta_t$). The two terms on the right-hand side of Equation (7.54) correspond to the incident and reflected waves, respectively. The term on the right-hand side of Equation (7.55) corresponds to the refracted wave. Substitution of Equations (7.54)–(7.55) into the governing differential equations, (7.50)–(7.52), yields

$$D_x(x, z, t) = \frac{\psi_i \cos \theta_i}{v_1} \cos(\omega t - n_1 k_0 \sin \theta_i \, x - n_1 k_0 \cos \theta_i \, z)$$
$$- \frac{\psi_r \cos \theta_i}{v_1} \cos(\omega t - n_1 k_0 \sin \theta_i \, x + n_1 k_0 \cos \theta_i \, z), \quad (7.56)$$

$$D_z(x, z, t) = -\frac{\psi_i \sin \theta_i}{v_1} \cos(\omega t - n_1 k_0 \sin \theta_i \, x - n_1 k_0 \cos \theta_i \, z)$$
$$- \frac{\psi_r \sin \theta_i}{v_1} \cos(\omega t - n_1 k_0 \sin \theta_i \, x + n_1 k_0 \cos \theta_i \, z) \quad (7.57)$$

in the region $z < 0$, and

$$D_x(x, z, t) = \frac{\psi_t \cos \theta_t}{v_2} \cos(\omega t - n_1 k_0 \sin \theta_i \, x - n_2 k_0 \cos \theta_t \, z), \quad (7.58)$$

$$D_z(x, z, t) = -\frac{\psi_t \sin \theta_i}{v_1} \cos(\omega t - n_1 k_0 \sin \theta_i \, x - n_2 k_0 \cos \theta_t \, z) \quad (7.59)$$

in the region $z > 0$.

According to standard electromagnetic theory (see Appendix C), both the normal and the tangential components of the magnetic intensity must be continuous at the interface. This implies that

$$[H_y]_{z=0_-}^{z=0_+} = 0, \quad (7.60)$$

which yields

$$\psi_i + \psi_r = \psi_t \quad (7.61)$$

Standard electromagnetic theory (ibid.) also implies that the normal component of the electric displacement, as well as the tangential component of the electric field, are continuous at the interface. In other words,

$$[D_z]_{z=0_-}^{z=0_+} = 0, \quad (7.62)$$

and

$$[E_x]_{z=0_-}^{z=0_+} = \left[\frac{D_x}{\epsilon \, \epsilon_0} \right]_{z=0_-}^{z=0_+} = 0. \quad (7.63)$$

The former of these conditions again gives Equation (7.61), whereas the latter yields

$$\psi_i - \psi_r = \frac{\alpha}{\beta} \psi_t, \quad (7.64)$$

where

$$\alpha = \frac{\cos\theta_t}{\cos\theta_i}, \tag{7.65}$$

$$\beta = \frac{v_1}{v_2} = \frac{n_2}{n_1}. \tag{7.66}$$

It follows that

$$\psi_r = \left(\frac{\beta - \alpha}{\beta + \alpha}\right)\psi_i, \tag{7.67}$$

$$\psi_t = \left(\frac{2\beta}{\beta + \alpha}\right)\psi_i. \tag{7.68}$$

The electromagnetic energy flux in the z-direction (i.e., normal to the interface) is (see Appendix C),

$$I_z = E_x H_y \tag{7.69}$$

[cf. Equation (6.127)]. Thus, the mean energy fluxes associated with the incident, reflected, and refracted waves are

$$\langle I_z \rangle_i = \frac{\psi_i^2 \cos\theta_i}{2\,\epsilon_0\,c\,n_1}, \tag{7.70}$$

$$\langle I_z \rangle_r = -\frac{\psi_r^2 \cos\theta_i}{2\,\epsilon_0\,c\,n_1}, \tag{7.71}$$

$$\langle I_z \rangle_t = \frac{\psi_t^2 \cos\theta_t}{2\,\epsilon_0\,c\,n_2}, \tag{7.72}$$

respectively. The coefficients of reflection and transmission are defined

$$R = \frac{-\langle I_z \rangle_r}{\langle I_z \rangle_i}, \tag{7.73}$$

$$T = \frac{\langle I_z \rangle_t}{\langle I_z \rangle_i}, \tag{7.74}$$

respectively. Hence, it follows that

$$R = \left(\frac{\beta - \alpha}{\beta + \alpha}\right)^2, \tag{7.75}$$

$$T = \frac{4\,\alpha\,\beta}{(\beta + \alpha)^2} = 1 - R. \tag{7.76}$$

These expressions are known as *Fresnel relations*, and are the generalizations of expressions (6.151) and (6.152) for the case of oblique incidence (with the polarization in which the magnetic field is parallel to the interface).

Let us now consider the second polarization, in which the electric components of the incident, reflected, and refracted waves are all parallel to the interface. In this case, the governing equations are (see Appendix C)

$$\frac{\partial H_x}{\partial t} = v^2 \frac{\partial D_y}{\partial z}, \tag{7.77}$$

$$\frac{\partial H_z}{\partial t} = -v^2 \frac{\partial D_y}{\partial x}, \tag{7.78}$$

$$\frac{\partial D_y}{\partial t} = -\frac{\partial H_z}{\partial x} + \frac{\partial H_x}{\partial z}. \tag{7.79}$$

If we make the transformations $H_y \rightarrow -v^2 D_y$, $D_x \rightarrow H_x$, $D_z \rightarrow H_z$, $\psi_{i,r,t} \rightarrow -v \psi_{i,r,t}$ then we can reuse the solutions that we derived for the other polarization. We find that

$$D_y(x, z, t) = \frac{\psi_i}{v_1} \cos(\omega t - n_1 k_0 \sin \theta_i x - n_1 k_0 \cos \theta_i z)$$

$$+ \frac{\psi_r}{v_1} \cos(\omega t - n_1 k_0 \sin \theta_i x + n_1 k_0 \cos \theta_i z), \tag{7.80}$$

$$H_x(x, z, t) = -\psi_i \cos \theta_i \cos(\omega t - n_1 k_0 \sin \theta_i x - n_1 k_0 \cos \theta_i z)$$

$$+ \psi_r \cos \theta_i \cos(\omega t - n_1 k_0 \sin \theta_i x + n_1 k_0 \cos \theta_i z), \tag{7.81}$$

$$H_z(x, z, t) = \psi_i \sin \theta_i \cos(\omega t - n_1 k_0 \sin \theta_i x - n_1 k_0 \cos \theta_i z)$$

$$+ \psi_r \sin \theta_i \cos(\omega t - n_1 k_0 \sin \theta_i x + n_1 k_0 \cos \theta_i z) \tag{7.82}$$

in the region $z < 0$, and

$$D_y(x, z, t) = \frac{\psi_t}{v_2} \cos(\omega t - n_1 k_0 \sin \theta_i x - n_2 k_0 \cos \theta_t z), \tag{7.83}$$

$$H_x(x, z, t) = -\psi_t \cos \theta_t \cos(\omega t - n_1 k_0 \sin \theta_i x - n_2 k_0 \cos \theta_t z), \tag{7.84}$$

$$H_z(x, z, t) = \psi_t \sin \theta_t \cos(\omega t - n_1 k_0 \sin \theta_i x - n_2 k_0 \cos \theta_t z) \tag{7.85}$$

in the region $z > 0$. The first two matching conditions at the interface are that the normal and tangential components of the magnetic intensity are continuous. (See Appendix C.) In other words,

$$[H_z]_{z=0_+}^{z=0_-} = 0, \tag{7.86}$$

$$[H_x]_{z=0_+}^{z=0_-} = 0. \tag{7.87}$$

The first of these conditions yields

$$\psi_i + \psi_r = \beta^{-1} \psi_t, \tag{7.88}$$

whereas the second gives

$$\psi_i - \psi_r = \alpha \psi_t. \tag{7.89}$$

The final matching condition at the interface is that the tangential component of the electric field is continuous. (See Appendix C.) In other words,

$$[E_y]_{z=0_-}^{z=0_+} = \left[\frac{D_y}{\epsilon \epsilon_0} \right]_{z=0_-}^{z=0_+} = 0, \tag{7.90}$$

which again yields Equation (7.88). It follows that

$$\psi_r = \left(\frac{1 - \alpha\beta}{1 + \alpha\beta} \right) \psi_i, \tag{7.91}$$

$$\psi_t = \left(\frac{2\beta}{1 + \alpha\beta} \right) \psi_i. \tag{7.92}$$

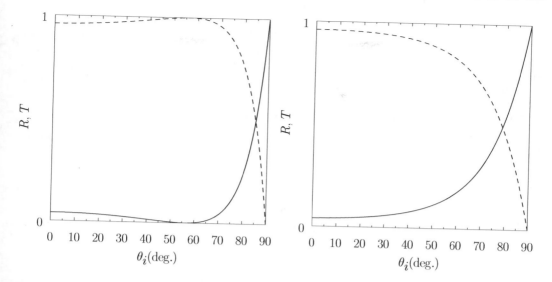

FIGURE 7.8 Coefficients of reflection (solid curves) and transmission (dashed curves) for oblique incidence from air ($n_1 = 1.0$) to glass ($n_2 = 1.5$). The left-hand panel shows the wave polarization for which the magnetic field is parallel to the interface, whereas the right-hand panel shows the wave polarization for which the electric field is parallel to the interface. The Brewster angle is 56.3°.

The electromagnetic energy flux in the z-direction is (see Appendix C),

$$I_z = -E_y H_x. \tag{7.93}$$

Thus, the mean energy fluxes associated with the incident, reflected, and refracted waves are

$$\langle I_z \rangle_i = \frac{\psi_i^2 \, \cos \theta_i}{2 \, \epsilon_0 \, c \, n_1}, \tag{7.94}$$

$$\langle I_z \rangle_r = -\frac{\psi_r^2 \, \cos \theta_i}{2 \, \epsilon_0 \, c \, n_1}, \tag{7.95}$$

$$\langle I_z \rangle_t = \frac{\psi_t^2 \, \cos \theta_t}{2 \, \epsilon_0 \, c \, n_2}, \tag{7.96}$$

respectively. Hence, the coefficients of reflection and transmission are

$$R = \left(\frac{1 - \alpha \beta}{1 + \alpha \beta} \right)^2, \tag{7.97}$$

$$T = \frac{4 \, \alpha \beta}{(1 + \alpha \beta)^2} = 1 - R, \tag{7.98}$$

respectively. These expressions are the Fresnel relations for the polarization in which the electric field is parallel to the interface.

It can be seen that, at oblique incidence, the Fresnel relations (7.75) and (7.76) for the polarization in which the magnetic field is parallel to the interface are different to the corresponding relations (7.97) and (7.98) for the polarization in which the electric field is parallel to the interface. This implies that the coefficients of reflection and transmission for these two polarizations are, in general, different.

Figure 7.8 shows the coefficients of reflection and transmission for oblique incidence from air ($n_1 = 1.0$) to glass ($n_2 = 1.5$). In general, it can be seen that the coefficient of reflection rises, and the coefficient of transmission falls, as the angle of incidence increases. However, for the polarization in which the magnetic field is parallel to the interface, there is a particular angle of incidence, known as the *Brewster angle*, at which the reflected intensity is zero. There is no similar behavior for the polarization in which the electric field is parallel to the interface.

It follows from Equation (7.75) that the Brewster angle corresponds to the condition

$$\alpha = \beta, \tag{7.99}$$

or

$$\beta^2 = \frac{\cos^2 \theta_t}{\cos^2 \theta_i} = \frac{1 - \sin^2 \theta_t}{1 - \sin^2 \theta_i} = \frac{1 - \sin^2 \theta_i / \beta^2}{1 - \sin^2 \theta_i}, \tag{7.100}$$

where use has been made of Snell's law. The previous expression reduces to

$$\sin \theta_i = \frac{\beta}{\sqrt{1 + \beta^2}}, \tag{7.101}$$

or $\tan \theta_i = \beta = n_2/n_1$. Hence, the Brewster angle corresponds to $\theta_i = \theta_B$, where

$$\theta_B = \tan^{-1} \left(\frac{n_2}{n_1} \right). \tag{7.102}$$

If unpolarized light is incident on an air/glass (say) interface at the Brewster angle then the reflected light is 100% linearly polarized. (See Section 7.7.)

The fact that the coefficient of reflection for the polarization in which the electric field is parallel to the interface is generally greater than that for the other polarization (see Figure 7.8) implies that sunlight reflected from a horizontal water or snow surface is partially linearly polarized, with the horizontal polarization predominating over the vertical one. Such reflected light may be so intense as to cause glare. Polaroid sunglasses help reduce this glare by blocking horizontally polarized light.

7.10 TOTAL INTERNAL REFLECTION

According to Equation (7.49), when light is obliquely incident at an interface between two dielectric media, the angle of refraction θ_t is related to the angle of incidence θ_i according to

$$\sin \theta_t = \frac{n_1}{n_2} \sin \theta_i. \tag{7.103}$$

This formula presents no problems when $n_1 < n_2$. However, if $n_1 > n_2$ then the formula predicts that $\sin \theta_t$ is greater than unity when the angle of incidence exceeds some *critical angle* given by

$$\theta_c = \sin^{-1} \left(\frac{n_2}{n_1} \right). \tag{7.104}$$

In this situation, the analysis of the previous section requires modification.

Consider the polarization in which the magnetic field is parallel to the interface. We can write

$$H_y(x, z, t) = \psi_i \cos(\omega t - n_1 k_0 \sin\theta_i\, x - n_1 k_0 \cos\theta_i\, z)$$
$$+ \psi_r \cos(\omega t - n_1 k_0 \sin\theta_i\, x + n_1 k_0 \cos\theta_i\, z + \phi_r), \tag{7.105}$$

$$D_x(x, z, t) = \frac{\psi_i \cos\theta_i}{v_1} \cos(\omega t - n_1 k_0 \sin\theta_i\, x - n_1 k_0 \cos\theta_i\, z)$$
$$- \frac{\psi_r \cos\theta_i}{v_1} \cos(\omega t - n_1 k_0 \sin\theta_i\, x + n_1 k_0 \cos\theta_i\, z + \phi_r), \tag{7.106}$$

$$D_z(x, z, t) = -\frac{\psi_i \sin\theta_i}{v_1} \cos(\omega t - n_1 k_0 \sin\theta_i\, x - n_1 k_0 \cos\theta_i\, z)$$
$$- \frac{\psi_r \sin\theta_i}{v_1} \cos(\omega t - n_1 k_0 \sin\theta_i\, x + n_1 k_0 \cos\theta_i\, z + \phi_r) \tag{7.107}$$

in the region $z < 0$, and

$$H_y(x, z, t) = \psi_t\, e^{-n_2 k_0 \sinh\theta_t\, z} \cos(\omega t - n_1 k_0 \sin\theta_i\, x + \phi_t), \tag{7.108}$$

$$D_x(x, z, t) = \frac{\psi_t \sinh\theta_t}{v_2}\, e^{-n_2 k_0 \sinh\theta_t\, z} \sin(\omega t - n_1 k_0 \sin\theta_i\, x + \phi_t), \tag{7.109}$$

$$D_z(x, z, t) = -\frac{\psi_t \sin\theta_i}{v_1}\, e^{-n_2 k_0 \sinh\theta_t\, z} \cos(\omega t - n_1 k_0 \sin\theta_i\, x + \phi_t) \tag{7.110}$$

in the region $z > 0$. Here,

$$\cosh\theta_t = \sin\theta_t = \frac{n_1}{n_2} \sin\theta_i. \tag{7.111}$$

The matching conditions (7.60) and (7.62) both yield

$$\psi_i + \cos\phi_r\, \psi_r = \cos\phi_t\, \psi_t, \tag{7.112}$$
$$\sin\phi_r\, \psi_r = \sin\phi_t\, \psi_t, \tag{7.113}$$

whereas the matching condition (7.63) gives

$$\psi_i - \cos\phi_r\, \psi_r = \frac{\hat{\alpha}}{\beta} \sin\phi_t\, \psi_t, \tag{7.114}$$

$$\sin\phi_r\, \psi_r = \frac{\hat{\alpha}}{\beta} \cos\phi_t\, \psi_t. \tag{7.115}$$

Here,

$$\hat{\alpha} = \frac{\sinh\theta_t}{\cos\theta_i}. \tag{7.116}$$

It follows that

$$\tan\phi_r = \frac{2\,\hat{\alpha}\,\beta}{\beta^2 - \hat{\alpha}^2}, \tag{7.117}$$

$$\tan\phi_t = \frac{\hat{\alpha}}{\beta}, \tag{7.118}$$

$$\psi_i = \psi_t, \tag{7.119}$$

$$\psi_t = \frac{2\beta}{(\beta^2 + \hat{\alpha}^2)^{1/2}}\, \psi_i. \tag{7.120}$$

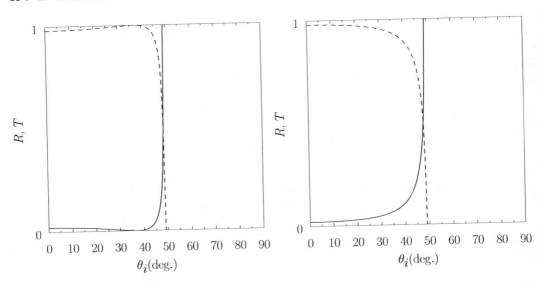

FIGURE 7.9 Coefficients of reflection (solid curves) and transmission (dashed curves) for oblique incidence from water ($n_1 = 1.33$) to air ($n_2 = 1.0$). The left-hand panel shows the wave polarization for which the magnetic field is parallel to the interface, whereas the right-hand panel shows the wave polarization for which the electric field is parallel to the interface. The critical angle is $48.8°$.

Moreover,

$$\langle I_z \rangle_i = -\langle I_z \rangle_r = \frac{\psi_i^2 \cos \theta_i}{2 \epsilon_0 c n_1}, \tag{7.121}$$

and

$$\langle I_z \rangle_t = 0. \tag{7.122}$$

The last result follows because H_y and D_x for the transmitted wave oscillate $\pi/2$ radians out of phase. Hence, when the angle of incidence exceeds the critical angle, the coefficient of reflection is unity, and the coefficient of transmission zero.

Consider the polarization in which the electric field is parallel to the interface. We can write

$$D_y(x, z, t) = \frac{\psi_i}{v_1} \cos(\omega t - n_1 k_0 \sin \theta_i\, x - n_1 k_0 \cos \theta_i\, z)$$

$$+ \frac{\psi_r}{v_1} \cos(\omega t - n_1 k_0 \sin \theta_i\, x + n_1 k_0 \cos \theta_i\, z + \phi_r), \tag{7.123}$$

$$H_x(x, z, t) = -\psi_i \cos \theta_i \cos(\omega t - n_1 k_0 \sin \theta_i\, x - n_1 k_0 \cos \theta_i\, z)$$

$$+ \psi_r \cos \theta_i \cos(\omega t - n_1 k_0 \sin \theta_i\, x + n_1 k_0 \cos \theta_i\, z + \phi_r), \tag{7.124}$$

$$H_z(x, z, t) = \psi_i \sin \theta_i \cos(\omega t - n_1 k_0 \sin \theta_i\, x - n_1 k_0 \cos \theta_i\, z)$$

$$+ \psi_r \sin \theta_i \cos(\omega t - n_1 k_0 \sin \theta_i\, x + n_1 k_0 \cos \theta_i\, z + \phi_r) \tag{7.125}$$

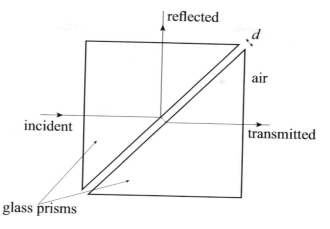

FIGURE 7.10 Frustrated total internal reflection. [From Fitzpatrick 2008.]

in the region $z < 0$, and

$$D_y(x, z, t) = \frac{\psi_t}{v_2} e^{-n_2 k_0 \sinh \theta_t z} \cos(\omega t - n_1 k_0 \sin \theta_i x + \phi_t), \qquad (7.126)$$

$$H_x(x, z, t) = -\psi_t \sinh \theta_t e^{-n_2 k_0 \sinh \theta_t z} \sin(\omega t - n_1 k_0 \sin \theta_i x + \phi_t), \qquad (7.127)$$

$$H_z(x, z, t) = \psi_t \cosh \theta_t e^{-n_2 k_0 \sinh \theta_t z} \cos(\omega t - n_1 k_0 \sin \theta_i x + \phi_t) \qquad (7.128)$$

in the region $z > 0$. The matching conditions (7.86) and (7.90) both yield

$$\psi_i + \cos \phi_r \psi_r = \beta^{-1} \cos \phi_t \psi_t \qquad (7.129)$$

$$\sin \phi_r \psi_r = \beta^{-1} \sin \phi_t \psi_t, \qquad (7.130)$$

whereas the matching condition (7.87) gives

$$\psi_i - \cos \phi_r \psi_r = \hat{\alpha} \sin \phi_t \psi_t, \qquad (7.131)$$

$$\sin \phi_r \psi_r = \hat{\alpha} \cos \phi_t \psi_t. \qquad (7.132)$$

It follows that

$$\tan \phi_r = \frac{2 \hat{\alpha} \beta}{1 - \hat{\alpha}^2 \beta^2}, \qquad (7.133)$$

$$\tan \phi_t = \hat{\alpha} \beta, \qquad (7.134)$$

$$\psi_i = \psi_t, \qquad (7.135)$$

$$\psi_t = \frac{2\beta}{(1 + \hat{\alpha}^2 \beta^2)^{1/2}} \psi_i. \qquad (7.136)$$

As before, if the angle of incidence exceeds the critical angle, the coefficient of reflection is unity, and the coefficient of transmission zero.

According to the previous analysis, when light is incident on an interface separating a medium of high refractive index from a medium of low refractive index, and the angle of incidence exceeds the critical angle, θ_c, the transmitted ray becomes evanescent (i.e., its amplitude decays exponentially), and all of the incident energy is reflected. This process is known as *total internal reflection*.

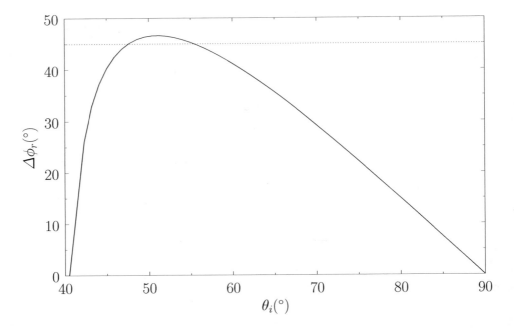

FIGURE 7.11 Phase advance introduced between the two different wave polarizations by total internal reflection at an interface between glass ($n_1 = 1.52$) and air ($n_2 = 1.0$).

Figure 7.9 shows the coefficients of reflection and transmission for oblique incidence from water ($n_1 = 1.33$) to air ($n_2 = 1.0$). In this case, the critical angle is $\theta_c = 48.8°$.

When total internal reflection takes place, the evanescent transmitted wave penetrates a few wavelengths into the lower refractive index medium. The existence of the evanescent wave can be demonstrated using the apparatus pictured in Figure 7.10. This shows two right-angled glass prisms separated by a small air gap of width d. Light incident on the internal surface of the first prism is internally reflected (assuming that $\theta_c < 45°$). However, if the spacing d is not too much larger than the wavelength of the light (in air) then the evanescent wave in the air gap still has a finite amplitude when it reaches the second prism. In this case, a detectable transmitted wave is excited in the second prism. The amplitude of this wave has an inverse exponential dependence on the width of the gap. This effect is called *frustrated total internal reflection*, and is analogous to the tunneling of wavefunctions through potential barriers in quantum mechanics. (See Section 11.12.)

According to Equations (7.117) and (7.133), total internal reflection produces a phase shift, ϕ_r, between the reflected and the incident waves. Moreover, this phase shift is different for the two possible wave polarizations. Hence, if unpolarized light is subject to total internal reflection then a phase advance, $\Delta\phi_r$, is introduced between the different polarizations. (The phase of the polarization in which the magnetic field is parallel to the interface is advanced with respect to that of the other polarization.) Figure 7.11 shows the phase advance due to total internal reflection at a glass/air interface, as a function of the angle of incidence. Here, the refractive indices of the glass and air are taken to be $n_1 = 1.52$ and $n_2 = 1.0$, respectively. It can be seen that there are two special values of the angle of incidence (i.e., $47.6°$ and $55.5°$) at which the phase advance is $\pi/4$ radians.

The aforementioned phase advance on total internal reflection is exploited in the so-called *Fresnel rhomb* to convert linearly polarized light into circular polarized light. A Fresnel rhomb is a prism-like device (usually in the form of a right-parallelepiped) that is shaped such that light entering one of the small faces is internally reflected twice (once from each of the two sloped faces)

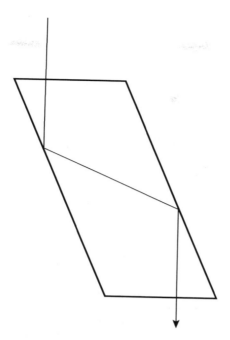

FIGURE 7.12   Path of light ray through Fresnel rhomb (schematic).

before exiting through the other small face. See Figure 7.12. The angle of internal reflection is the same in each case, and is designed to produces a $\pi/4$ phase difference between the two wave polarizations. For the case of a prism made up of glass of refractive index 1.52, this is achieved by ensuring that the reflection angle is either 47.6° or 55.5°. The net result of sending light though the device is thus to introduce a $\pi/2$ phase difference between the two polarizations. If the incoming light is linearly polarized at 45° to the plane of the incident and reflected waves then the amplitudes of the two wave polarizations are the same. This ensures that the $\pi/2$ phase difference introduced by the rhomb produces circularly (rather than elliptically) polarized light. (See Section 7.7.)

7.11   BIREFRINGENCE

Up until now, our treatment of electromagnetic wave propagation through transparent dielectrics has been restricted to isotropic media in which the refractive index is independent of either the direction of propagation or the polarization of the wave. However, there exists a certain class of optically anisotropic materials (e.g., crystals with non-cubic lattices, and plastics under mechanical stress) which are such that the refractive index varies with both the direction of propagation and the polarization. Such materials are said to be *birefringent*. Let us investigate the propagation of electromagnetic waves through birefringent media.

In a birefringent medium, the constitutive relation (7.53) generalizes to give

$$\mathbf{D} = \epsilon_0\, \boldsymbol{\epsilon} \cdot \mathbf{E}. \tag{7.137}$$

Here, $\mathbf{E}$ is the electric field-strength, $\mathbf{D} = \epsilon_0\, \mathbf{E} + \mathbf{P}$ the electric displacement, $\mathbf{P}$ the electric dipole moment per unit volume, and ϵ is termed the *dielectric tensor*. In a general Cartesian coordinate system, the dielectric tensor takes the form of a real 3 by 3 matrix. (However, the components of this matrix transform under rotation of the coordinate axes in an analogous manner to that in which the components of a vector transform.) Thus, in component form, the previous equation becomes

$$D_x = \epsilon_0 \left(\epsilon_{xx} E_x + \epsilon_{xy} E_y + \epsilon_{xz} E_z \right), \tag{7.138}$$

$$D_y = \epsilon_0 \left(\epsilon_{yx} E_x + \epsilon_{yy} E_y + \epsilon_{yz} E_z \right), \tag{7.139}$$

$$D_z = \epsilon_0 \left(\epsilon_{zx} E_x + \epsilon_{zy} E_y + \epsilon_{zz} E_z \right). \tag{7.140}$$

Note that $\epsilon_{yx} = \epsilon_{xy}$, $\epsilon_{zx} = \epsilon_{xz}$, and $\epsilon_{zy} = \epsilon_{yz}$ (i.e., the dielectric tensor is symmetric) in a lossless birefringent medium (i.e., one that does not absorb wave energy). Furthermore, the **D** and **E** vectors are not necessarily parallel to one another in such a medium.

It is a well-known fact that it is always possible to find a particular orientation of the Cartesian coordinate axes that diagonalizes a symmetric tensor (Riley 1974). For the case of the dielectric tensor, these special axes are called the *principal axes* of the dielectric medium in question. When the Cartesian axes are aligned along the principal axes, the dielectric tensor takes the form

$$\epsilon = \begin{pmatrix} \epsilon_{xx}, & 0, & 0 \\ 0, & \epsilon_{yy}, & 0 \\ 0, & 0, & \epsilon_{zz} \end{pmatrix}. \tag{7.141}$$

Here, ϵ_{xx}, ϵ_{yy}, and ϵ_{zz} are known as the *principal values* of the dielectric tensor. If the principal values are all equal to one another then the medium is termed *isotropic*. If one of the principal values is different from the other two then the medium is termed *monaxial*. This nomenclature arises because the medium possesses a single optic axis (i.e., a direction of wave propagation in which the phase velocity is independent of the wave polarization). Finally, if all of the principal values are different from one another then the medium is termed *biaxial* (because it possesses two optic axes). (Obviously, an isotropic medium possesses an infinite number of optic axes corresponding to all the possible directions of wave propagation.)

The equations that govern electromagnetic wave propagation through dielectric media can be written (see Appendix C)

$$\frac{\partial D_x}{\partial t} = \frac{\partial H_z}{\partial y} - \frac{\partial H_y}{\partial z}, \tag{7.142}$$

$$\frac{\partial D_y}{\partial t} = \frac{\partial H_x}{\partial z} - \frac{\partial H_z}{\partial x}, \tag{7.143}$$

$$\frac{\partial D_z}{\partial t} = \frac{\partial H_y}{\partial x} - \frac{\partial H_x}{\partial y}, \tag{7.144}$$

$$\frac{\partial B_x}{\partial t} = \frac{\partial E_y}{\partial z} - \frac{\partial E_z}{\partial y}, \tag{7.145}$$

$$\frac{\partial B_y}{\partial t} = \frac{\partial E_z}{\partial x} - \frac{\partial E_x}{\partial z}, \tag{7.146}$$

$$\frac{\partial B_z}{\partial t} = \frac{\partial E_x}{\partial y} - \frac{\partial E_y}{\partial x}, \tag{7.147}$$

where $\mathbf{B} = \mu_0 \mathbf{H}$. Here, **B** is the magnetic field-strength, and **H** the magnetic intensity. Taking E_x times Equation (7.142) plus E_y times Equation (7.143) plus E_z times Equation (7.144) plus H_x times Equation (7.145) plus H_y times Equation (7.146) plus H_z times Equation (7.147), and rearranging, we obtain

$$0 = E_x \frac{\partial D_x}{\partial t} + E_y \frac{\partial D_y}{\partial t} + E_z \frac{\partial D_z}{\partial t} + H_x \frac{\partial B_x}{\partial t} + H_y \frac{\partial B_y}{\partial t} + H_z \frac{\partial B_z}{\partial t}$$

$$+ \frac{\partial}{\partial x} (E_y H_z - E_z H_y) + \frac{\partial}{\partial y} (E_z H_x - E_x H_z) + \frac{\partial}{\partial z} (E_x H_y - E_y H_z). \tag{7.148}$$

However, assuming that the coordinate axes are aligned with the principal axes,

$$E_x \frac{\partial D_x}{\partial t} + E_y \frac{\partial D_y}{\partial t} + E_z \frac{\partial D_z}{\partial t} = \epsilon_0 \left(\epsilon_{xx} E_x \frac{\partial E_x}{\partial t} + \epsilon_{yy} E_y \frac{\partial E_y}{\partial t} + \epsilon_{zz} E_z \frac{\partial E_z}{\partial t} \right)$$

$$= \frac{1}{2} \epsilon_0 \frac{\partial}{\partial t} \left(\epsilon_{xx} E_x^2 + \epsilon_{yy} E_y^2 + \epsilon_{zz} E_z^2 \right)$$

$$= \frac{1}{2} \frac{\partial}{\partial t} \left(E_x D_x + E_y D_y + E_z D_z \right). \tag{7.149}$$

Here, we are also assuming that ϵ_{xx}, ϵ_{yy}, and ϵ_{zz} are time independent. Likewise,

$$H_x \frac{\partial B_x}{\partial t} + H_y \frac{\partial B_y}{\partial t} + H_z \frac{\partial B_z}{\partial z} = \frac{1}{2} \frac{\partial}{\partial t} \left(B_x H_x + B_y H_y + B_z H_z \right). \tag{7.150}$$

Hence, Equation (7.148) reduces to

$$\frac{\partial \mathcal{E}}{\partial t} + \frac{\partial \mathcal{I}_x}{\partial x} + \frac{\partial \mathcal{I}_y}{\partial y} + \frac{\partial \mathcal{I}_z}{\partial z} = 0, \tag{7.151}$$

where

$$\mathcal{E} = \frac{1}{2} \mathbf{E} \cdot \mathbf{D} + \frac{1}{2} \mathbf{B} \cdot \mathbf{H}, \tag{7.152}$$

$$\mathcal{I} = \mathbf{E} \times \mathbf{H}. \tag{7.153}$$

We can recognize Equation (7.151) as a three-dimensional generalization of the energy conservation equation (6.125). It follows that $\mathcal{E}$ is the electromagnetic *energy density* (i.e., the electromagnetic energy per unit volume), whereas $\mathcal{I}$ is the electromagnetic *energy flux* (i.e., electromagnetic energy flows at the rate $|\mathcal{I}|$ joules per unit area per unit time in the direction of the vector $\mathcal{I}$).

Let us search for wave-like solutions of Equations (7.142)–(7.147) of the form

$$\mathbf{E}(x, y, z, t) = \hat{\mathbf{E}} \cos [k (v t - \mathbf{n} \cdot \mathbf{r})], \tag{7.154}$$

$$\mathbf{D}(x, y, z, t) = \hat{\mathbf{D}} \cos [k (v t - \mathbf{n} \cdot \mathbf{r})], \tag{7.155}$$

$$\mathbf{H}(x, y, z, t) = \hat{\mathbf{H}} \cos [k (v t - \mathbf{n} \cdot \mathbf{r})], \tag{7.156}$$

with $\mathbf{B} = \mu_0 \mathbf{H}$. Here, the wavevector is $\mathbf{k} = k \mathbf{n}$, and the phase velocity is $\mathbf{v}_p = v \mathbf{n}$, where $\mathbf{n}$ is a unit vector. Equations (7.142)–(7.144) yield

$$v \hat{\mathbf{D}} = \hat{\mathbf{H}} \times \mathbf{n}, \tag{7.157}$$

whereas Equations (7.145)–(7.147) give

$$v \hat{\mathbf{H}} = \epsilon_0 c^2 \mathbf{n} \times \hat{\mathbf{E}}. \tag{7.158}$$

The previous two equations immediately imply that

$$\mathbf{n} \cdot \hat{\mathbf{D}} = \mathbf{n} \cdot \hat{\mathbf{H}} = 0, \tag{7.159}$$

and

$$\hat{\mathbf{H}} \cdot \hat{\mathbf{D}} = \hat{\mathbf{H}} \cdot \hat{\mathbf{E}} = 0. \tag{7.160}$$

In other words, the electric displacement and the magnetic intensity are both perpendicular to the wavevector. Furthermore, the magnetic intensity is perpendicular to both the electric displacement and the electric field-strength. Equations (7.157) and (7.158) can also be combined to give

$$v^2 \hat{\mathbf{D}} = \epsilon_0 c^2 \left[\hat{\mathbf{E}} - (\mathbf{n} \cdot \hat{\mathbf{E}}) \mathbf{n} \right], \tag{7.161}$$

where use has been made of a standard vector identity. The quantities $v_x = c/\sqrt{\epsilon_{xx}}$, $v_y = c/\sqrt{\epsilon_{yy}}$, and $v_z = c/\sqrt{\epsilon_{zz}}$, where ϵ_{xx}, ϵ_{yy}, ϵ_{zz} are the principal components of the dielectric tensor, are termed the *principal velocities* of the dielectric medium in question. Now, given that $\epsilon_0 c^2 \hat{E}_x = c^2 \hat{D}_x/\epsilon_{xx} = v_x^2 \hat{D}_x$, et cetera, when the Cartesian axes are aligned along the principal axes, the three Cartesian components of the previous equation can be written

$$\left(v^2 - v_x^2 \right) \hat{D}_x = -\epsilon_0 c^2 (\mathbf{n} \cdot \hat{\mathbf{E}}) n_x, \tag{7.162}$$

$$\left(v^2 - v_y^2 \right) \hat{D}_x = -\epsilon_0 c^2 (\mathbf{n} \cdot \hat{\mathbf{E}}) n_y, \tag{7.163}$$

$$\left(v^2 - v_z^2 \right) \hat{D}_x = -\epsilon_0 c^2 (\mathbf{n} \cdot \hat{\mathbf{E}}) n_z. \tag{7.164}$$

Finally, because [see Equation (7.159)]

$$\mathbf{n} \cdot \hat{\mathbf{D}} = n_x \hat{D}_x + n_y \hat{D}_y + n_z \hat{D}_z = 0, \tag{7.165}$$

we obtain

$$\left(\frac{n_x^2}{v^2 - v_x^2} + \frac{n_y^2}{v^2 - v_y^2} + \frac{n_z^2}{v^2 - v_z^2} \right) \epsilon_0 c^2 (\mathbf{n} \cdot \hat{\mathbf{E}}) = 0. \tag{7.166}$$

Assuming that $\mathbf{n} \cdot \hat{\mathbf{E}} \neq 0$, the previous equation yields

$$\frac{n_x^2}{v^2 - v_x^2} + \frac{n_y^2}{v^2 - v_y^2} + \frac{n_z^2}{v^2 - v_z^2} = 0, \tag{7.167}$$

which is known as the *Fresnel equation*.

The Fresnel equation is a quadratic equation for v^2 that specifies the phase speeds, v, of the two independent electromagnetic wave polarizations that can propagate through a birefringent medium in a particular direction, $\mathbf{n}$. In general, these two speeds are different. Let v_a be the first speed, $\hat{\mathbf{D}}_a$ the associated electric displacement, and $\hat{\mathbf{E}}_a$ the associated electric field-strength. Likewise, let v_b be the second speed, $\hat{\mathbf{D}}_b$ the associated electric displacement, and $\hat{\mathbf{E}}_b$ the associated electric field-strength It follows from Equations (7.162)–(7.164) that

$$\hat{D}_{ax} = -\frac{\epsilon_0 c^2 (\mathbf{n} \cdot \hat{\mathbf{E}}_a) n_x}{v_a^2 - v_x^2}, \tag{7.168}$$

$$\hat{D}_{ay} = -\frac{\epsilon_0 c^2 (\mathbf{n} \cdot \hat{\mathbf{E}}_a) n_y}{v_a^2 - v_y^2}, \tag{7.169}$$

$$\hat{D}_{az} = -\frac{\epsilon_0 c^2 (\mathbf{n} \cdot \hat{\mathbf{E}}_a) n_z}{v_a^2 - v_z^2}, \tag{7.170}$$

$$\hat{D}_{bx} = -\frac{\epsilon_0\, c^2\, (\mathbf{n}\cdot\hat{\mathbf{E}}_b)\, n_x}{v_b^2 - v_x^2},\tag{7.171}$$

$$\hat{D}_{by} = -\frac{\epsilon_0\, c^2\, (\mathbf{n}\cdot\hat{\mathbf{E}}_b)\, n_y}{v_b^2 - v_y^2},\tag{7.172}$$

$$\hat{D}_{bz} = -\frac{\epsilon_0\, c^2\, (\mathbf{n}\cdot\hat{\mathbf{E}}_b)\, n_z}{v_b^2 - v_z^2}.\tag{7.173}$$

Hence,

$$\hat{\mathbf{D}}_a \cdot \hat{\mathbf{D}}_b = D_{ax}\, D_{bx} + D_{ay}\, D_{by} + D_{az}\, D_{bz}\tag{7.174}$$

$$= \left[\frac{n_x^2}{(v_a^2 - v_x^2)(v_b^2 - v_x^2)} + \frac{n_y^2}{(v_a^2 - v_y^2)(v_b^2 - v_y^2)} + \frac{n_z^2}{(v_a^2 - v_z^2)(v_b^2 - v_z^2)} \right] (\epsilon_0\, c^2)^2\, (\mathbf{n}\cdot\hat{\mathbf{E}}_a)\, (\mathbf{n}\cdot\hat{\mathbf{E}}_b)$$

$$= \left[\frac{n_x^2}{v_a^2 - v_x^2} + \frac{n_y^2}{v_a^2 - v_y^2} + \frac{n_z^2}{v_a^2 - v_z^2} - \frac{n_x^2}{v_b^2 - v_x^2} - \frac{n_y^2}{v_b^2 - v_y^2} - \frac{n_z^2}{v_b^2 - v_z^2} \right] \frac{(\epsilon_0\, c^2)^2\, (\mathbf{n}\cdot\hat{\mathbf{E}}_a)\, (\mathbf{n}\cdot\hat{\mathbf{E}}_b)}{v_b^2 - v_a^2}.$$

However, according to the Fresnel equation, (7.167),

$$\frac{n_x^2}{v_a^2 - v_x^2} + \frac{n_y^2}{v_a^2 - v_y^2} + \frac{n_z^2}{v_a^2 - v_z^2} = \frac{n_x^2}{v_b^2 - v_x^2} + \frac{n_y^2}{v_b^2 - v_y^2} + \frac{n_z^2}{v_b^2 - v_z^2} = 0.\tag{7.175}$$

Hence, we deduce that

$$\hat{\mathbf{D}}_a \cdot \hat{\mathbf{D}}_b = 0.\tag{7.176}$$

In other words, the two independent wave polarizations have mutually orthogonal electric displacements.

As an example, consider the propagation of an electromagnetic wave through a monaxial material whose principal velocities are $v_x = v_y = v_\perp$ and $v_z = v_\|$, where $v_\perp \neq v_\|$. The corresponding principal components of the dielectric tensor are $\epsilon_{xx} = \epsilon_{yy} = \epsilon_\perp$ and $\epsilon_{zz} = \epsilon_\|$. Of course, $v_\perp = c/\sqrt{\epsilon_\perp}$ and $v_\| = c/\sqrt{\epsilon_\|}$. In this case, the optic axis corresponds to the z-axis. It is convenient to specify the direction of wave propagation in terms of standard spherical angles,

$$\mathbf{n} = (\sin\theta\, \cos\phi,\ \sin\theta\, \sin\phi,\ \cos\theta).\tag{7.177}$$

In particular, θ is the angle subtended between the direction of wave propagation and the optic axis.

The two independent electromagnetic wave polarizations that can propagate through a monaxial material are termed the ordinary wave and the extraordinary wave. The *ordinary wave* is such that $\mathbf{n}\cdot\hat{\mathbf{E}} = 0$, which is one way of satisfying Equation (7.166). Assuming that the Cartesian axes correspond to the principal axes, it is easily demonstrated that

$$\hat{\mathbf{D}} = \hat{D}\,(\sin\phi,\ -\cos\phi,\ 0),\tag{7.178}$$

$$\hat{\mathbf{E}} = \frac{\hat{D}}{\epsilon_0\, \epsilon_\perp}\,(\sin\phi,\ -\cos\phi,\ 0) = \frac{\hat{\mathbf{D}}}{\epsilon_0\, \epsilon_\perp},\tag{7.179}$$

$$\hat{\mathbf{H}} = v_\perp\, \hat{D}\,(\cos\theta\, \cos\phi,\ \cos\theta\, \sin\phi,\ -\sin\theta),\tag{7.180}$$

$$\langle I \rangle = \frac{v_\perp\, \hat{D}^2}{2\, \epsilon_0\, \epsilon_\perp}\,(\sin\theta\, \cos\phi,\ \sin\theta\, \sin\phi,\ \cos\theta) = \frac{v_\perp\, \hat{D}^2}{2\, \epsilon_0\, \epsilon_\perp}\,\mathbf{n}.\tag{7.181}$$

Here, $\langle\cdots\rangle$ denotes an average over a wave period. Substitution of Equations (7.178) and (7.179) into Equation (7.161) reveals that

$$v = v_\perp.\tag{7.182}$$

In other words, the ordinary wave propagates at the fixed phase speed $v_\perp$, irrespective of its direction of propagation. Furthermore, the electric field-strength is parallel to the electric displacement, and the electromagnetic energy flux is parallel to the wavevector.

The phase speed of the *extraordinary wave* is obtained directly from the Fresnel equation, (7.167), which yields

$$\frac{\sin^2 \theta \cos^2 \phi}{v^2 - v_\perp^2} + \frac{\sin^2 \theta \sin^2 \phi}{v^2 - v_\perp^2} + \frac{\cos^2 \theta}{v^2 - v_\parallel^2} = 0, \tag{7.183}$$

or

$$v^2 = \cos^2 \theta \, v_\perp^2 + \sin^2 \theta \, v_\parallel^2. \tag{7.184}$$

It follows that the phase speed of the extraordinary wave varies with its direction of propagation. The phase speed matches that of the ordinary wave when the wave propagates along the optic axis (i.e., when $\theta = 0$); otherwise, it is different. It is easily demonstrated that

$$\hat{\mathbf{D}} = \hat{D}(\cos \theta \cos \phi, \cos \theta \sin \phi, -\sin \theta), \tag{7.185}$$

$$\hat{\mathbf{E}} = \frac{\hat{D}}{\epsilon_0 \epsilon} \left(\frac{\epsilon}{\epsilon_\perp} \cos \theta \cos \phi, \frac{\epsilon}{\epsilon_\perp} \cos \theta \sin \phi, -\frac{\epsilon}{\epsilon_\parallel} \sin \theta \right), \tag{7.186}$$

$$\hat{\mathbf{H}} = v \hat{D}(-\sin \phi, \cos \phi, 0), \tag{7.187}$$

$$\langle \mathbf{I} \rangle = \frac{v \hat{D}^2}{2 \epsilon_0 \epsilon} \left(\frac{\epsilon}{\epsilon_\parallel} \sin \theta \cos \phi, \frac{\epsilon}{\epsilon_\parallel} \sin \theta \sin \phi, \frac{\epsilon}{\epsilon_\perp} \cos \theta \right). \tag{7.188}$$

Here,

$$\frac{1}{\epsilon^2} = \frac{\cos^2 \theta}{\epsilon_\perp^2} + \frac{\sin^2 \theta}{\epsilon_\parallel^2}. \tag{7.189}$$

It can be seen that the $\hat{\mathbf{D}}$ and $\hat{\mathbf{E}}$ vectors are not parallel to one another. Moreover, the electromagnetic energy flux is not parallel to the wavevector. If α is the angle subtended between the directions of the $\hat{\mathbf{D}}$ and $\hat{\mathbf{E}}$ vectors then

$$\tan \alpha = \frac{|\hat{\mathbf{D}} \times \hat{\mathbf{E}}|}{\hat{\mathbf{D}} \cdot \hat{\mathbf{E}}} = \frac{\sin \theta \cos \theta |1/\epsilon_\parallel - 1/\epsilon_\perp|}{\cos^2 \theta / \epsilon_\perp + \sin^2 \theta / \epsilon_\parallel}. \tag{7.190}$$

Likewise, if β is the angle subtended between the directions of the electromagnetic energy flux and the wavevector then

$$\tan \beta = \frac{|\langle \mathbf{I} \rangle \times \mathbf{n}|}{\langle \mathbf{I} \rangle \cdot \mathbf{n}} = \frac{\sin \theta \cos \theta |1/\epsilon_\parallel - 1/\epsilon_\perp|}{\cos^2 \theta / \epsilon_\perp + \sin^2 \theta / \epsilon_\parallel}. \tag{7.191}$$

It follows that $\alpha = \beta$. Moreover, these two angles are only zero when the extraordinary wave propagates parallel (i.e., $\theta = 0, \pi$) or perpendicular (i.e., $\theta = \pi/2$) to the optic axis.

Figure 7.13 illustrates what happens when unpolarized light is normally incident on a slab of birefringent material in such a manner that the incident light is neither parallel nor perpendicular to the optic axis. Because of the normal incidence, the wavevectors of both the ordinary and the extraordinary rays are not refracted, and remain parallel to the direction of incidence. However, the ray path is actually coincident with the associated electromagnetic energy flux. For the ordinary ray, the electromagnetic energy flux is parallel to the wavevector. However, for the extraordinary ray, the electromagnetic energy flux subtends a finite angle with the wavevector. Consequently, although the ordinary ray passes through the slab without changing direction, the direction of the extraordinary ray suffers a sideways deviation. This type of double refraction is known as *birefringence*.

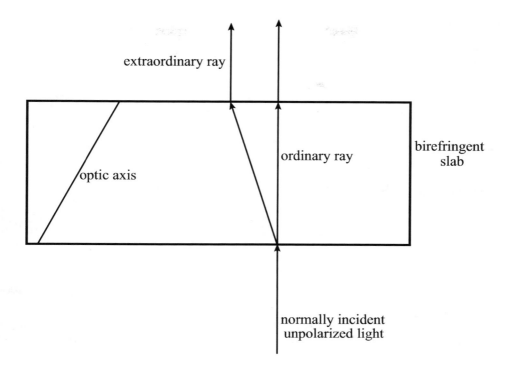

extraordinary ray

ordinary ray

optic axis

birefringent slab

normally incident
unpolarized light

FIGURE 7.13 Birefringence.

7.12 SOUND WAVES IN FLUIDS

Consider a uniform fluid (at rest) whose equilibrium density and pressure are ρ and p, respectively. Suppose that a sound wave propagates though the fluid, and that the density and pressure perturbations produced by the wave are $\tilde{\rho}$ and $\tilde{p}$, respectively. Incidentally, the quantity $\tilde{p}$ is often referred to as the *acoustic pressure*. In two dimensions (i.e., neglecting any variation of perturbed quantities in the y-direction), the equations governing the propagation of a sound wave though a uniform fluid are (Landau and Lifshitz 1959)

$$\rho \frac{\partial v_x}{\partial t} = -\frac{\partial \tilde{p}}{\partial x}, \tag{7.192}$$

$$\rho \frac{\partial v_z}{\partial t} = -\frac{\partial \tilde{p}}{\partial z}, \tag{7.193}$$

$$\frac{\partial \tilde{p}}{\partial t} = \frac{K}{\rho} \frac{\partial \tilde{\rho}}{\partial t} = -K \left(\frac{\partial v_x}{\partial x} + \frac{\partial v_z}{\partial z} \right). \tag{7.194}$$

(See Section 9.12 for a more complete discussion of fluid equations.) Here, $\mathbf{v} = (v_x, 0, v_z)$ is the perturbation to the fluid velocity produced by the wave, and

$$K = \rho \frac{dp}{d\rho} \tag{7.195}$$

is the *bulk modulus*. The bulk modulus is a quantity with the units of pressure that measures a given substance's resistance to uniform compression. The bulk modulus of an ideal gas is $\gamma\,p$, where γ is the ratio of specific heats. On the other hand, the bulk modulus of water at $20°\,\mathrm{C}$ is $K = 2.2 \times 10^9\,\mathrm{N\,m^{-2}}$ (Wikipedia contributors 2018).

It is helpful to write

$$v_x = \frac{\partial \phi}{\partial x}, \tag{7.196}$$

$$v_z = \frac{\partial \phi}{\partial z}, \tag{7.197}$$

where $\phi(x, z, t)$ is conventionally referred to as a *velocity potential*. It follows from Equations (7.192) and (7.193) that

$$\tilde{p} = -\rho \frac{\partial \phi}{\partial t}. \tag{7.198}$$

Moreover, substitution into Equation (7.194) yields the wave equation

$$\frac{\partial^2 \phi}{\partial t^2} = v^2 \left(\frac{\partial^2 \phi}{\partial x^2} + \frac{\partial^2 \phi}{\partial z^2} \right), \tag{7.199}$$

where the characteristic wave speed is

$$v = \sqrt{\frac{K}{\rho}}. \tag{7.200}$$

For the case of an ideal gas, for which $K = \gamma p$, we obtain $v = \sqrt{\gamma p / \rho}$. (See Section 5.4.) On the other hand, for the case of water at $20°$ C, for which $K = 2.2 \times 10^9$ N m^{-2} and $\rho = 1.0 \times 10^3$ kg m^{-3}, we get $v = 1483$ m s^{-1}. This prediction is in good agreement with the measured sound speed in water at $20°$ C, which is 1481 m s^{-1} (Wikipedia contributors 2018).

Forming the sum of v_x times Equation (7.192), v_z times Equation (7.193), and $K^{-1} \tilde{p}$ times Equation (7.194), we obtain

$$\frac{\partial \mathcal{E}}{\partial t} + \frac{\partial \mathcal{I}_x}{\partial x} + \frac{\partial \mathcal{I}_z}{\partial z} = 0, \tag{7.201}$$

where

$$\mathcal{E} = \frac{1}{2} \rho v_x^2 + \frac{1}{2} \rho v_z^2 + \frac{1}{2} \frac{\tilde{p}^2}{K}, \tag{7.202}$$

$$\mathcal{I}_x = \tilde{p} v_x, \tag{7.203}$$

$$\mathcal{I}_z = \tilde{p} v_z. \tag{7.204}$$

Equation (7.201) can be recognized as a two-dimensional energy conservation equation. (See Section 6.5.) Here, $\mathcal{E}$ is the *acoustic energy density*, and $\mathcal{I}_x$ and $\mathcal{I}_z$ are the *acoustic energy fluxes* in the x- and z-directions, respectively.

Consider a situation (analogous to that illustrated in Figure 7.7) in which a sound wave is incident at an interface between two uniform immiscible fluids. Let the region $z < 0$ be occupied by a fluid of equilibrium density ρ_1 and sound speed v_1, and let the region $z > 0$ be occupied by a fluid of equilibrium density ρ_2 and sound speed v_2. We can write the wavevectors of the incident, reflected, and refracted waves as

$$\mathbf{k}_i = \frac{\omega}{v_1} (\sin \theta_i, 0, \cos \theta_i), \tag{7.205}$$

$$\mathbf{k}_r = \frac{\omega}{v_1} (\sin \theta_r, 0, -\cos \theta_r), \tag{7.206}$$

$$\mathbf{k}_t = \frac{\omega}{v_2} (\sin \theta_t, 0, \cos \theta_t), \tag{7.207}$$

respectively. Here, for the sake of simplicity, we have assumed that all three wavevectors lie in the same plane (as is readily demonstrated; see Section 7.8.) Moreover, in order to be valid solutions of the wave equation, (7.199), all three waves must satisfy the dispersion relation $\omega = k\,v$, where ω is the common wave frequency. Finally, θ_i, θ_r, and θ_t are the angles of incidence, reflection, and refraction, respectively. (See Section 7.8.)

The velocity potential in the region $z < 0$ is written

$$\phi(x, z, t) = \phi_i \cos[\omega\,(t - \sin\theta_i\,x/v_1 - \cos\theta_i\,z/v_1)]$$
$$+\,\phi_r \cos[\omega\,(t - \sin\theta_r\,x/v_1 + \cos\theta_r\,z/v_1)], \tag{7.208}$$

where the first and second terms on the right-hand side specify the incident and reflected waves, respectively. The velocity potential in the region $z > 0$ takes the form

$$\phi(x, z, t) = \phi_t \cos[\omega\,(t - \sin\theta_t\,x/v_2 - \cos\theta_t\,z/v_2)], \tag{7.209}$$

where the term on the right-hand side specifies the refracted wave. The first physical matching constraint that must be satisfied at the interface is continuity of the acoustic pressure; that is,

$$[\tilde{p}]_{z=0_-}^{z=0_+} = -\left[\rho\,\frac{\partial\phi}{\partial t}\right]_{z=0_-}^{z=0_+} = 0. \tag{7.210}$$

This contraint yields

$$\rho_1\,\omega\,\phi_i\,\sin[\omega\,(t - \sin\theta_i\,x/v_1)] + \rho_1\,\omega\,\phi_r\,\sin[\omega\,(t - \sin\theta_r\,x/v_1)]$$
$$= \rho_2\,\omega\,\phi_t\,\sin[\omega\,(t - \sin\theta_t\,x/v_2)]. \tag{7.211}$$

The previous equation holds at all values of x. This is only possible if

$$\sin\theta_i = \sin\theta_r \tag{7.212}$$

$$\frac{\sin\theta_i}{v_1} = \frac{\sin\theta_t}{v_2}. \tag{7.213}$$

These two expressions are analogous to the laws of reflection and refraction, respectively, of geometric optics. (See Section 7.8.) This suggests that these laws are of universal validity, rather than being restricted to light waves. Equation (7.211) reduces to

$$\rho_1\,(\phi_i + \phi_r) = \rho_2\,\phi_t. \tag{7.214}$$

The second physical matching constraint that must be satisfied at the interface is continuity of the normal velocity; that is,

$$[v_z]_{z=0_-}^{z=0_+} = \left[\frac{\partial\phi}{\partial z}\right]_{z=0_-}^{z=0_+} = 0. \tag{7.215}$$

This constraint yields

$$\tan\theta_t\,(\phi_i - \phi_r) = \tan\theta_i\,\phi_t, \tag{7.216}$$

where use has been made of Equation (7.213). Equations (7.214) and (7.216) can be combined to give

$$\phi_r = \left(\frac{\rho_2\,\tan\theta_t - \rho_1\,\tan\theta_i}{\rho_2\,\tan\theta_t + \rho_1\,\tan\theta_i}\right)\phi_i, \tag{7.217}$$

$$\phi_t = \left(\frac{2\,\rho_1\,\tan\theta_t}{\rho_2\,\tan\theta_t + \rho_1\,\tan\theta_i}\right)\phi_i. \tag{7.218}$$

Equations (7.197), (7.198), and (7.204) reveal that the mean acoustic energy fluxes, normal to the interface, associated with the incident, reflected, and refracted waves are

$$\langle I_z \rangle_i = \frac{\rho_1 \, \omega \, \cos \theta_i}{2 \, v_1} \, \phi_i^2, \tag{7.219}$$

$$\langle I_z \rangle_r = -\frac{\rho_1 \, \omega \, \cos \theta_i}{2 \, v_1} \, \phi_r^2, \tag{7.220}$$

$$\langle I_z \rangle_t = \frac{\rho_2 \, \omega \, \cos \theta_t}{2 \, v_2} \, \phi_t^2, \tag{7.221}$$

respectively. Thus, it follows that the coefficients of reflection and transmission at the interface are

$$R = \frac{-\langle I_z \rangle_r}{\langle I_z \rangle_i} = \left(\frac{\rho_2 \, \tan \theta_t - \rho_1 \, \tan \theta_i}{\rho_2 \, \tan \theta_t + \rho_1 \, \tan \theta_i} \right)^2, \tag{7.222}$$

$$T = \frac{\langle I_z \rangle_t}{\langle I_z \rangle_i} = \frac{4 \, \rho_1 \, \rho_2 \, \tan \theta_i \, \tan \theta_t}{(\rho_2 \, \tan \theta_t + \rho_1 \, \tan \theta_i)^2}, \tag{7.223}$$

respectively. It is actually possible for there to be no reflection at the interface (i.e., $R = 0$), provided that $\rho_2 \, \tan \theta_t = \rho_1 \, \tan \theta_i$. This criterion yields

$$\tan^2 \theta_i = \frac{\rho_2^2 \, v_2^2 - \rho_1^2 \, v_1^2}{\rho_1^2 \, (v_1^2 - v_2^2)}, \tag{7.224}$$

which can only be satisfied if $\rho_2 \, v_2 > \rho_1 \, v_1$ and $v_1 > v_2$, or if $\rho_2 \, v_2 < \rho_1 \, v_1$ and $v_1 < v_2$. The critical angle of incidence at which there is no reflection is sometimes called the *angle of intromission*. However, not every pair of immiscible fluids possesses such an angle.

EXERCISES

7.1 (a) Show that the one-dimensional plane wave, (7.1), is a solution of the one-dimensional wave equation, (7.8), provided that

$$\omega = k \, v.$$

(b) Demonstrate that the three-dimensional plane wave, (7.5), is a solution of the three-dimensional wave equation, (7.9), as long as

$$\omega = |\mathbf{k}| \, v.$$

7.2 (a) Demonstrate that for a cylindrically symmetric wavefunction $\psi(\rho, t)$, where $\rho = (x^2 + y^2)^{1/2}$, the three-dimensional wave equation (7.9) can be rewritten

$$\frac{\partial^2 \psi}{\partial t^2} = v^2 \left(\frac{\partial^2 \psi}{\partial \rho^2} + \frac{1}{\rho} \frac{\partial \psi}{\partial \rho} \right).$$

(b) Show that

$$\psi(\rho, t) \simeq \frac{\psi_0}{\rho^{1/2}} \, \cos(\omega t - k \rho - \phi)$$

is an approximate solution of this equation in the limit $k\rho \gg 1$, where $\omega/k = v$.

7.3 (a) Demonstrate that for a spherically symmetric wavefunction $\psi(r,t)$, where $r = (x^2 + y^2 + z^2)^{1/2}$, the three-dimensional wave equation (7.9) can be rewritten

$$\frac{\partial^2 \psi}{\partial t^2} = v^2 \left(\frac{\partial^2 \psi}{\partial r^2} + \frac{2}{r} \frac{\partial \psi}{\partial r} \right).$$

(b) Show that

$$\psi(r,t) = \frac{\psi_0}{r} \cos(\omega t - k r - \phi)$$

is a solution of this equation, where $\omega/k = v$.

7.4 Consider an elastic sheet stretched over a rectangular frame that extends from $x = 0$ to $x = a$, and from $y = 0$ to $y = b$. Suppose that

$$\psi(x,y,0) = F(x,y),$$
$$\dot{\psi}(x,y,0) = G(x,y).$$

Show that the amplitudes and phase angles in the normal mode expansion (7.28) are given by

$$A_{m,n} = \left(I_{m,n}^2 + J_{m,n}^2 \right)^{1/2},$$

$$\phi_{m,n} = \tan^{-1} \left(\frac{J_{m,n}}{I_{m,n}} \right),$$

where

$$I_{m,n} = \frac{4}{ab} \int_0^a \int_0^b F(x,y) \sin\left(m\pi \frac{x}{a} \right) \sin\left(n\pi \frac{y}{b} \right) dx\,dy,$$

$$J_{m,n} = \frac{4}{ab\,\omega_{m,n}} \int_0^a \int_0^b G(x,y) \sin\left(m\pi \frac{x}{a} \right) \sin\left(n\pi \frac{y}{b} \right) dx\,dy.$$

7.5 The radial oscillations of an ideal gas in a spherical cavity of radius a are governed by the spherical wave equation

$$\frac{\partial^2 \psi}{\partial t^2} = v^2 \left(\frac{\partial^2 \psi}{\partial r^2} + \frac{2}{r} \frac{\partial \psi}{\partial r} \right),$$

subject to the boundary condition $\psi(a,t) = 0$. Here, $r = (x^2 + y^2 + z^2)^{1/2}$ is a spherical coordinate, $\psi(r,t)$ is the radial displacement, and v is the speed of sound. Show that the general solution of this equation is written

$$\psi(r,t) = \sum_{j=1,\infty} A_j \operatorname{sinc}\left(j\pi \frac{r}{a} \right) \cos(\omega_j t - \phi_j),$$

where $\operatorname{sinc}(x) \equiv \sin x / x$,

$$\omega_j = j\pi \frac{v}{a},$$

and A_j, ϕ_j are arbitrary constants.

7.6 Show that a light-ray entering a planar transparent plate of thickness d and refractive index n emerges parallel to its original direction. Show that the lateral displacement of the ray is

$$s = \frac{d\,\sin(\theta_1 - \theta_2)}{\cos\theta_2},$$

where θ_1 and θ_2 are the angles of incidence and refraction, respectively, at the front side of the plate.

7.7 Suppose that a light-ray is incident on the front (air/glass) interface of a uniform pane of glass of refractive index n at the Brewster angle. Demonstrate that the refracted ray is also incident on the rear (glass/air) interface of the pane at the Brewster angle. [From Fitzpatrick 2008.]

7.8 (a) Show that the Fresnel relations, (7.75) and (7.76), for the polarization in which the magnetic intensities of all three waves are parallel to the interface can be written

$$R = \left(\frac{Z_1\,\cos\theta_i - Z_2\,\cos\theta_t}{Z_1\,\cos\theta_i + Z_2\,\cos\theta_t}\right)^2,$$

$$T = \frac{4\,Z_1\,Z_2\,\cos\theta_i\,\cos\theta_t}{(Z_1\,\cos\theta_i + Z_2\,\cos\theta_t)^2},$$

where $Z = Z_0/n$ represents impedance. (Here, Z_0 is the impedance of free space, and n the refractive index.)

(b) Demonstrate that the Fresnel relations, (7.97) and (7.98), for the other polarization take the form

$$R = \left(\frac{Z_2\,\cos\theta_i - Z_1\,\cos\theta_t}{Z_2\,\cos\theta_i + Z_1\,\cos\theta_t}\right)^2,$$

$$T = \frac{4\,Z_2\,Z_1\,\cos\theta_i\,\cos\theta_t}{(Z_2\,\cos\theta_i + Z_1\,\cos\theta_t)^2}.$$

7.9 (a) Show that the expressions, (7.222) and (7.223), for the coefficients of reflection and transmission for a sound wave obliquely incident at an interface between two immiscible fluids can be written

$$R = \left(\frac{Z_2\,\cos\theta_i - Z_1\,\cos\theta_t}{Z_2\,\cos\theta_i + Z_1\,\cos\theta_t}\right)^2,$$

$$T = \frac{4\,Z_2\,Z_1\,\cos\theta_i\,\cos\theta_t}{(Z_2\,\cos\theta_i + Z_1\,\cos\theta_t)^2},$$

where $Z_1 = \rho_1 v_1$ and $Z_2 = \rho_2 v_2$ are the *acoustic impedances* of the two fluids.

(b) Show that the expression, (7.224), for the angle of intromission can be written

$$\tan^2\theta_i = \frac{(Z_2/Z_1)^2 - 1}{1 - (v_2/v_1)^2}.$$

CHAPTER 8

Wave Pulses

8.1 INTRODUCTION

In this chapter, we shall discuss how arbitrarily-shaped wave pulses can be built up as superpositions of sinusoidal waves with a range of different frequencies and wavelengths. Conversely, we shall examine how wave pulses can be resolved into their component sinusoidal waves.

8.2 FOURIER TRANSFORMS

Consider a function $F(x)$ that is periodic in x with period L. In other words,

$$F(x + L) = F(x) \tag{8.1}$$

for all x. Recall, from Section 5.5, that we can represent such a function as a Fourier series; that is,

$$F(x) = \sum_{n=1,\infty} [C_n \cos(n \, \delta k \, x) + S_n \sin(n \, \delta k \, x)], \tag{8.2}$$

where

$$\delta k = \frac{2\pi}{L}. \tag{8.3}$$

[We have neglected the $n = 0$ term in Equation (8.2), for the sake of convenience.] Equation (8.2) automatically satisfies the periodicity constraint (8.1), because $\cos(\theta + n \, 2\pi) = \cos \theta$ and $\sin(\theta + n \, 2\pi) = \sin \theta$ for all θ and n (with the proviso that n is an integer). The so-called Fourier coefficients, C_n and S_n, appearing in Equation (8.2), can be determined from the function $F(x)$ by means of the following readily demonstrated (see Exercise 8.1) results:

$$\frac{2}{L} \int_{-L/2}^{L/2} \cos(n \, \delta k \, x) \, \cos(n' \, \delta k \, x) \, dx = \delta_{n,n'}, \tag{8.4}$$

$$\frac{2}{L} \int_{-L/2}^{L/2} \sin(n \, \delta k \, x) \, \sin(n' \, \delta k \, x) \, dx = \delta_{n,n'}, \tag{8.5}$$

$$\frac{2}{L} \int_{-L/2}^{L/2} \cos(n \, \delta k \, x) \, \sin(n' \, \delta k \, x) \, dx = 0, \tag{8.6}$$

where n, n' are positive integers. Here, $\delta_{n,n'}$ is a Kronecker delta function. In fact,

$$C_n = \frac{2}{L} \int_{-L/2}^{L/2} F(x) \cos(n \, \delta k \, x) \, dx, \tag{8.7}$$

$$S_n = \frac{2}{L} \int_{-L/2}^{L/2} F(x) \sin(n \, \delta k \, x) \, dx. \tag{8.8}$$

(See Exercise 8.1.) Incidentally, any periodic function of x can be represented as a Fourier series.

Suppose, however, that we are dealing with a function $F(x)$ that is not periodic in x. We can think of such a function as one that is periodic in x with a period L that tends to infinity. Does this mean that we can still represent $F(x)$ as a Fourier series? Consider what happens to the series (8.2) in the limit $L \to \infty$, or, equivalently, $\delta k \to 0$. The series is basically a weighted sum of sinusoidal functions whose wavenumbers take the quantized values $k_n = n \, \delta k$. Moreover, as $\delta k \to 0$, these values become more and more closely spaced. In fact, we can write

$$F(x) = \sum_{n=1,\infty} \frac{C_n}{\delta k} \cos(n \, \delta k \, x) \, \delta k + \sum_{n=1,\infty} \frac{S_n}{\delta k} \sin(n \, \delta k \, x) \, \delta k. \tag{8.9}$$

In the continuum limit, $\delta k \to 0$, the summations in the previous expression become integrals, and we obtain

$$F(x) = \int_{-\infty}^{\infty} C(k) \cos(k \, x) \, dk + \int_{-\infty}^{\infty} S(k) \sin(k \, x) \, dk, \tag{8.10}$$

where $k = n \, \delta k$, $C(k) = C(-k) = C_n/(2 \, \delta k)$, and $S(k) = -S(-k) = S_n/(2 \, \delta k)$. Thus, for the case of an aperiodic function, the Fourier series (8.2) morphs into the so-called *Fourier transform* (8.10). This transform can be inverted using the continuum limits (i.e., the limit $\delta k \to 0$) of Equations (8.7) and (8.8), which are readily shown to be

$$C(k) = \frac{1}{2\pi} \int_{-\infty}^{\infty} F(x) \cos(k \, x) \, dx, \tag{8.11}$$

$$S(k) = \frac{1}{2\pi} \int_{-\infty}^{\infty} F(x) \sin(k \, x) \, dx, \tag{8.12}$$

respectively. (See Exercise 8.5.) The previous equations confirm that $C(-k) = C(k)$ and $S(-k) = -S(k)$. The Fourier-space (i.e., k-space) functions $C(k)$ and $S(k)$ are known as the *cosine Fourier transform* and the *sine Fourier transform* of the real-space (i.e., x-space) function $F(x)$, respectively. Furthermore, because we already know that any periodic function can be represented as a Fourier series, it seems plausible that any aperiodic function can be represented as a Fourier transform. This is indeed the case.

When sinusoidal waves of different amplitudes, phases, and wavelengths are superposed, they interfere with one another. In some regions of space, the interference is constructive, and the resulting wave amplitude is comparatively large. In other regions, the interference is destructive, and the resulting wave amplitude is comparatively small, or even zero. Equations (8.10)–(8.12) essentially allow us to construct an interference pattern that mimics any given function of position (in one dimension). Alternatively, they allow us to decompose any given function of position into sinusoidal waves that, when superposed, reconstruct the function. Let us consider some examples.

Consider the "top-hat" function,

$$F(x) = \begin{cases} 1 & |x| \le l/2 \\ 0 & |x| > l/2 \end{cases}. \tag{8.13}$$

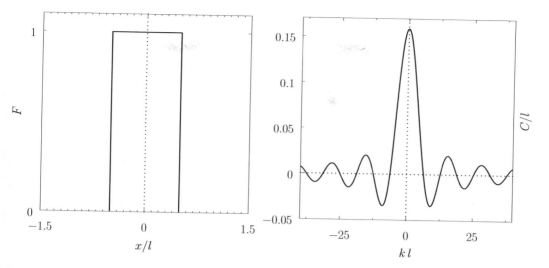

FIGURE 8.1   Fourier transform of a top-hat function.

See Figure 8.1. Given that $\cos(-k\,x) = \cos(k\,x)$ and $\sin(-k\,x) = -\sin(k\,x)$, it follows from Equations (8.11) and (8.12) that if $F(x)$ is even in x, so that $F(-x) = F(x)$, then $S(k) = 0$, and if $F(x)$ is odd in x, so that $F(-x) = -F(x)$, then $C(k) = 0$. Hence, because the top-hat function (8.13) is even in x, its sine Fourier transform is automatically zero. On the other hand, its cosine Fourier transform takes the form

$$C(k) = \frac{1}{2\pi} \int_{-l/2}^{l/2} \cos(k\,x)\, dx = \frac{l}{2\pi} \frac{\sin(k\,l/2)}{k\,l/2}. \tag{8.14}$$

Figure 8.1 shows the function $F(x)$, together with its associated cosine transform, $C(k)$.

As a second example, consider the so-called Gaussian function,

$$F(x) = \exp\left(-\frac{x^2}{2\sigma_x^2}\right). \tag{8.15}$$

As illustrated in Figure 8.2, this is a smoothly-varying even function of x that attains its peak value 1 at $x = 0$, and becomes completely negligible when $|x| \gtrsim 3\,\sigma_x$. Thus, σ_x is a measure of the "width" of the function in real (as opposed to Fourier) space. By symmetry, the sine Fourier transform of the preceding function is zero. On the other hand, the cosine Fourier transform is readily shown to be

$$C(k) = \frac{1}{(2\pi\,\sigma_k^2)^{1/2}} \exp\left(-\frac{k^2}{2\sigma_k^2}\right), \tag{8.16}$$

where

$$\sigma_k = \frac{1}{\sigma_x}. \tag{8.17}$$

(See Exercise 8.2.) This function is a Gaussian in Fourier space of characteristic width $\sigma_k = 1/\sigma_x$. The original function $F(x)$ can be reconstructed from its Fourier transform using

$$F(x) = \int_{-\infty}^{\infty} C(k)\, \cos(k\,x)\, dk. \tag{8.18}$$

This reconstruction is simply a linear superposition of cosine waves of differing wavenumbers.

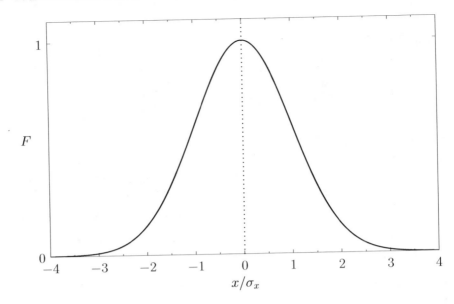

FIGURE 8.2 A Gaussian function.

Moreover, $C(k)$ can be interpreted as the amplitude of waves of wavenumber k within this superposition. The fact that $C(k)$ is a Gaussian of characteristic width $\sigma_k = 1/\sigma_x$ [which means that $C(k)$ is negligible for $|k| \gtrsim 3\,\sigma_k$] implies that in order to reconstruct a real-space function whose width in real space is approximately σ_x it is necessary to combine sinusoidal functions with a range of different wavenumbers that is approximately $\sigma_k = 1/\sigma_x$ in extent. To be slightly more exact, the real-space Gaussian function $F(x)$ falls to half of its peak value when $|x| \simeq \sqrt{\pi/2}\,\sigma_x$. Hence, the full width at half maximum of the function is $\Delta x \simeq 2\sqrt{\pi/2}\,\sigma_x = \sqrt{2\pi}\,\sigma_x$. Likewise, the full width at half maximum of the Fourier-space Gaussian function $C(k)$ is $\Delta k \simeq \sqrt{2\pi}\,\sigma_k$. Thus,

$$\Delta x\,\Delta k \simeq 2\pi, \tag{8.19}$$

because $\sigma_k\,\sigma_x = 1$. We conclude that a function that is highly localized in real space has a transform that is highly delocalized in Fourier space, and vice versa. Finally,

$$\int_{-\infty}^{\infty} \frac{1}{(2\pi\,\sigma_k^2)^{1/2}} \exp\left(-\frac{k^2}{2\,\sigma_k^2}\right) dk = 1. \tag{8.20}$$

(See Exercise 8.3.) In other words, a Gaussian function in real space, of unit height and characteristic width σ_x, has a cosine Fourier transform that is a Gaussian in Fourier space, of characteristic width $\sigma_k = 1/\sigma_x$, and whose integral over all k-space is unity.

8.3 DIRAC DELTA FUNCTION

Consider what happens to the previously mentioned real-space Gaussian, and its Fourier transform, in the limit $\sigma_x \to \infty$, or, equivalently, $\sigma_k \to 0$. There is no difficulty in seeing, from Equation (8.15), that

$$F(x) \to 1. \tag{8.21}$$

In other words, the real-space Gaussian morphs into a function that takes the constant value unity everywhere. The Fourier transform is more problematic. In the limit $\sigma_k \to 0$, Equation (8.16) yields a k-space function that is zero everywhere apart from $k = 0$ (because the function is negligible for $|k| \gtrsim \sigma_k$), where it is infinite [because the function takes the value $(2\pi \sigma_k)^{-1/2}$ at $k = 0$]. Moreover, according to Equation (8.20), the integral of the function over all k remains unity. Thus, the Fourier transform of the uniform function $F(x) = 1$ is a sort of integrable "spike" located at $k = 0$. This unusual function is known as the *Dirac delta function*, and is denoted $\delta(k)$. Thus, one definition of a delta function is

$$\delta(k) = \lim_{\sigma_k \to 0} \frac{1}{(2\pi \sigma_k^2)^{1/2}} \exp\left(-\frac{k^2}{2\sigma_k^2}\right).$$ (8.22)

As has already been mentioned, $\delta(k) = 0$ for $k \neq 0$, and $\delta(0) = \infty$. Moreover,

$$\int_{-\infty}^{\infty} \delta(k)\, dk = 1.$$ (8.23)

Consider the integral

$$\int_{-\infty}^{\infty} F(k)\, \delta(k)\, dk,$$ (8.24)

where $F(k)$ is an arbitrary function. Because of the peculiar properties of the delta function, the only contribution to the previous integral comes from the region in k-space in the immediate vicinity of $k = 0$. Furthermore, provided $F(k)$ is well behaved in this region, we can write

$$\int_{-\infty}^{\infty} F(k)\, \delta(k)\, dk = \int_{-\infty}^{\infty} F(0)\, \delta(k)\, dk = F(0) \int_{-\infty}^{\infty} \delta(k)\, dk = F(0),$$ (8.25)

where use has been made of Equation (8.23).

A change of variables allows us to define $\delta(k - k')$, which is a "spike" function centered on $k = k'$. The previous result can be generalized to give

$$\int_{-\infty}^{\infty} F(k)\, \delta(k - k')\, dk = F(k'),$$ (8.26)

for all $F(k)$ that are well behaved at $k = k'$. Indeed, this expression can be thought of as an alternative definition of a delta function. Incidentally, a Dirac delta function is sometimes referred to as a *generalized function*. This nomenclature arises because the actual value of the function is ill defined (for instance, it would be impossible to graph the function), whereas its integral is perfectly well defined (as evidenced by the previous equation).

We have seen that the delta function $\delta(k)$ is the cosine Fourier transform of the uniform function $F(x) = 1$. It, thus, follows from Equation (8.11) that

$$\delta(k) = \frac{1}{2\pi} \int_{-\infty}^{\infty} \cos(k\, x)\, dx.$$ (8.27)

This result represents yet another definition of the delta function. By symmetry, we also have

$$0 = \frac{1}{2\pi} \int_{-\infty}^{\infty} \sin(k\, x)\, dx.$$ (8.28)

It follows that

$$\frac{1}{2\pi} \int_{-\infty}^{\infty} \cos(k\, x)\, \cos(k'\, x)\, dx = \frac{1}{4\pi} \int_{-\infty}^{\infty} \{\cos[(k - k')\, x] + \cos[(k + k')\, x]\}\, dx,$$ (8.29)

which yields

$$\frac{1}{2\pi} \int_{-\infty}^{\infty} \cos(k\, x)\, \cos(k'\, x)\, dx = \frac{1}{2} [\delta(k - k') + \delta(k + k')],$$ (8.30)

where use has been made of Equation (8.27), and a standard trigonometric identity. (See Appendix B.) Likewise,

$$\frac{1}{2\pi} \int_{-\infty}^{\infty} \sin(k\,x) \sin(k'\,x)\,dx = \frac{1}{2}\left[\delta(k-k') - \delta(k+k')\right], \tag{8.31}$$

$$\frac{1}{2\pi} \int_{-\infty}^{\infty} \cos(k\,x) \sin(k'\,x)\,dx = 0. \tag{8.32}$$

(See Exercise 8.4.) Incidentally, Equations (8.30)–(8.32) can be used to derive Equations (8.11) and (8.12) directly from Equation (8.10). (See Exercise 8.5.)

8.4 GENERAL SOLUTION OF 1D WAVE EQUATION

Consider the one-dimensional wave equation,

$$\frac{\partial^2 \psi}{\partial t^2} = v^2 \frac{\partial^2 \psi}{\partial x^2}, \tag{8.33}$$

where $\psi(x, t)$ is the wavefunction, and v the characteristic phase velocity. We have seen a number of particular solutions of this equation. For instance,

$$\psi(x, t) = A \cos(\omega t - k x - \phi) \tag{8.34}$$

represents a traveling wave of amplitude A, angular frequency ω, wavenumber k, and phase angle ϕ, that propagates in the positive x-direction. The previous expression is a solution of the one-dimensional wave equation, (8.33), provided that it satisfies the dispersion relation

$$\omega = k\,v; \tag{8.35}$$

that is, provided the wave propagates at the fixed phase velocity v. We can also write the wavefunction (8.34) in the form

$$\psi(x, t) = C_+ \cos[k\,(v\,t - x)] + S_+ \sin[k\,(v\,t - x)], \tag{8.36}$$

where $C_+ = A \cos\phi$, $S_+ = A \sin\phi$, and we have explicitly incorporated the dispersion relation $\omega = k\,v$ into the solution. The previous expression can be regarded as the most general form for a traveling wave of wavenumber k propagating in the positive x-direction. Likewise, the most general form for a traveling wave of wavenumber k propagating in the negative x-direction is

$$\psi(x, t) = C_- \cos[k\,(v\,t + x)] + S_- \sin[k\,(v\,t + x)]. \tag{8.37}$$

We have also encountered standing wave solutions of Equation (8.33). However, as we have seen, these can be regarded as linear superpositions of traveling waves, of equal amplitude and wavenumber, propagating in opposite directions. (See Section 6.4.) In other words, standing waves are not fundamentally different to traveling waves.

The wave equation, (8.33), is linear. This suggests that its most general solution can be written as a linear superposition of all of its valid wavelike solutions. In the absence of specific boundary conditions, there is no restriction on the possible wavenumbers of such solutions. Thus, it is plausible that the most general solution of Equation (8.33) can be written

$$\psi(x, t) = \int_{-\infty}^{\infty} C_+(k) \cos[k\,(v\,t - x)]\,dk + \int_{-\infty}^{\infty} S_+(k) \sin[k\,(v\,t - x)]\,dk$$

$$+ \int_{-\infty}^{\infty} C_-(k) \cos[k\,(v\,t + x)]\,dk + \int_{-\infty}^{\infty} S_-(k) \sin[k\,(v\,t + x)]\,dk. \tag{8.38}$$

In other words, the general solution is a linear superposition of traveling waves propagating to the right (i.e., in the positive x-direction), and to the left. Here, $C_+(k)$ represents the amplitude of right-propagating cosine waves of wavenumber k in this superposition. Moreover, $S_+(k)$ represents the amplitude of right-propagating sine waves of wavenumber k, $C_-(k)$ the amplitude of left-propagating cosine waves, and $S_-(k)$ the amplitude of left-propagating sine waves. Because each of these waves is individually a solution of Equation (8.33), we are guaranteed, from the linear nature of this equation, that the previous superposition is also a solution.

But, how can we prove that Equation (8.38) is the *most general* solution of the wave equation, (8.33)? Our understanding of Newtonian dynamics tells us that if we know the initial wave amplitude $\psi(x, 0)$, and its time derivative $\dot\psi(x, 0)$, then this should constitute sufficient information to uniquely specify the solution at all subsequent times. Hence, if Equation (8.38) is the most general solution of Equation (8.33) then it must be consistent with any initial wave amplitude, and any initial wave velocity. In other words, given any $\psi(x, 0)$ and $\dot\psi(x, 0)$, we should be able to uniquely determine the functions $C_+(k)$, $S_+(k)$, $C_-(k)$, and $S_-(k)$ appearing in Equation (8.38). Let us see if this is the case.

From Equation (8.38),

$$\psi(x, 0) = \int_{-\infty}^{\infty} [C_+(k) + C_-(k)] \cos(k\, x)\, dk + \int_{-\infty}^{\infty} [-S_+(k) + S_-(k)] \sin(k\, x)\, dk. \qquad (8.39)$$

However, this is a Fourier transform of the form (8.10). Moreover, Equations (8.11) and (8.12) allow us to uniquely invert this transform. In fact,

$$C_+(k) + C_-(k) = \frac{1}{2\pi} \int_{-\infty}^{\infty} \psi(x, 0) \cos(k\, x)\, dx, \qquad (8.40)$$

$$-S_+(k) + S_-(k) = \frac{1}{2\pi} \int_{-\infty}^{\infty} \psi(x, 0) \sin(k\, x)\, dx. \qquad (8.41)$$

Equation (8.38) also yields

$$\dot\psi(x, 0) = \int_{-\infty}^{\infty} k\, v\, [C_+(k) - C_-(k)] \sin(k\, x)\, dk + \int_{-\infty}^{\infty} k\, v\, [S_+(k) + S_-(k)] \cos(k\, x)\, dk. \qquad (8.42)$$

This is, again, a Fourier transform, and can be inverted to give

$$k\, v\, [C_+(k) - C_-(k)] = \frac{1}{2\pi} \int_{-\infty}^{\infty} \dot\psi(x, 0) \sin(k\, x)\, dx, \qquad (8.43)$$

$$k\, v\, [S_+(k) + S_-(k)] = \frac{1}{2\pi} \int_{-\infty}^{\infty} \dot\psi(x, 0) \cos(k\, x)\, dx. \qquad (8.44)$$

Hence,

$$C_+(k) = \frac{1}{4\pi} \left[\int_{-\infty}^{\infty} \psi(x, 0) \cos(k\, x)\, dx + \int_{-\infty}^{\infty} \frac{\dot\psi(x, 0)}{k\, v} \sin(k\, x)\, dx \right], \qquad (8.45)$$

$$C_-(k) = \frac{1}{4\pi} \left[\int_{-\infty}^{\infty} \psi(x, 0) \cos(k\, x)\, dx - \int_{-\infty}^{\infty} \frac{\dot\psi(x, 0)}{k\, v} \sin(k\, x)\, dx \right], \qquad (8.46)$$

$$S_+(k) = \frac{1}{4\pi} \left[-\int_{-\infty}^{\infty} \psi(x, 0) \sin(k\, x)\, dx + \int_{-\infty}^{\infty} \frac{\dot\psi(x, 0)}{k\, v} \cos(k\, x)\, dx \right], \qquad (8.47)$$

$$S_-(k) = \frac{1}{4\pi} \left[\int_{-\infty}^{\infty} \psi(x, 0) \sin(k\, x)\, dx + \int_{-\infty}^{\infty} \frac{\dot\psi(x, 0)}{k\, v} \cos(k\, x)\, dx \right]. \qquad (8.48)$$

It follows that we can uniquely determine the functions $C_+(k)$, $C_-(k)$, $S_+(k)$, and $S_-(k)$, appearing in Equation (8.38), for any $\psi(x,0)$ and $\dot\psi(x,0)$. This proves that Equation (8.38) is indeed the most general solution of the wave equation, (8.33).

Let us examine our solution in more detail. Equation (8.38) can be written

$$\psi(x,t) = F(v\,t - x) + G(v\,t + x), \tag{8.49}$$

where

$$F(x) = \int_{-\infty}^{\infty} [C_+(k)\cos(k\,x) + S_+(k)\sin(k\,x)]\,dk, \tag{8.50}$$

$$G(x) = \int_{-\infty}^{\infty} [C_-(k)\cos(k\,x) + S_-(k)\sin(k\,x)]\,dk. \tag{8.51}$$

(See Exercise 8.6.) What is the significance of Equation (8.49)? Actually, $F(v\,t-x)$ represents a wave disturbance of arbitrary shape that propagates in the positive x-direction, at the fixed speed v, without changing shape. This follows because a point with a given amplitude on the wave, $F(v\,t-x) = c$, has an equation of motion $v\,t - x = F^{-1}(c) = $ constant, and thus propagates in the positive x-direction at the speed v. Moreover, because all points on the wave propagate in the same direction at the same speed, it follows that the wave does not change shape as it moves. By analogy, $G(v\,t+x)$ represents a wave disturbance of arbitrary shape that propagates in the negative x-direction, at the fixed speed v, without changing shape. We conclude that the most general solution to the wave equation, (8.33), is a superposition of two wave disturbances of arbitrary shapes that propagate in opposite directions, at the fixed speed v, without changing shape. Such solutions are generally termed *wave pulses*. What is the relationship between a general wave pulse and the sinusoidal traveling wave solutions to the wave equation that we found previously? As is apparent from Equations (8.50) and (8.51), a wave pulse is a superposition of sinusoidal traveling waves propagating in the same direction as the pulse. Moreover, the amplitude of cosine waves of wavenumber k in this superposition is the cosine Fourier transform of the pulse shape, evaluated at wavenumber k. Likewise, the amplitude of sine waves of wavenumber k in the superposition is the sine Fourier transform of the pulse shape, evaluated at wavenumber k.

For instance, suppose that we have a triangular wave pulse of the form

$$F(x) = \begin{cases} 1 - 2\,|x|/l & |x| \le l/2 \\ 0 & |x| > l/2 \end{cases}. \tag{8.52}$$

The sine Fourier transform of this pulse shape is zero by symmetry. However, the cosine Fourier transform is

$$C(k) = \frac{1}{2\pi}\int_{-\infty}^{\infty} F(x)\cos(k\,x)\,dx = \frac{l}{4\pi}\frac{\sin^2(k\,l/4)}{(k\,l/4)^2}. \tag{8.53}$$

(See Exercise 8.7.) The functions $F(x)$ and $C(k)$ are shown in Figure 8.3. It follows that the right-propagating triangular wave pulse

$$\psi(x,t) = \begin{cases} 1 - 2\,|v\,t - x|/l & |v\,t - x| \le l/2 \\ 0 & |v\,t - x| > l/2 \end{cases} \tag{8.54}$$

can be written as the following superposition of right-propagating cosine waves:

$$\psi(x,t) = \frac{1}{4\pi}\int_{-\infty}^{\infty} \frac{\sin^2(k\,l/4)}{(k\,l/4)^2}\cos[k\,(v\,t - x)]\,l\,dk. \tag{8.55}$$

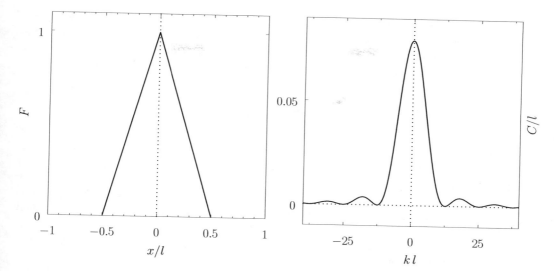

FIGURE 8.3 Fourier transform of a triangular wave pulse.

Likewise, the left-propagating triangular wave pulse

$$\psi(x, t) = \begin{cases} 1 - 2\,|v\,t + x|/l & |v\,t + x| \le l/2 \\ 0 & |v\,t + x| > l/2 \end{cases} \tag{8.56}$$

becomes

$$\psi(x, t) = \frac{1}{4\pi} \int_{-\infty}^{\infty} \frac{\sin^2(k\,l/4)}{(k\,l/4)^2} \cos[k\,(v\,t + x)]\,l\,dk. \tag{8.57}$$

The ideas developed in this section can be extended to multi-dimensional waves in a straightforward fashion. (See Exercises 8.12 and 8.13.)

8.5 BANDWIDTH

It is possible to Fourier transform in time, as well as in space. Thus, a general temporal waveform $F(t)$ can be written as a superposition of sinusoidal waveforms of various different angular frequencies, ω. In other words,

$$F(t) = \int_{-\infty}^{\infty} C(\omega)\,\cos(\omega\,t)\,d\omega + \int_{-\infty}^{\infty} S(\omega)\,\sin(\omega\,t)\,d\omega, \tag{8.58}$$

where $C(\omega)$ and $S(\omega)$ are the temporal cosine and sine Fourier transforms of the waveform, respectively. By analogy with Equations (8.10)–(8.12), we can invert the previous expression to give

$$C(\omega) = \frac{1}{2\pi} \int_{-\infty}^{\infty} F(t)\,\cos(\omega\,t)\,dt, \tag{8.59}$$

$$S(\omega) = \frac{1}{2\pi} \int_{-\infty}^{\infty} F(t)\,\sin(\omega\,t)\,dt. \tag{8.60}$$

These equations make it manifest that $C(-\omega) = C(\omega)$, and $S(-\omega) = -S(\omega)$. Moreover, it is apparent that if $F(t)$ is an even function of t then $S(\omega) = 0$, but if it is an odd function then $C(\omega) = 0$.

The current flowing in the antenna of an amplitude-modulated (AM) radio transmitter is driven by a voltage signal that oscillates sinusoidally at a frequency, ω_0, which is known as the *carrier frequency*. In commercial (medium wave) AM radio, each station is assigned a single carrier frequency that lies somewhere between about 500 kHz and 1600 kHz. However, the voltage signal fed to the antenna does not have a constant amplitude. Rather, it has a modulated amplitude that can be expressed, somewhat schematically, as a Fourier series:

$$A(t) = A_0 + \sum_{n>0} A_n \cos(\omega_n t - \phi_n), \tag{8.61}$$

where $A(t) - A_0$ represents the information being transmitted. Typically, this information is speech or music that is picked up by a microphone, and converted into an electrical signal. The constant amplitude A_0 is present even when the transmitter is sending no information. The remaining terms in the previous expression are due to the signal picked up by the microphone. The modulation frequencies, ω_n, are thus the frequencies of audible sound waves. In other words, they are so-called *audio frequencies* lying between about 20 Hz and 20 kHz. This implies that the modulation frequencies are much smaller than the carrier frequency; that is, $\omega_n \ll \omega_0$ for all $n > 0$. Furthermore, the modulation amplitudes A_n are all generally smaller than the carrier amplitude A_0.

The signal transmitted by an AM station, and received by an AM receiver, is an amplitude-modulated sinusoidal oscillation of the form

$$\psi(t) = A(t) \cos(\omega_0 t)$$
$$= A_0 \cos(\omega_0 t) + \sum_{n>0} A_n \cos(\omega_n t - \phi_n) \cos(\omega_0 t), \tag{8.62}$$

which, with the help of some standard trigonometric identities (see Appendix B), can also be written

$$\psi(t) = A_0 \cos(\omega_0 t) + \frac{1}{2}\sum_{n>0} A_n \cos[(\omega_0 + \omega_n)t - \phi_n)] + \frac{1}{2}\sum_{n>0} A_n \cos[(\omega_0 - \omega_n)t + \phi_n)]$$
$$= A_0 \cos(\omega_0 t) + \frac{1}{2}\sum_{n>0} A_n \cos\phi_n \cos[(\omega_0 + \omega_n)t]$$
$$+ \frac{1}{2}\sum_{n>0} A_n \sin\phi_n \sin[(\omega_0 + \omega_n)t] + \frac{1}{2}\sum_{n>0} A_n \cos\phi_n \cos[(\omega_0 - \omega_n)t]$$
$$- \frac{1}{2}\sum_{n>0} A_n \sin\phi_n \sin[(\omega_0 - \omega_n)t]. \tag{8.63}$$

We can calculate the cosine and sine Fourier transforms of the signal,

$$C(\omega) = \frac{1}{2\pi}\int_{-\infty}^{\infty} \psi(t)\cos(\omega t)\,dt, \tag{8.64}$$

$$S(\omega) = \frac{1}{2\pi}\int_{-\infty}^{\infty} \psi(t)\sin(\omega t)\,dt, \tag{8.65}$$

by making use of Equation (8.30)–(8.32). It follows that

$$C(\omega > 0) = \frac{1}{2}A_0\,\delta(\omega - \omega_0) + \frac{1}{4}\sum_{n>0} A_n\cos\phi_n\,[\delta(\omega - \omega_0 - \omega_n) + \delta(\omega - \omega_0 + \omega_n)], \tag{8.66}$$

$$S(\omega > 0) = \frac{1}{4}\sum_{n>0} A_n\sin\phi_n\,[\delta(\omega - \omega_0 - \omega_n) - \delta(\omega - \omega_0 + \omega_n)]. \tag{8.67}$$

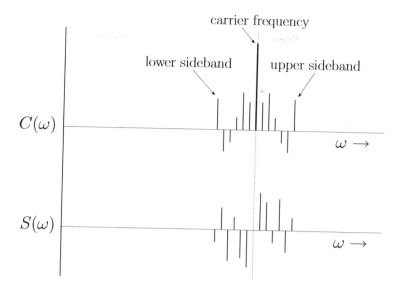

FIGURE 8.4   Frequency spectrum of an AM radio signal.

Here, we have only shown the positive frequency components of $C(\omega)$ and $S(\omega)$, because we know that $C(-\omega) = C(\omega)$ and $S(-\omega) = -S(\omega)$.

The AM frequency spectrum specified in Equations (8.66) and (8.67) is shown, somewhat schematically, in Figure 8.4. The spectrum consists of a series of delta function spikes. The largest spike corresponds to the carrier frequency, ω_0. However, this spike carries no information. Indeed, the signal information is carried in so-called *sideband frequencies* which are equally spaced on either side of the carrier frequency. The upper sidebands correspond to the frequencies $\omega_0 + \omega_n$, whereas the lower sidebands correspond to the frequencies $\omega_0 - \omega_n$. Thus, in order for an AM radio signal to carry all of the information present in audible sound, for which the appropriate modulation frequencies, ω_n, range from about 0 Hz to about 20 kHz, the signal would have to consist of a superposition of sinusoidal oscillations with frequencies that range from the carrier frequency minus 20 kHz to the carrier frequency plus 20 kHz. In other words, the signal would have to occupy a range of frequencies from $\omega_0 - \omega_N$ to $\omega_0 + \omega_N$, where ω_N is the largest modulation frequency. This is an important result. An AM radio signal that only consists of a single frequency, such as the carrier frequency, transmits no information. Only a signal that occupies a finite range of frequencies, centered on the carrier frequency, is capable of transmitting useful information. The difference between the highest and the lowest frequency components of an AM radio signal, which is twice the maximum modulation frequency, is called the *bandwidth* of the signal. Thus, to transmit all of the information present in audible sound an AM signal would need to have a bandwidth of 40 kHz. In fact, commercial AM radio signals are only allowed to broadcast a bandwidth of 10 kHz, in order to maximize the number of available stations. (Two different stations obviously cannot broadcast in frequency ranges that overlap.) This means that commercial AM radio can only carry audible information in the range 0 to about 5 kHz. This is perfectly adequate for ordinary speech, but only barely adequate for music.

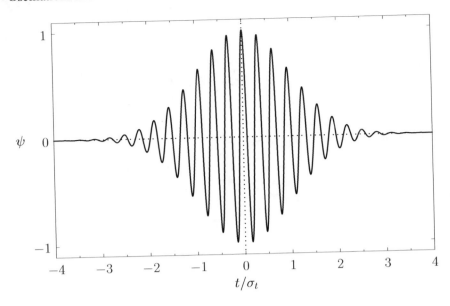

FIGURE 8.5   A digital bit transmitted over AM radio.

8.6   BANDWIDTH THEOREM

Let us now consider how we might transmit a digital signal over AM radio. Suppose that each data "bit" in the signal takes the form of a Gaussian envelope, of characteristic duration σ_t, superimposed on a carrier wave whose frequency is ω_0; that is,

$$\psi(t) = \exp\left(-\frac{t^2}{2\,\sigma_t^2}\right)\cos(\omega_0\,t). \tag{8.68}$$

Let us assume that $\omega_0\,\sigma_t \gg 1$. In other words, the period of the carrier wave is much less than the duration of the bit. Figure 8.5 illustrates a digital bit calculated for $\omega_0\,\sigma_t = 20$.

The sine Fourier transform of the signal (8.68) is zero by symmetry. However, its cosine Fourier transform takes the form

$$C(\omega) = \frac{1}{2\pi}\int_{-\infty}^{\infty}\exp\left(-\frac{t^2}{2\,\sigma_t^2}\right)\cos(\omega_0\,t)\,\cos(\omega\,t)\,dt,$$

$$= \frac{1}{4\pi}\int_{-\infty}^{\infty}\exp\left(-\frac{t^2}{2\,\sigma_t^2}\right)\{\cos[(\omega-\omega_0)\,t] + \cos[(\omega+\omega_0)\,t]\}\,dt. \tag{8.69}$$

A comparison with Equations (8.15)–(8.18) reveals that

$$C(\omega > 0) = \frac{1}{2\,(2\pi\,\sigma_\omega^2)^{1/2}}\exp\left[-\frac{(\omega-\omega_0)^2}{2\,\sigma_\omega^2}\right], \tag{8.70}$$

where

$$\sigma_\omega = \frac{1}{\sigma_t}. \tag{8.71}$$

In other words, the Fourier transform of the signal takes the form of a Gaussian in ω-space that is centered on the carrier frequency, ω_0, and is of characteristic width $\sigma_\omega = 1/\sigma_t$. Thus, the bandwidth of the signal is of order σ_ω. The shorter the signal duration, the higher the bandwidth. This is a general rule. A signal of full width at half maximum temporal duration $\Delta t \simeq \sqrt{2\pi}\,\sigma_t$ generally has a Fourier transform of full width at half maximum bandwidth $\Delta\omega \simeq \sqrt{2\pi}\,\sigma_\omega$, so that

$$\Delta\omega\,\Delta t \simeq 2\pi. \tag{8.72}$$

This can also be written

$$\Delta f\,\Delta t \simeq 1, \tag{8.73}$$

where $\Delta f = \Delta\omega/2\pi$ is the bandwidth in hertz. The previous result is known as the *bandwidth theorem*. The duration of a digital bit is closely related to the maximum rate at which information can be transmitted by a digital signal. Because the individual bits cannot overlap in time, the maximum number of bits per second that can be transmitted by a digital signal is of order $1/\Delta t$. In other words, it is of order the bandwidth of the signal. Thus, digital signals that transmit information at a rapid rate require large bandwidths, and, consequently occupy a wide range of frequency space.

An old-fashioned black and white TV screen consists of a rectangular grid of black and white spots scanned by an electron beam that can be switched on or off rapidly. A given spot is "white" if the phosphorescent TV screen was recently (i.e., within about 1/50 th of a second) struck by the energized electron beam at that location. The spot separation is about 1 mm. A typical screen is 50 cm × 50 cm, and thus has 500 lines with 500 spots per line, or 2.5×10^5 spots. Each spot is renewed every 1/30 th of a second. (Every other horizontal line is skipped during a given traversal of the electron beam over the screen. The skipped lines are renewed on the next traversal. This technique is known as *interlacing*. Consequently, a given region of the screen, that includes many horizontal lines, has a flicker rate of 60 Hz.) Thus, the rate at which the instructions "turn on" and "turn off" must be sent to the electron beam is $30 \times 2.5 \times 10^5$ or 8×10^6 times a second. The transmitted TV signal must therefore have about 10^7 on-off instruction blips per second. If temporal overlap is to be avoided, each blip can be no longer than $\Delta t \sim 10^{-7}$ seconds in duration. Thus, the required bandwidth is $\Delta f \sim 1/\Delta t \sim 10^7\,\text{Hz} = 10\,\text{MHz}$. The carrier wave frequencies used for conventional broadcast TV lie in the so-called VHF band, and range from about 55 to 210 MHz. Our previous discussion of AM radio might lead us to think that the 10 MHz bandwidth represents the combined extents of an upper and a lower sideband of modulation frequencies. In practice, the carrier wave and one of the sidebands are suppressed. That is, they are filtered out, and never applied to the antenna. However, they are regenerated in the receiver from the information contained in the single sideband that is broadcast. This technique, which is called *single sideband transmission*, halves the bandwidth requirement to about 5 MHz. (Incidentally, the lower sideband carries the same information as the upper one, and thus can be used to completely regenerate the upper sideband, and vice versa.) Thus, between 55 and 210 MHz there is room for about 30 TV channels, each using a 5 MHz bandwidth. (Actually, there are far fewer TV channels than this in the VHF band, because part of this band is reserved for FM radio, air traffic control, air navigation beacons, marine communications, etc.)

EXERCISES

8.1 (a) Verify Equations (8.4)–(8.6). (b) Derive Equations (8.7) and (8.8) from Equation (8.2) and Equations (8.4)–(8.6).

8.2 Suppose that

$$F(x) = \exp\left(-\frac{x^2}{2\sigma_x^2}\right).$$

Demonstrate that

$$\bar{F}(k) \equiv \frac{1}{2\pi}\int_{-\infty}^{\infty} F(x)\,e^{ikx}\,dx = \frac{1}{\sqrt{2\pi\sigma_k^2}}\exp\left(-\frac{k^2}{2\sigma_k^2}\right),$$

where i is the square-root of minus one, and $\sigma_k = 1/\sigma_x$. [Hint: You will need to complete the square of the exponent of e, transform the variable of integration, and then make use of the standard result that $\int_{-\infty}^{\infty} e^{-y^2} dy = \sqrt{\pi}$.] Hence, show from *Euler's theorem*, $\exp(i\theta) \equiv \cos\theta + i\sin\theta$, that

$$C(k) = \frac{1}{2\pi} \int_{-\infty}^{\infty} F(x)\cos(kx)\,dx = \frac{1}{\sqrt{2\pi\,\sigma_k^2}}\exp\left(-\frac{k^2}{2\sigma_k^2}\right),$$

$$S(k) = \frac{1}{2\pi} \int_{-\infty}^{\infty} F(x)\sin(kx)\,dx = 0.$$

8.3 Demonstrate that

$$\int_{-\infty}^{\infty} \frac{1}{(2\pi\,\sigma_k^2)^{1/2}}\exp\left(-\frac{k^2}{2\sigma_k^2}\right)dk = 1.$$

8.4 Verify Equations (8.31) and (8.32).

8.5 Derive Equations (8.11) and (8.12) directly from Equation (8.10) using the results (8.30)–(8.32).

8.6 Verify directly that Equation (8.49) is a solution of the wave equation, (8.33), for arbitrary pulse shapes $F(x)$ and $G(x)$.

8.7 Verify Equation (8.53).

8.8 Consider a function $F(t)$ that is zero for negative t, and takes the value $\exp(-t/2\tau)$ for $t \geq 0$. Find its Fourier transforms, $C(\omega)$ and $S(\omega)$, defined by

$$F(t) = \int_{-\infty}^{\infty} C(\omega)\cos(\omega t)\,d\omega + \int_{-\infty}^{\infty} S(\omega)\sin(\omega t)\,d\omega.$$

[Hint: Use Euler's theorem.]

8.9 Let $F(t)$ be zero, except in the interval from $t = -\Delta t/2$ to $t = \Delta t/2$. Suppose that in this interval $F(t)$ makes exactly one sinusoidal oscillation at the angular frequency $\omega_0 = 2\pi/\Delta t$, starting and ending with the value zero. Find the previously defined Fourier transforms $C(\omega)$ and $S(\omega)$.

8.10 Demonstrate that

$$\int_{-\infty}^{\infty} F^2(t)\,dt = 2\pi \int_{-\infty}^{\infty} \left[C^2(\omega) + S^2(\omega)\right]d\omega,$$

where the relation between $F(t)$, $C(\omega)$, and $S(\omega)$ is as previously defined. This result is known as *Parseval's theorem*.

8.11 Suppose that $F(t)$ and $G(t)$ are both even functions of t with the cosine transforms $\bar{F}(\omega)$ and $\bar{G}(\omega)$, so that

$$F(t) = \int_{-\infty}^{\infty} \bar{F}(\omega)\cos(\omega t)\,d\omega,$$

$$G(t) = \int_{-\infty}^{\infty} \bar{G}(\omega)\cos(\omega t)\,d\omega.$$

Let $H(t) = F(t)\,G(t)$, and let $\bar{H}(\omega)$ be the cosine transform of this even function, so that

$$H(t) = \int_{-\infty}^{\infty} \bar{H}(\omega)\cos(\omega t)\,d\omega.$$

(a) Demonstrate that

$$\bar{H}(\omega) = \frac{1}{2} \int_{-\infty}^{\infty} \bar{F}(\omega') \left[\bar{G}(\omega' + \omega) + \bar{G}(\omega' - \omega) \right] d\omega'.$$

This result is known as the *convolution theorem*, because the previous type of integral is known as a convolution integral.

(b) Suppose that $F(t) = \cos(\omega_0 t)$. Show that

$$\bar{H}(\omega) = \frac{1}{2} \left[\bar{G}(\omega - \omega_0) + \bar{G}(\omega + \omega_0) \right].$$

8.12 Demonstrate that

$$\psi(\mathbf{r}, t) = F(v\,t - \mathbf{n} \cdot \mathbf{r}),$$

where F is an arbitrary function, and $\mathbf{n}$ a constant unit vector, is a solution of the three-dimensional wave equation, (7.9). How would you interpret this solution?

8.13 Demonstrate that

$$\psi(\mathbf{r}, t) = \frac{F(v\,t - r)}{r},$$

where F is an arbitrary function, is a solution of the spherical wave equation, (7.12). How would you interpret this solution?

CHAPTER 9

Dispersive Waves

9.1 INTRODUCTION

This chapter is devoted to the investigation of waves whose dispersion relations are nonlinear in nature.

9.2 PULSE PROPAGATION

Consider a one-dimensional wave pulse,

$$\psi(x, t) = \int_{-\infty}^{\infty} C(k) \cos(\omega t - k x) \, dk, \qquad (9.1)$$

made up of a linear superposition of cosine waves, with a range of different wavenumbers, all traveling in the positive x-direction. The angular frequency, ω, of each of these waves is related to its wavenumber, k, via the so-called *dispersion relation*, which can be written schematically as

$$\omega = \omega(k). \qquad (9.2)$$

In general, this relation is derivable from the wave disturbance's equation of motion. Up to now, we have only considered sinusoidal waves that have linear dispersion relations of the form

$$\omega = k\, v, \qquad (9.3)$$

where v is a constant. The previous expression immediately implies that such waves have the same phase velocity,

$$v_p = \frac{\omega}{k} = v, \qquad (9.4)$$

irrespective of their frequencies. Substituting Equation (9.3) into Equation (9.1), we obtain

$$\psi(x, t) = \int_{-\infty}^{\infty} C(k) \cos[k\,(v\,t - x)] \, dk, \qquad (9.5)$$

which is the equation of a wave pulse that propagates in the positive x-direction, at the fixed speed v, without changing shape. (See Chapter 8.) The previous analysis seems to suggest that arbitrarily-shaped wave pulses generally propagate at the same speed as sinusoidal waves, and do so without dispersing or, otherwise, changing shape. In fact, these statements are only true of pulses made up of superpositions of sinusoidal waves with linear dispersion relations. There are, however, many types of sinusoidal wave whose dispersion relations are nonlinear. For instance, the dispersion relation of sinusoidal electromagnetic waves propagating through an unmagnetized plasma is (see Section 9.3)

$$\omega = \sqrt{k^2 c^2 + \omega_{pe}^2}, \tag{9.6}$$

where c is the speed of light in vacuum, and ω_{pe} is a constant, known as the (electron) plasma frequency, that depends on the properties of the plasma. [See Equation (9.28).] Moreover, the dispersion relation of sinusoidal surface waves in deep water is

$$\omega = \sqrt{gk + \frac{T}{\rho} k^3}, \tag{9.7}$$

where g is the acceleration due to gravity, T the surface tension of water, and ρ the mass density. Sinusoidal waves that satisfy nonlinear dispersion relations, such as (9.6) or (9.7), are known as *dispersive waves*, as opposed to waves that satisfy linear dispersion relations, such as (9.3), which are called *non-dispersive* waves. As we saw previously, a wave pulse made up of a linear superposition of non-dispersive sinusoidal waves, all traveling in the same direction, propagates at the common phase velocity of these waves, without changing shape. How does a wave pulse made up of a linear superposition of dispersive sinusoidal waves evolve in time?

Suppose that

$$C(k) = \frac{1}{\sqrt{2\pi \sigma_k^2}} \exp\left[-\frac{(k - k_0)^2}{2\sigma_k^2}\right]. \tag{9.8}$$

In other words, the function $C(k)$ in Equation (9.1) is a Gaussian, of characteristic width σ_k, centered on wavenumber $k = k_0$. It follows, from the properties of the Gaussian function, that $C(k)$ is negligible for $|k - k_0| \gtrsim 3\sigma_k$. Thus, the only significant contributions to the wave integral

$$\psi(x, t) = \int_{-\infty}^{\infty} \frac{1}{\sqrt{2\pi \sigma_k^2}} \exp\left[-\frac{(k - k_0)^2}{2\sigma_k^2}\right] \cos(\omega t - k x) \, dk \tag{9.9}$$

come from a small region of k-space centered on $k = k_0$. Let us Taylor expand the dispersion relation, $\omega = \omega(k)$, about $k = k_0$. Neglecting second-order terms in the expansion, we obtain

$$\omega \simeq \omega(k_0) + (k - k_0) \frac{d\omega(k_0)}{dk}. \tag{9.10}$$

It follows that

$$\omega t - k x \simeq \omega_0 t - k_0 x + (k - k_0)(v_g t - x), \tag{9.11}$$

where $\omega_0 = \omega(k_0)$, and

$$v_g = \frac{d\omega(k_0)}{dk} \tag{9.12}$$

is a constant with the dimensions of velocity. If σ_k is sufficiently small then the neglect of second-order terms in the expansion (9.11) is a good approximation, and Equation (9.9) becomes

$$\psi(x, t) \simeq \frac{\cos(\omega_0 t - k_0 x)}{\sqrt{2\pi \sigma_k^2}} \int_{-\infty}^{\infty} \exp\left[-\frac{(k - k_0)^2}{2\sigma_k^2}\right] \cos[(k - k_0)(v_g t - x)] \, dk$$

$$- \frac{\sin(\omega_0 t - k_0 x)}{\sqrt{2\pi \sigma_k^2}} \int_{-\infty}^{\infty} \exp\left[-\frac{(k - k_0)^2}{2\sigma_k^2}\right] \sin[(k - k_0)(v_g t - x)] \, dk, \tag{9.13}$$

where use has been made of a standard trigonometric identity. (See Appendix B.) The integral involving $\sin[(k-k_0)(v_g t - x)]$ is zero, by symmetry. Moreover, an examination of Equations (8.15)–(8.18) reveals that

$$\frac{1}{\sqrt{2\pi\,\sigma_k^2}} \int_{-\infty}^{\infty} \exp\left(-\frac{k^2}{2\,\sigma_k^2}\right) \cos(k\,x)\,dk = \exp\left(-\frac{x^2}{2\,\sigma_x^2}\right), \tag{9.14}$$

where $\sigma_x = 1/\sigma_k$. Hence, by analogy with this expression, Equation (9.13) reduces to

$$\psi(x,t) \simeq \exp\left[-\frac{(v_g t - x)^2}{2\,\sigma_x^2}\right] \cos(\omega_0 t - k_0\,x). \tag{9.15}$$

This is the equation of a wave pulse, of wavenumber k_0, and angular frequency ω_0, with a Gaussian envelope, of characteristic width σ_x, whose peak (which is located by setting the argument of the exponential to zero) has the equation of motion

$$x = v_g\,t. \tag{9.16}$$

In other words, the pulse peak—and, hence, the pulse itself—propagates at the velocity v_g, which is known as the *group velocity*. In the case of non-dispersive waves, the group velocity is the same as the phase velocity (because, if $\omega = k\,v$ then $\omega/k = d\omega/dk = v$). However, for the case of dispersive waves, the two velocities are, in general, different.

Equation (9.15) indicates that, as the wave pulse propagates, its envelope remains the same shape. Actually, this result is misleading, and is only obtained because of the neglect of second-order terms in the expansion (9.11). If we keep more terms in this expansion then we can show that the wave pulse does actually change shape as it propagates. However, this demonstration is most readily effected by means of the following simple argument. The pulse extends in Fourier space from $k_0 - \Delta k/2$ to $k_0 + \Delta k/2$, where $\Delta k \sim \sigma_k$. Thus, part of the pulse propagates at the velocity $v_g(k_0 - \Delta k/2)$, and part at the velocity $v_g(k_0 + \Delta k/2)$. Consequently, the pulse spreads out as it propagates, because some parts of it move faster than others. Roughly speaking, the spatial extent of the pulse in real space grows as

$$\Delta x \sim (\Delta x)_0 + \left[v_g(k_0 + \Delta k/2) - v_g(k_0 - \Delta k/2)\right] t \sim (\Delta x)_0 + \frac{dv_g(k_0)}{dk}\,\Delta k\,t, \tag{9.17}$$

where $(\Delta x)_0 \sim \sigma_x = \sigma_k^{-1}$ is the extent of the pulse at $t = 0$. Hence, from Equation (9.12),

$$\Delta x \sim (\Delta x)_0 + \frac{d^2\omega(k_0)}{dk^2}\,\frac{t}{(\Delta x)_0}. \tag{9.18}$$

We, thus, conclude that the spatial extent of the pulse grows linearly in time, at a rate proportional to the second derivative of the dispersion relation with respect to k (evaluated at the pulse's central wavenumber). This effect is known as *pulse dispersion*. Incidentally, there is no pulse dispersion in a non-dispersive medium, because, by definition, $d^2\omega/dk^2 = 0$ in such a medium. In summary, a wave pulse made up of a linear superposition of dispersive sinusoidal waves, with a range of different wavenumbers, propagates at the group velocity, and also gradually disperses as time progresses.

9.3 ELECTROMAGNETIC WAVES IN UNMAGNETIZED PLASMAS

Consider a point particle of mass m and electric charge q interacting with a linearly polarized, sinusoidal, electromagnetic plane wave that propagates in the z-direction. Provided that the wave amplitude is not sufficiently large to cause the particle to move at relativistic speeds, the electric component of the wave exerts a much greater force on the particle than the magnetic component. [This follows, from standard electrodynamics, because the ratio of the magnetic to the electric force is of order $B_0 v / E_0$, where E_0 is the amplitude of the wave electric field-strength, $B_0 = E_0/c$ the amplitude of the wave magnetic field-strength, v the particle velocity, and c the velocity of light in vacuum. Hence, the ratio of the forces is approximately v/c (Fitzpatrick 2008).] Suppose that the electric component of the wave oscillates in the x-direction, and takes the form

$$E_x(z, t) = E_0 \cos(\omega t - k z), \tag{9.19}$$

where k is the wavenumber, and ω the angular frequency. The equation of motion of the particle is thus (see Appendix C)

$$m \frac{d^2 x}{dt^2} = q E_x, \tag{9.20}$$

where x measures its wave-induced displacement in the x-direction. The previous equation can be solved to give

$$x = -\frac{q E_0}{m \omega^2} \cos(\omega t - k z). \tag{9.21}$$

Thus, the wave causes the particle to execute sympathetic simple harmonic oscillations, in the x-direction, with an amplitude that is directly proportional to its charge, and inversely proportional to its mass.

Suppose that the wave is actually propagating through an unmagnetized, electrically neutral, plasma consisting of free electrons, of mass m_e and charge $-e$, and free ions, of mass m_i and charge $+e$. Because the plasma is assumed to be electrically neutral, each species must have the same equilibrium number density, n_e. Given that the electrons are much less massive than the ions (i.e., $m_e \ll m_i$), but have the same charge (modulo a sign), it follows from Equation (9.21) that the wave-induced oscillations of the electrons are of much higher amplitude than those of the ions. In fact, to a first approximation, we can say that the electrons oscillate while the ions remain stationary. Assuming that the electrons and ions are evenly distributed throughout the plasma, the wave-induced displacement of an individual electron generates an effective electric dipole moment in the x-direction of the form $p_x = -e x$ (the other component of the dipole is a stationary ion of charge $+e$ located at $x = 0$). Hence, the x-directed electric dipole moment per unit volume is

$$P_x = n_e p_x = -n_e e x. \tag{9.22}$$

Given that all of the electrons oscillate according to Equation (9.21) (with $q = -e$ and $m = m_e$), we obtain

$$P_x(z, t) = -\frac{n_e e^2 E_0}{m_e \omega^2} \cos(\omega t - k z). \tag{9.23}$$

We saw earlier, in Section 6.7, that the z-directed propagation of a plane electromagnetic wave, linearly polarized in the x-direction, through a dielectric medium is governed by (see Appendix C)

$$\frac{\partial E_x}{\partial t} = -\frac{1}{\epsilon_0} \left(\frac{\partial P_x}{\partial t} + \frac{\partial H_y}{\partial z} \right), \tag{9.24}$$

$$\frac{\partial H_y}{\partial t} = -\frac{1}{\mu_0} \frac{\partial E_x}{\partial z}. \tag{9.25}$$

Thus, writing E_x in the form (9.19), H_y in the form

$$H_y(z, t) = Z^{-1} E_0 \cos(\omega t - k z), \tag{9.26}$$

where Z is the effective impedance of the plasma, and P_x in the form (9.23), Equations (9.24) and (9.25) yield the nonlinear dispersion relation (see Exercise 9.3)

$$\omega^2 = k^2 c^2 + \omega_{pe}^2, \tag{9.27}$$

where $c = 1/\sqrt{\epsilon_0 \mu_0}$ is the velocity of light in vacuum, and the so-called (electron) *plasma frequency*,

$$\omega_{pe} = \left(\frac{n_e e^2}{\epsilon_0 m_e}\right)^{1/2}, \tag{9.28}$$

is the characteristic frequency of collective electron oscillations in the plasma (Stix 1962). Equations (9.24) and (9.25) also yield

$$Z = \frac{Z_0}{n}, \tag{9.29}$$

where $Z_0 = \sqrt{\mu_0/\epsilon_0}$ is the impedance of free space, and

$$n = \frac{kc}{\omega} = \left(1 - \frac{\omega_{pe}^2}{\omega^2}\right)^{1/2} \tag{9.30}$$

the effective refractive index of the plasma. We, thus, conclude that sinusoidal electromagnetic waves propagating through an unmagnetized plasma have a nonlinear dispersion relation. Moreover, this nonlinearity arises because the effective refractive index of the plasma is frequency dependent.

The expression (9.30) for the refractive index of a plasma has some rather unusual properties. For wave frequencies lying above the plasma frequency (i.e., $\omega > \omega_{pe}$), it yields a real refractive index that is less than unity. On the other hand, for wave frequencies lying below the plasma frequency (i.e., $\omega < \omega_{pe}$), it yields an imaginary refractive index. Neither of these results makes much sense. The former result is problematic because if the refractive index is less than unity then the phase velocity of the wave, $v_p = \omega/k = c/n$, becomes superluminal (i.e., $v_p > c$), and superluminal velocities are generally thought to be unphysical. The latter result is problematic because an imaginary refractive index implies an imaginary phase velocity, which seems utterly meaningless. Let us investigate further.

Consider, first of all, the high-frequency limit, $\omega > \omega_{pe}$. According to Equation (9.30), a sinusoidal electromagnetic wave of angular frequency $\omega > \omega_{pe}$ propagates through the plasma at the superluminal phase velocity

$$v_p = \frac{\omega}{k} = \frac{c}{n} = \frac{c}{(1 - \omega_{pe}^2/\omega^2)^{1/2}}. \tag{9.31}$$

Is this really unphysical? As is well known, Einstein's special theory of relativity forbids information from traveling faster than the velocity of light in vacuum, because this would violate causality (i.e., it would be possible to transform to a valid frame of reference in which an effect occurs prior to its cause) (Rindler 1997). However, a sinusoidal wave with a unique frequency, and an infinite spatial extent, does not transmit any information. (Recall, for instance, from Section 8.5, that the carrier wave in an AM radio signal transmits no information.) At what speed do electromagnetic waves propagating through a plasma transmit information? The most obvious way of using such waves to transmit information would be to send a message via Morse code. In other words, we could transmit a message by means of short wave pulses, of varying lengths and inter-pulse spacings, that are made to propagate through the plasma. The pulses in question would definitely transmit information, so the velocity of information propagation must be the same as that of the pulses; that is, the group velocity, $v_g = d\omega/dk$. Differentiating the dispersion relation (9.27) with respect to k, we obtain

$$2\omega \frac{d\omega}{dk} = 2kc^2, \tag{9.32}$$

or

$$\frac{\omega}{k}\frac{d\omega}{dk} = v_p\, v_g = c^2. \tag{9.33}$$

Thus, it follows, from Equation (9.31), that the group velocity of high-frequency electromagnetic waves in a plasma is

$$v_g = n\, c = (1 - \omega_{pe}^2/\omega^2)^{1/2}\, c. \tag{9.34}$$

The group velocity is sub-luminal (i.e., $v_g < c$). Hence, as long as we accept that high-frequency electromagnetic waves transmit information through a plasma at the group velocity, rather than the phase velocity, then there is no problem with causality. Incidentally, it follows, from this discussion, that the phase velocity of dispersive waves has very little physical significance. It is the group velocity that matters. For instance, according to Equations (6.128), (9.29), (9.30), and (9.34), the mean flux of electromagnetic energy in the z-direction due to a high-frequency sinusoidal wave propagating through a plasma is given by

$$\langle I_z\rangle = \frac{1}{2}\,\epsilon_0\, E_0^2\, n\, c = \frac{1}{2}\,\epsilon_0\, E_0^2\, v_g, \tag{9.35}$$

because $Z_0 = \sqrt{\mu_0/\epsilon_0}$ and $c = 1/\sqrt{\epsilon_0\mu_0}$. Thus, if the group velocity is zero, as is the case when $\omega = \omega_{pe}$, then there is zero energy flux associated with the wave.

The fact that the energy flux and the group velocity of a sinusoidal wave propagating through a plasma both go to zero when $\omega = \omega_{pe}$ suggests that the wave ceases to propagate at all in the low-frequency limit, $\omega < \omega_{pe}$. This observation leads us to search for spatially decaying, standing wave solutions to Equations (9.24) and (9.25) of the form,

$$E_x(z, t) = E_0\, e^{-kz}\, \cos(\omega t), \tag{9.36}$$

$$H_y(z, t) = Z^{-1}\, E_0\, e^{-kz}\, \sin(\omega t). \tag{9.37}$$

It follows from Equations (9.20) and (9.22) that

$$P_x(z, t) = -\frac{n_e\, e^2\, E_0}{m_e\, \omega^2}\, e^{-kz}\, \cos(\omega t). \tag{9.38}$$

Substitution into Equations (9.24) and (9.25) reveals that (9.36) and (9.37) are indeed the correct solutions when $\omega < \omega_{pe}$, and also yields

$$k\, c = \sqrt{\omega_{pe}^2 - \omega^2}, \tag{9.39}$$

as well as

$$Z = Z_0\,\frac{\omega}{k\, c} = Z_0\left(\frac{\omega_{pe}^2}{\omega^2} - 1\right)^{-1/2}. \tag{9.40}$$

(See Exercise 9.4.) Furthermore, the mean z-directed electromagnetic energy flux becomes

$$\langle I_z\rangle = \langle E_x\, H_y\rangle = E_0^2\, Z^{-1}\, e^{-2kz}\, \langle\cos(\omega t)\, \sin(\omega t)\rangle = 0. \tag{9.41}$$

The previous analysis demonstrates that a sinusoidal electromagnetic wave cannot propagate through a plasma when its frequency lies below the plasma frequency. Instead, the amplitude of the wave decays exponentially into the plasma. Moreover, the electric and magnetic components of the wave oscillate in phase quadrature (i.e., $\pi/2$ radians out of phase), and the wave consequentially has zero associated net energy flux. This suggests that a plasma reflects, rather than absorbs, an incident electromagnetic wave whose frequency is less than the plasma frequency (because if the wave were absorbed then there would be a net flux of energy into the plasma). Let us investigate what happens when a low-frequency electromagnetic wave is normally incident on a plasma in more detail.

Suppose that the region $z < 0$ is a vacuum, and the region $z > 0$ is occupied by a plasma of plasma frequency ω_{pe}. Let the wave electric and magnetic fields in the vacuum region take the form

$$E_x(z, t) = E_i \cos[k_0 (c t - z)] + E_r \cos[k_0 (c t + z) + \phi_r],$$ (9.42)

$$H_y(z, t) = E_i Z_0^{-1} \cos[k_0 (c t - z)] - E_r Z_0^{-1} \cos[k_0 (c t + z) + \phi_r],$$ (9.43)

where $k_0 = \omega/c$ is the vacuum wavenumber. Here, E_i is the amplitude of an electromagnetic wave of frequency $\omega < \omega_{pe}$ that is normally incident on the plasma, whereas E_r is the amplitude of the reflected wave, and ϕ_r the phase of this wave with respect to the incident wave. The wave electric and magnetic fields in the plasma are written

$$E_x(z, t) = E_t e^{-k_0 \alpha z} \cos(\omega t + \phi_t),$$ (9.44)

$$H_y(z, t) = E_t Z_0^{-1} \alpha e^{-k_0 \alpha z} \sin(\omega t + \phi_t),$$ (9.45)

where E_t is the amplitude of the evanescent wave that penetrates into the plasma, ϕ_t is the phase of this wave with respect to the incident wave, and

$$\alpha = \frac{k c}{\omega} = \left(\frac{\omega_{pe}^2}{\omega^2} - 1 \right)^{1/2}.$$ (9.46)

The appropriate matching conditions are the continuity of E_x and H_y at the vacuum/plasma interface ($z = 0$). (See Appendix C.) In other words,

$$E_i \cos(\omega t) + E_r \cos(\omega t + \phi_r) = E_t \cos(\omega t + \phi_t),$$ (9.47)

$$E_i \cos(\omega t) - E_r \cos(\omega t + \phi_r) = E_t \alpha \sin(\omega t + \phi_t).$$ (9.48)

These two equations, which must be satisfied at all times, can be solved to give

$$E_r = E_i,$$ (9.49)

$$\tan \phi_r = \frac{2 \alpha}{1 - \alpha^2},$$ (9.50)

$$E_t = \frac{2 E_i}{(1 + \alpha^2)^{1/2}},$$ (9.51)

$$\tan \phi_t = \alpha.$$ (9.52)

(See Exercise 9.5.) Thus, the coefficient of reflection,

$$R = \left(\frac{E_r}{E_i} \right)^2 = 1,$$ (9.53)

is unity, which implies that all of the incident wave energy is reflected by the plasma, and there is no energy absorption. The relative phase of the reflected wave varies from 0 (when $\omega = \omega_{pe}$) to π (when $\omega \ll \omega_{pe}$) radians.

The outer regions of the Earth's atmosphere consist of a tenuous gas that is partially ionized by ultraviolet and X-ray radiation from the Sun, as well as by cosmic rays incident from outer space. This region, which is known as the *ionosphere*, acts like a plasma as far as its interaction with radio

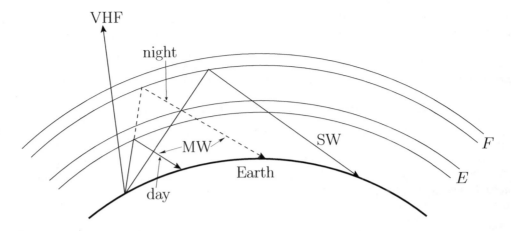

FIGURE 9.1   Reflection and transmission of radio waves by the ionosphere.

waves is concerned. The ionosphere consists of many layers. The two most important, as far as radio wave propagation is concerned, are the *E layer*, which lies at an altitude of about 90 to 120 km above the Earth's surface, and the *F layer*, which lies at an altitude of about 120 to 400 km (Pain 1999). The plasma frequency in the F layer is generally larger than that in the E layer, because of the greater density of free electrons in the former (recall that $\omega_{pe} \propto \sqrt{n_e}$). The free electron number density in the E layer drops steeply after sunset, due to the lack of solar ionization combined with the gradual recombination of free electrons and ions. Consequently, the plasma frequency in the E layer also drops steeply after sunset. Recombination in the F layer occurs at a much slower rate, so there is nothing like as great a reduction in the plasma frequency of this layer at night. Very High Frequency (VHF) radio signals (i.e., signals with frequencies greater than 30 MHz), which include FM radio and TV signals, have frequencies well in excess of the plasma frequencies of both the E and the F layers, and thus pass straight through the ionosphere. Short Wave (SW) radio signals (i.e., signals with frequencies in the range 3 to 30 MHz) have frequencies in excess of the plasma frequency of the E layer, but not of the F layer. Hence, SW signals pass through the E layer, but are reflected by the F layer. Finally, Medium Wave (MW) radio signals (i.e., signals with frequencies in the range 0.5 to 3 MHz) have frequencies that lie below the plasma frequency of the F layer, and also lie below the plasma frequency of the E layer during daytime, but not during nighttime. Thus, MW signals are reflected by the E layer during the day, but pass through the E layer, and are reflected by the F layer, during the night.

The reflection and transmission of the various different types of radio wave by the ionosphere is shown schematically in Figure 9.1. This diagram explains many of the characteristic features of radio reception. For instance, because of the curvature of the Earth's surface, VHF reception is only possible when the receiving antenna lies in the line of sight of the transmitting antenna, and is consequently fairly local in nature. MW reception is possible over much larger distances, because the signal is reflected by the ionosphere back toward the Earth's surface. Moreover, long range MW reception improves at night, because the signal is reflected at a higher altitude. Finally, SW radio reception is possible over very large distances, because the signal is reflected at extremely high altitudes.

9.4 FARADAY ROTATION

Consider a high-frequency, circularly polarized, electromagnetic wave propagating, along the z-axis, through a plasma with a longitudinal equilibrium magnetic field of strength, $\mathbf{B} = B_0\,\mathbf{e}_z$. The equations of motion of an individual electron making up the plasma take the form (see Appendix C)

$$m_e \frac{d^2x}{dt^2} = -e\,E_x - e\,B_0\,\frac{dy}{dt}, \tag{9.54}$$

$$m_e \frac{d^2y}{dt^2} = -e\,E_y + e\,B_0\,\frac{dx}{dt}, \tag{9.55}$$

where m_e is the electron mass, and $-e$ the electron charge. Here, x and y are the wave-induced displacements of the electron in the x- and y-directions, respectively. (As before, it is a good approximation to neglect the wave-induced displacements of the ions, because of their relatively large mass.) The former terms on the right-hand sides of the previous equations represent the x- and y-directed forces exerted on the electron by the wave electric field, $\mathbf{E}$, whereas the latter terms represent the forces exerted by the equilibrium magnetic field when the electron moves (Fitzpatrick 2008). (As before, we can neglect the forces due to the wave magnetic field, provided that the electron motion remains non-relativistic.)

Consider a right-hand circularly polarized wave, of angular frequency ω, whose electric field takes the form

$$E_x(z, t) = E_R \cos(\omega t - k_R z), \tag{9.56}$$

$$E_y(z, t) = E_R \sin(\omega t - k_R z). \tag{9.57}$$

(See Section 7.7.) Let us search for solutions of Equations (9.54) and (9.55) of the form

$$x = x_0 \cos(\omega t - k_R z), \tag{9.58}$$

$$y = y_0 \sin(\omega t - k_R z). \tag{9.59}$$

It is readily demonstrated (see Exercise 9.6) that

$$x_0 = \frac{e\,E_R}{m_e\,\omega\,(\omega - \Omega_e)}, \tag{9.60}$$

$$y_0 = \frac{e\,E_R}{m_e\,\omega\,(\omega - \Omega_e)}, \tag{9.61}$$

where

$$\Omega_e = \frac{e\,B_0}{m_e} \tag{9.62}$$

is the so-called *electron cyclotron frequency* at which the electrons gyrate in the equilibrium magnetic field (Stix 1962). If n_e is the number density of electrons in the plasma then it follows that the x- and y-components of the electric dipole moment per unit volume are

$$P_x(z, t) = -e\,n_e\,x = -\frac{n_e\,e^2\,E_R}{m_e\,\omega\,(\omega - \Omega_e)}\,\cos(\omega t - k_R z), \tag{9.63}$$

$$P_y(z, t) = -e\,n_e\,y = -\frac{n_e\,e^2\,E_R}{m_e\,\omega\,(\omega - \Omega_e)}\,\sin(\omega t - k_R z), \tag{9.64}$$

respectively. For the case of a circularly polarized wave, Equations (9.24) and (9.25) generalize to give (see Appendix C)

$$\frac{\partial E_x}{\partial t} = -\frac{1}{\epsilon_0}\left(\frac{\partial P_x}{\partial t} + \frac{\partial H_y}{\partial z}\right), \tag{9.65}$$

$$\frac{\partial E_y}{\partial t} = -\frac{1}{\epsilon_0}\left(\frac{\partial P_y}{\partial t} - \frac{\partial H_x}{\partial z}\right), \tag{9.66}$$

$$\frac{\partial H_x}{\partial t} = \frac{1}{\mu_0}\frac{\partial E_y}{\partial z}, \tag{9.67}$$

$$\frac{\partial H_y}{\partial t} = -\frac{1}{\mu_0}\frac{\partial E_x}{\partial z}. \tag{9.68}$$

Thus, writing E_x and E_y in the form (9.56) and (9.57), respectively, the components of the magnetic intensity in the form

$$H_x(z,t) = -Z_R^{-1}\,E_R\,\sin(\omega t - k_R z), \tag{9.69}$$

$$H_y(z,t) = Z_R^{-1}\,E_R\,\cos(\omega t - k_R z), \tag{9.70}$$

and P_x and P_y in the form (9.63) and (9.64), respectively, Equations (9.65)–(9.68) yield the nonlinear dispersion relation

$$\omega^2 = k_R^2 c^2 + \frac{\omega\,\omega_{pe}^2}{\omega - \Omega_e}, \tag{9.71}$$

where ω_{pe} is the (electron) plasma frequency. [See Equation (9.28).] It follows that the refractive index of the plasma for right-hand circularly polarized waves is

$$n_R = \frac{k_R c}{\omega} = \left[1 - \frac{\omega_{pe}^2}{\omega(\omega - \Omega_e)}\right]^{1/2}, \tag{9.72}$$

whereas the effective impedance becomes

$$Z_R = \frac{Z_0}{n_R}. \tag{9.73}$$

Here, Z_0 is the impedance of free space.

Consider a left-hand circularly polarized wave whose electric field takes the form

$$E_x(z,t) = E_L\,\cos(\omega t - k_L z), \tag{9.74}$$

$$E_y(z,t) = -E_L\,\sin(\omega t - k_L z). \tag{9.75}$$

(See Section 7.7.) Repeating the previous analysis (with suitable modifications), we deduce that the dispersion relation for left-hand circularly polarized waves is

$$\omega^2 = k_L^2 c^2 + \frac{\omega\,\omega_{pe}^2}{\omega + \Omega_e}. \tag{9.76}$$

It follows that the refractive index of the plasma for such waves is

$$n_L = \frac{k_L c}{\omega} = \left[1 - \frac{\omega_{pe}^2}{\omega(\omega + \Omega_e)}\right]^{1/2}, \tag{9.77}$$

whereas the effective impedance becomes

$$Z_L = \frac{Z_0}{n_L}.$$
(9.78)

According to the previous analysis, in the presence of a longitudinal equilibrium magnetic field, the refractive indices of right-hand and left-hand circularly polarized electromagnetic waves propagating through a plasma are slightly different. Consider what happens when a linearly polarized electromagnetic wave, whose electric field is initially of the form

$$E_x(z, t) = E_0 \cos(\omega t - k_0 z),$$
(9.79)

$$E_y(z, t) = 0,$$
(9.80)

propagates through the plasma. We can represent a linearly polarized wave as a superposition of right- and left-hand circularly polarized waves of equal amplitudes. In other words,

$$E_x(z, t) = E_R \cos(\omega t - k_R z) + E_L \cos(\omega t - k_L z),$$
(9.81)

$$E_y(z, t) = E_R \sin(\omega t - k_R z) - E_L \sin(\omega t - k_L z),$$
(9.82)

where $E_R = E_L = E_0/2$. Furthermore, in the high-frequency limit $\omega \gg \omega_{pe}, \Omega_e$, Equations (9.72) and (9.77) yield

$$k_R \simeq k_0 - \Delta k,$$
(9.83)

$$k_L \simeq k_0 + \Delta k,$$
(9.84)

where $k_0 = (\omega/c)[1 - (1/2)(\omega_{pe}/\omega)^2]$, and

$$\Delta k = \frac{1}{2}\left(\frac{\omega_{pe}}{\omega}\right)^2\left(\frac{\Omega_e}{c}\right).$$
(9.85)

Equations (9.81)–(9.84), in combination with some standard trigonometric identities (see Appendix B), give

$$E_x(z, r) = E_0 \cos(\omega t - k_0 z) \cos(\Delta k z),$$
(9.86)

$$E_y(z, r) = E_0 \cos(\omega t - k_0 z) \sin(\Delta k z).$$
(9.87)

It can be seen that the presence of the longitudinal magnetic field (whose strength is parameterized by Δk) causes the plane of polarization of the wave to rotate as it propagates through the plasma. This effect is known as *Faraday rotation*. Defining the angle of polarization,

$$\varphi = \tan^{-1}\left(\frac{E_y}{E_x}\right) = \Delta k z,$$
(9.88)

the rate at which this angle advances as the wave propagates is

$$\frac{d\varphi}{dz} = \Delta k = \frac{\omega_{pe}^2 \Omega_e}{2\omega^2 c} = \frac{e^3}{2\epsilon_0 m_e^2 c \omega^2} n_e B_0.$$
(9.89)

Thus, a linearly polarized electromagnetic wave that propagates through a plasma with a slowly varying electron number density, $n_e(z)$, and longitudinal magnetic field, $B_0(z)$, has its plane of rotation rotated through a net angle

$$\Delta\varphi = \phi(z) - \phi(0) = \frac{e^3}{2\epsilon_0 m_e^2 c \omega^2}\int_0^z n_e(z') B_0(z') dz'.$$
(9.90)

Observe the very strong inverse depedence of $\Delta\varphi$ with the wave frequency, ω.

Pulsars are rapidly rotating neutron stars that emit regular blips of highly polarized radio waves (Longair 2011). Hundreds of such objects have been found in our galaxy since the first was discovered in 1967. By measuring the variation of the angle of polarization, φ, of radio emission from a pulsar with frequency, ω, astronomers can effectively determine the line integral of $n_e B_0$ along the straight line joining the pulsar to the Earth using formula (9.90) (ibid.). Here, n_e is the number density of free electrons in the interstellar medium, whereas B_0 is the parallel (to the line joining the pulsar to the Earth) component of the galactic magnetic field. In order to perform this calculation, astronomers must make the reasonable assumption that the radiation was emitted by the pulsar with a common angle of polarization, φ_0, over a wide range of different frequencies. By fitting Equation (9.90) to the data, and then extrapolating to large ω, it is possible to determine φ_0, and, hence, the amount, $\Delta\varphi(\omega)$, through which the polarization angle of the radiation has rotated, at a given frequency, during its passage to Earth.

Astronomers can also determine the line integral of n_e by looking at the variation of the arrival time of the various components of a pulsar radio blip with frequency (Longair 2011). This calculation depends on the reasonable assumption that the components were emitted simultaneously, and then traveled through interstellar space at the frequency dependent group velocity $v_g = (1 - \omega_{pe}^2/\omega^2)^{1/2} c$. [See Equation (9.34).] It follows that the arrival time can be written

$$t \simeq t_0 + \frac{e^2}{2\,\epsilon_0\,m_e\,c\,\omega^2} \int_0^z n_e(z')\,dz'. \tag{9.91}$$

By fitting Equation (9.91) to the data, and then extrapolating to large ω, it is possible to determine t_0, and, hence, $t - t_0$ at a given frequency. Finally, once the line integrals of $n_e B_0$ and n_e have been independently determined, estimates can be made of the mean electron number density, and the mean galactic magnetic field, along the straight line joining the pulsar to the Earth.

9.5 ELECTROMAGNETIC WAVES IN MAGNETIZED PLASMAS

Let us extend the analysis of the previous section to consider a general electromagnetic wave propagating through a uniform plasma with an equilibrium magnetic field of strength, $\mathbf{B} = B_0\,\mathbf{e}_z$. The plasma is assumed to consist of two species: electrons of mass m_e and electric charge $-e$, and ions of mass m_i and electric charge $+e$. The plasma is also assumed to be electrically neutral, so that the equilibrium number density of the ions is the same as that of the electrons; namely, n_e (Stix 1962). The equations of motion of a constituent ion of the plasma are written

$$m_i \frac{d^2 x_i}{dt^2} = e\,E_x + e\,B_0\,\frac{dy_i}{dt}, \tag{9.92}$$

$$m_i \frac{d^2 y_i}{dt^2} = e\,E_y - e\,B_0\,\frac{dx_i}{dt}, \tag{9.93}$$

$$m_i \frac{d^2 z_i}{dt^2} = e\,E_z, \tag{9.94}$$

where x_i, y_i, and z_i are the wave-induced displacements of the ion along the three Cartesian axes. (Here, we are including ion motion in our analysis because such motion is important in certain frequency ranges.) As before, the former terms on the right-hand sides of the previous equations represent the forces exerted on the ion by the wave electric field, $\mathbf{E}$, whereas the latter terms represent the forces exerted by the equilibrium magnetic field when the ion moves (Fitzpatrick 2008). (As before, we can neglect any forces due to the wave magnetic field, as long as the particle motion remains non-relativistic.) The equations of motion of a constituent electron take the form

$$m_e \frac{d^2 x_e}{dt^2} = -e\, E_x - e\, B_0 \frac{dy_e}{dt}, \tag{9.95}$$

$$m_e \frac{d^2 y_e}{dt^2} = -e\, E_y + e\, B_0 \frac{dx_e}{dt}, \tag{9.96}$$

$$m_e \frac{d^2 z_e}{dt^2} = -e\, E_z. \tag{9.97}$$

The Cartesian components of the electric dipole moment per unit volume are

$$P_x = e\, n_e\, (x_i - x_e), \tag{9.98}$$

$$P_y = e\, n_e\, (y_i - y_e), \tag{9.99}$$

$$P_z = e\, n_e\, (z_i - z_e). \tag{9.100}$$

Finally, the electric displacement is written

$$\mathbf{D} = \epsilon_0\, \mathbf{E} + \mathbf{P}. \tag{9.101}$$

Consider a right-hand circularly polarized (with respect to the direction of the equilibrium magnetic field) wave whose electric field takes the form

$$E_x(\mathbf{r}, t) = E_R\, \cos(\omega t - \mathbf{k} \cdot \mathbf{r}), \tag{9.102}$$

$$E_y(\mathbf{r}, t) = E_R\, \sin(\omega t - \mathbf{k} \cdot \mathbf{r}), \tag{9.103}$$

$$E_z(\mathbf{r}, t) = 0. \tag{9.104}$$

Let us write

$$x_i(\mathbf{r}, t) = \hat{x}_i\, \cos(\omega t - \mathbf{k} \cdot \mathbf{r}), \tag{9.105}$$

$$y_i(\mathbf{r}, t) = \hat{y}_i\, \sin(\omega t - \mathbf{k} \cdot \mathbf{r}), \tag{9.106}$$

$$z_i(\mathbf{r}, t) = 0, \tag{9.107}$$

$$x_e(\mathbf{r}, t) = \hat{x}_e\, \cos(\omega t - \mathbf{k} \cdot \mathbf{r}), \tag{9.108}$$

$$y_e(\mathbf{r}, t) = \hat{y}_e\, \sin(\omega t - \mathbf{k} \cdot \mathbf{r}), \tag{9.109}$$

$$z_e(\mathbf{r}, t) = 0, \tag{9.110}$$

$$P_x(\mathbf{r}, t) = \hat{P}_x\, \cos(\omega t - \mathbf{k} \cdot \mathbf{r}), \tag{9.111}$$

$$P_y(\mathbf{r}, t) = \hat{P}_y\, \sin(\omega t - \mathbf{k} \cdot \mathbf{r}), \tag{9.112}$$

$$P_z(\mathbf{r}, t) = 0. \tag{9.113}$$

Equations (9.92)–(9.113) yield

$$\hat{x}_i = \hat{y}_i = -\frac{e\, E_R}{m_i\, \omega\, (\omega + \Omega_i)}, \tag{9.114}$$

$$\hat{x}_e = \hat{y}_e = \frac{e\, E_R}{m_e\, \omega\, (\omega - \Omega_e)}, \tag{9.115}$$

$$\hat{P}_x = \hat{P}_y = -\epsilon_0 \left[\frac{\omega_{pi}^2}{\omega\, (\omega + \Omega_i)} + \frac{\omega_{pe}^2}{\omega\, (\omega - \Omega_e)} \right], \tag{9.116}$$

where

$$\Omega_i = \frac{e\,B_0}{m_i}, \tag{9.117}$$

$$\Omega_e = \frac{e\,B_0}{m_e}, \tag{9.118}$$

$$\omega_{pe} = \left(\frac{n_e\,e^2}{\epsilon_0\,m_e}\right)^{1/2}, \tag{9.119}$$

$$\omega_{pi} = \left(\frac{n_e\,e^2}{\epsilon_0\,m_i}\right)^{1/2}. \tag{9.120}$$

Here, Ω_i is termed the *ion cyclotron frequency*, and is the frequency at which ions gyrate in the plane perpendicular to the equilibrium magnetic field (Stix 1962). Moreover, Ω_e is the *electron cyclotron frequency*, and is the frequency at which electrons gyrate in the plane perpendicular to the equilibrium magnetic field (ibid). Finally, ω_{pi} and ω_{pe} are termed the *ion plasma frequency*, and the *electron plasma frequency*, respectively (ibid). Of course, $\Omega_e \gg \Omega_i$ and $\omega_{pe} \gg \omega_{pi}$, because $m_i \gg m_e$. Finally, it follows from Equations (9.101)–(9.104), (9.113), and (9.116), that the electric displacement of a right-hand circularly polarized wave propagating through a magnetized plasma has the components

$$D_x(\mathbf{r}, t) = \epsilon_0\,R\,E_R\,\cos(\omega\,t - \mathbf{k} \cdot \mathbf{r}), \tag{9.121}$$

$$D_y(\mathbf{r}, t) = \epsilon_0\,R\,E_R\,\sin(\omega\,t - \mathbf{k} \cdot \mathbf{r}), \tag{9.122}$$

$$D_z(\mathbf{r}, t) = 0, \tag{9.123}$$

where

$$R = 1 - \frac{\omega_{pi}^2}{\omega^2}\left(\frac{\omega}{\omega + \Omega_i}\right) - \frac{\omega_{pe}^2}{\omega^2}\left(\frac{\omega}{\omega - \Omega_e}\right). \tag{9.124}$$

Consider a left-hand circularly polarized (with respect to the direction of the equilibrium magnetic field) wave whose electric field takes the form

$$E_x(\mathbf{r}, t) = E_L\,\cos(\omega\,t - \mathbf{k} \cdot \mathbf{r}), \tag{9.125}$$

$$E_y(\mathbf{r}, t) = -E_L\,\sin(\omega\,t - \mathbf{k} \cdot \mathbf{r}), \tag{9.126}$$

$$E_z(\mathbf{r}, t) = 0. \tag{9.127}$$

By repeating the previously described analysis (with appropriate modifications), we deduce that

$$D_x(\mathbf{r}, t) = \epsilon_0\,L\,D_L\,\cos(\omega\,t - \mathbf{k} \cdot \mathbf{r}), \tag{9.128}$$

$$D_y(\mathbf{r}, t) = -\epsilon_0\,L\,D_L\,\sin(\omega\,t - \mathbf{k} \cdot \mathbf{r}), \tag{9.129}$$

$$D_z(\mathbf{r}, t) = 0, \tag{9.130}$$

where

$$L = 1 - \frac{\omega_{pi}^2}{\omega^2}\left(\frac{\omega}{\omega - \Omega_i}\right) - \frac{\omega_{pe}^2}{\omega^2}\left(\frac{\omega}{\omega + \Omega_e}\right). \tag{9.131}$$

Finally, consider a wave whose electric field is polarized parallel to the equilibrium magnetic field, so that

$$E_x(\mathbf{r}, t) = 0, \tag{9.132}$$

$$E_y(\mathbf{r}, t) = 0, \tag{9.133}$$

$$E_z(\mathbf{r}, t) = E_P\,\cos(\omega\,t - \mathbf{k} \cdot \mathbf{r}). \tag{9.134}$$

Again, repeating the previous analysis (with suitable modifications), we obtain

$$D_x(\mathbf{r}, t) = 0, \tag{9.135}$$

$$D_y(\mathbf{r}, t) = 0, \tag{9.136}$$

$$D_z(\mathbf{r}, t) = \epsilon_0 \, P \, \cos(\omega t - \mathbf{k} \cdot \mathbf{r}), \tag{9.137}$$

where

$$P = 1 - \frac{\omega_{pi}^2}{\omega^2} - \frac{\omega_{pe}^2}{\omega^2}. \tag{9.138}$$

Now, the equations that govern electromagnetic wave propagation through a dielectric media are (see Appendix C)

$$\frac{\partial D_x}{\partial t} = \frac{\partial H_z}{\partial y} - \frac{\partial H_y}{\partial z}, \tag{9.139}$$

$$\frac{\partial D_y}{\partial t} = \frac{\partial H_x}{\partial z} - \frac{\partial H_z}{\partial x}, \tag{9.140}$$

$$\frac{\partial D_z}{\partial t} = \frac{\partial H_y}{\partial x} - \frac{\partial H_x}{\partial y}, \tag{9.141}$$

$$\frac{\partial B_x}{\partial t} = \frac{\partial E_y}{\partial z} - \frac{\partial E_z}{\partial y}, \tag{9.142}$$

$$\frac{\partial B_y}{\partial t} = \frac{\partial E_z}{\partial x} - \frac{\partial E_x}{\partial z}, \tag{9.143}$$

$$\frac{\partial B_z}{\partial t} = \frac{\partial E_x}{\partial y} - \frac{\partial E_y}{\partial x}. \tag{9.144}$$

Consider an electromagnetic wave with a general polarization (with respect to the equilibrium magnetic field). Such a wave can be written as a linear combination of a right-hand circularly polarized wave, a left-hand circularly polarized wave, and a wave with parallel polarization. In other words,

$$E_x(\mathbf{r}, t) = (E_R + E_L) \, \cos(\omega t - \mathbf{k} \cdot \mathbf{r}), \tag{9.145}$$

$$E_y(\mathbf{r}, t) = (E_R - E_L) \, \sin(\omega t - \mathbf{k} \cdot \mathbf{r}), \tag{9.146}$$

$$E_z(\mathbf{r}, t) = E_P \, \cos(\omega t - \mathbf{k} \cdot \mathbf{r}). \tag{9.147}$$

[See Equations (9.102)–(9.104), (9.125)–(9.127), and (9.132)–(9.134).] It follows, from the previous analysis, that

$$D_x(\mathbf{r}, t) = \epsilon_0 \, (R \, E_R + L \, E_L) \, \cos(\omega t - \mathbf{k} \cdot \mathbf{r}), \tag{9.148}$$

$$D_y(\mathbf{r}, t) = \epsilon_0 \, (R \, E_R - L \, E_L) \, \sin(\omega t - \mathbf{k} \cdot \mathbf{r}), \tag{9.149}$$

$$D_z(\mathbf{r}, t) = \epsilon_0 \, P \, E_P \, \cos(\omega t - \mathbf{k} \cdot \mathbf{r}). \tag{9.150}$$

[See Equations (9.121)–(9.123), (9.128)–(9.130), and (9.135)–(9.137).] Suppose that

$$H_x(\mathbf{r}, t) = \hat{H}_x \, \sin(\omega t - \mathbf{k} \cdot \mathbf{r}), \tag{9.151}$$

$$H_y(\mathbf{r}, t) = \hat{H}_y \, \cos(\omega t - \mathbf{k} \cdot \mathbf{r}), \tag{9.152}$$

$$H_z(\mathbf{r}, t) = \hat{H}_z \, \sin(\omega t - \mathbf{k} \cdot \mathbf{r}), \tag{9.153}$$

which implies that

$$B_x(\mathbf{r}, t) = \mu_0 \, \hat{H}_x \, \sin(\omega t - \mathbf{k} \cdot \mathbf{r}), \tag{9.154}$$

$$B_y(\mathbf{r}, t) = \mu_0 \, \hat{H}_y \, \cos(\omega t - \mathbf{k} \cdot \mathbf{r}), \tag{9.155}$$

$$B_z(\mathbf{r}, t) = \mu_0 \, \hat{H}_z \, \sin(\omega t - \mathbf{k} \cdot \mathbf{r}). \tag{9.156}$$

Finally, let

$$\mathbf{k} = k \, (\sin\theta, \, 0, \, \cos\theta), \tag{9.157}$$

which means that the wavevector lies in the x-z plane, and subtends an angle θ with the equilibrium magnetic field.

Equations (9.139)–(9.157) yield

$$\epsilon_0 \, \omega \, (R \, E_R + L \, E_L) = k \, \cos\theta \, \hat{H}_y, \tag{9.158}$$

$$\epsilon_0 \, \omega \, (R \, E_R - L \, E_L) = -k \, \cos\theta \, \hat{H}_x + k \, \sin\theta \, \hat{H}_z, \tag{9.159}$$

$$\epsilon_0 \, \omega \, P \, E_P = -k \, \sin\theta \, \hat{H}_y, \tag{9.160}$$

$$\mu_0 \, \omega \, \hat{H}_x = -k \, \cos\theta \, (E_R - E_L), \tag{9.161}$$

$$\mu_0 \, \omega \, \hat{H}_y = -k \, \sin\theta \, E_P + k \, \cos\theta \, (E_R + E_L), \tag{9.162}$$

$$\mu_0 \, \omega \, \hat{H}_z = k \, \sin\theta \, (E_R - E_L), \tag{9.163}$$

which can be combined to give

$$\begin{pmatrix} (\omega/c)^2 R - k^2 \cos^2\theta, & (\omega/c)^2 L - k^2 \cos^2\theta, & k^2 \cos\theta \sin\theta \\ (\omega/c)^2 R - k^2, & -(\omega/c)^2 L + k^2, & 0 \\ k^2 \cos\theta \sin\theta, & k^2 \cos\theta \sin\theta, & (\omega/c)^2 P - k^2 \sin^2\theta \end{pmatrix} \begin{pmatrix} E_R \\ E_L \\ E_P \end{pmatrix} = \begin{pmatrix} 0 \\ 0 \\ 0 \end{pmatrix}. \tag{9.164}$$

The previous equation determines the frequencies and polarizations of an electromagnetic wave of wavenumber k that propagates through a magnetized plasma, and whose direction of propagation subtends an angle θ with the magnetic field.

Suppose, finally, that

$$E_x(x, z, t) = \hat{E}_x \, \cos(\omega t - k \, \sin\theta \, x - k \, \cos\theta \, z), \tag{9.165}$$

$$E_y(x, z, t) = \hat{E}_y \, \sin(\omega t - k \, \sin\theta \, x - k \, \cos\theta \, z), \tag{9.166}$$

$$E_z(x, z, t) = \hat{E}_z \, \cos(\omega t - k \, \sin\theta \, x - k \, \cos\theta \, z). \tag{9.167}$$

It follows, from Equations (9.145)–(9.147), that $E_R = (\hat{E}_x + \hat{E}_y)/2$, $E_L = (\hat{E}_x - \hat{E}_y)/2$, and $E_P = \hat{E}_z$. Hence, Equation (9.164) transforms to give the eigenmode equation

$$\begin{pmatrix} S - n^2 \cos^2\theta, & D, & n^2 \cos\theta \sin\theta \\ D, & S - n^2, & 0 \\ n^2 \cos\theta \sin\theta, & 0, & P - n^2 \sin^2\theta \end{pmatrix} \begin{pmatrix} \hat{E}_x \\ \hat{E}_y \\ \hat{E}_z \end{pmatrix} = \begin{pmatrix} 0 \\ 0 \\ 0 \end{pmatrix}. \tag{9.168}$$

Here,

$$n = \frac{c \, k}{\omega} \tag{9.169}$$

is the effective *refractive index* of the plasma, whereas

$$S = \frac{R + L}{2}, \qquad (9.170)$$

$$D = \frac{R - L}{2}. \qquad (9.171)$$

9.6 LOW-FREQUENCY EM WAVES IN MAGNETIZED PLASMAS

Consider electromagnetic wave propagation through a magnetized plasma at frequencies far below the ion cyclotron or plasma frequencies, which are, in turn, well below the corresponding electron frequencies. In the low-frequency limit (i.e., $\omega \ll \Omega_i, \omega_{pi}$), we have [see Equations (9.124), (9.131), (9.138), (9.170), and (9.171)]

$$S \simeq 1 + \frac{\omega_{pi}^2}{\Omega_i^2}, \qquad (9.172)$$

$$D \simeq 0, \qquad (9.173)$$

$$P \simeq -\frac{\omega_{pe}^2}{\omega^2}. \qquad (9.174)$$

Here, use has been made of $\omega_{pe}^2/(\Omega_e \, \Omega_i) = \omega_{pi}^2/\Omega_i^2$. Thus, the eigenmode equation (9.168) reduces to

$$\begin{pmatrix} 1 + \omega_{pi}^2/\Omega_i^2 - n^2 \cos^2\theta, & 0, & n^2 \cos\theta \sin\theta \\ 0, & 1 + \omega_{pi}^2/\Omega_i^2 - n^2, & 0 \\ n^2 \cos\theta \sin\theta, & 0, & -\omega_{pe}^2/\omega^2 - n^2 \sin^2\theta \end{pmatrix} \begin{pmatrix} \hat{E}_x \\ \hat{E}_y \\ \hat{E}_z \end{pmatrix} = \begin{pmatrix} 0 \\ 0 \\ 0 \end{pmatrix}. \qquad (9.175)$$

The solubility condition (Riley 1974) for the homogeneous matrix equation (9.175) yields the dispersion relation

$$\begin{vmatrix} 1 + \omega_{pi}^2/\Omega_i^2 - n^2 \cos^2\theta, & 0, & n^2 \cos\theta \sin\theta \\ 0, & 1 + \omega_{pi}^2/\Omega_i^2 - n^2, & 0 \\ n^2 \cos\theta \sin\theta, & 0, & -\omega_{pe}^2/\omega^2 - n^2 \sin^2\theta \end{vmatrix} = 0. \qquad (9.176)$$

Now, in the low-frequency limit, $\omega_{pe}^2/\omega^2 \gg 1, \omega_{pi}^2/\Omega_i^2$. Thus, we can see that the bottom right-hand element of the previous determinant is far larger than any of the other elements. Hence, to a good approximation, the roots of the dispersion relation are obtained by equating the term multiplying this large factor to zero. In this manner, we obtain two roots:

$$n^2 \cos^2\theta = 1 + \frac{\omega_{pi}^2}{\Omega_i^2}, \qquad (9.177)$$

and

$$n^2 = 1 + \frac{\omega_{pi}^2}{\Omega_i^2}. \qquad (9.178)$$

It is fairly easy to show, from the definitions of the plasma and cyclotron frequencies [see Equations (9.117)–(9.120)], that

$$\frac{\omega_{pi}^2}{\Omega_i^2} = \frac{c^2}{B_0^2/(\mu_0 \rho)} = \frac{c^2}{V_A^2}. \qquad (9.179)$$

Here, $\rho \simeq n_e \, m_i$ is the plasma mass density, and

$$V_A = \sqrt{\frac{B_0^2}{\mu_0 \rho}} \tag{9.180}$$

is known as the *Alfvén speed*. Thus, the dispersion relations (9.177) and (9.178) can be written

$$\omega = \frac{k \, V_A \, \cos \theta}{\sqrt{1 + V_A^2/c^2}} \simeq k \, V_A \, \cos \theta, \tag{9.181}$$

and

$$\omega = \frac{k \, V_A}{\sqrt{1 + V_A^2/c^2}} \simeq k \, V_A, \tag{9.182}$$

respectively. Here, we have made use of the fact that $V_A \ll c$ in a conventional plasma.

The dispersion relation (9.181) corresponds to the so-called *shear-Alfvén wave*, whereas the dispersion relation (9.182) corresponds to the *compressional-Alfvén wave*. The shear-Alfvén wave bends magnetic field-lines without compressing them, whereas the compressional-Alfvén wave compresses magnetic field-lines without bending them. Likewise, the shear-Alfvén wave does not compress the plasma, whereas the compressional-Alfvén wave does (Hazeltine and Waelbroeck 2004).

The shear-Alfvén wave is analogous to a wave on a string in tension, which propagates at the phase velocity $v = (T/\rho)^{1/2}$, where T is the tension, and ρ the linear mass density. (See Section 6.3.) At low frequencies, the plasma and the magnetic field are "tied" (i.e., if one moves then so must the other), so it is possible to consider a magnetic field-line to be "loaded" with a plasma of density ρ (Fitzpatrick 2015). Furthermore, in terms of the Maxwell stress tensor, the field-line is under a tension B_0^2/μ_0 (Fitzpatrick 2008). Hence, $v = (B_0^2/\mu_0 \, \rho)^{1/2} = V_A$. We, thus, obtain the correct result for waves propagating along the magnetic field. The compressional-Alfvén wave is similar to a conventional sound wave (see Section 5.4), except that the restoring force emanates from magnetic pressure, rather than the thermal pressure of the plasma (which has actually been neglected in the present analysis) (Fitzpatrick 2015).

9.7 PARALLEL EM WAVES IN MAGNETIZED PLASMAS

Consider electromagnetic wave propagation, at arbitrary frequencies, parallel to the equilibrium magnetic field of a magnetized plasma. When $\theta = 0$, the eigenmode equation (9.168) simplifies to

$$\begin{pmatrix} S - n^2, & D, & 0 \\ D, & S - n^2, & 0 \\ 0, & 0, & P \end{pmatrix} \begin{pmatrix} \hat{E}_x \\ \hat{E}_y \\ \hat{E}_z \end{pmatrix} = \begin{pmatrix} 0 \\ 0 \\ 0 \end{pmatrix}. \tag{9.183}$$

One obvious way of solving this equation is to have

$$P \simeq 1 - \frac{\omega_{pe}^2}{\omega^2} = 0, \tag{9.184}$$

with the eigenvector $(0, 0, \hat{E}_z)$. This solution corresponds to an electrostatic plasma oscillation that does not propagate (i.e., ω is independent of k, so $v_g = d\omega/dk = 0$). The mode is longitudinal in nature, and, therefore, causes particles to oscillate parallel to $\mathbf{B}_0$. It follows that the particles experience zero Lorentz force due to the presence of the equilibrium magnetic field, with the result that this field has no effect on the mode dynamics.

The other two solutions to Equation (9.183) are obtained by setting the 2×2 determinant involving the x- and y-components of the electric field to zero. The first wave has the dispersion relation

$$n^2 = R \simeq 1 - \frac{\omega_{pe}^2}{(\omega - \Omega_e)(\omega + \Omega_i)}, \tag{9.185}$$

and the eigenvector $(\hat{E}_x, \hat{E}_x, 0)$. This is evidently a right-handed circularly polarized wave. (See Section 7.7.) The second wave has the dispersion relation

$$n^2 = L \simeq 1 - \frac{\omega_{pe}^2}{(\omega + \Omega_e)(\omega - \Omega_i)}, \tag{9.186}$$

and the eigenvector $(\hat{E}_x, -\hat{E}_x, 0)$. This is evidently a left-handed circularly polarized wave. (See Section 7.7.) At low frequencies (i.e., $\omega \ll \Omega_i$), both waves convert into the Alfvén wave discussed in the previous section. (The shear and compressional Alfvén waves are indistinguishable for parallel propagation.) Let us now examine the high-frequency behavior of the right- and left-handed waves.

For the right-handed wave, it is evident that $n^2 \to \infty$ as $\omega \to \Omega_e$. This so-called resonance, which corresponds to $R \to \infty$, is termed the *electron cyclotron resonance*. At the electron cyclotron resonance, the transverse electric field associated with a right-handed wave rotates at the same velocity, and in the same direction, as electrons gyrating around the equilibrium magnetic field. Thus, the electrons experience a continuous acceleration from the electric field, which tends to increase their perpendicular energy. It is, therefore, not surprising that right-handed waves, propagating parallel to the equilibrium magnetic field, and oscillating at the frequency Ω_e, are absorbed by the electrons. In fact, it is a general rule that an electromagnetic wave propagating through a magnetized plasma at a *resonant frequency*, at which $n^2 \to \infty$, is absorbed by the plasma.

When ω lies just above Ω_e, we find that n^2 is negative, and so there is no wave propagation. However, for frequencies much greater than the electron cyclotron or plasma frequencies, the solution to Equation (9.185) is approximately $n^2 = 1$. In other words, $\omega^2 = k^2 c^2$, which is the dispersion relation of a right-handed vacuum electromagnetic wave. Evidently, at some frequency above Ω_e, the solution for n^2 must pass through zero, and become positive again. Putting $n^2 = 0$ in Equation (9.185), we find that the equation reduces to

$$\omega^2 - \Omega_e \omega - \omega_{pe}^2 \simeq 0, \tag{9.187}$$

assuming that $V_A \ll c$. The previous equation has only one positive root, at $\omega = \omega_1$, where

$$\omega_1 \simeq \Omega_e/2 + \sqrt{\Omega_e^2/4 + \omega_{pe}^2} > \Omega_e. \tag{9.188}$$

Above this frequency, the wave propagates once again. Incidentally, a frequency at which $n^2 = 0$ is termed a *cutoff frequency*.

The dispersion curve for a right-handed wave propagating parallel to the equilibrium magnetic field is sketched in Figure 9.2. The continuation of the Alfvén wave above the ion cyclotron frequency is called the *electron cyclotron wave*, or, sometimes, the *whistler wave*. The latter terminology is prevalent in ionospheric and space plasma physics contexts. The wave that propagates above the cutoff frequency, ω_1, is a standard right-handed circularly polarized electromagnetic wave, somewhat modified by the presence of the plasma. The low-frequency branch of the dispersion curve differs fundamentally from the high-frequency branch, because the former branch corresponds to a wave that can only propagate through the plasma in the presence of an equilibrium magnetic field, whereas the latter branch corresponds to a wave that can propagate in the absence of an equilibrium field.

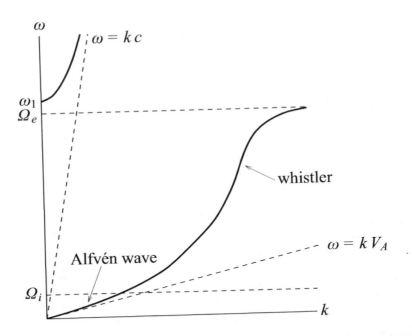

FIGURE 9.2 Schematic diagram showing the dispersion relation for a right-hand circularly polarized electromagnetic wave propagating parallel to the magnetic field in a magnetized plasma.

The curious name "whistler wave" for the branch of the dispersion relation lying between the ion and electron cyclotron frequencies is originally derived from ionospheric physics. Whistler waves are a very characteristic type of audio-frequency radio interference, most commonly encountered at high latitudes, which take the form of brief, intermittent pulses, starting at high frequencies, and rapidly descending in pitch.

Whistlers were discovered in the early days of radio communication, but were not explained until much later (Storey 1953). Whistler waves start off as "instantaneous" radio pulses, generated by lightning flashes at high latitudes. The pulses are channeled along the Earth's dipolar magnetic field, and eventually return to ground level in the opposite hemisphere. Now, in the frequency range $\Omega_i \ll \omega \ll \Omega_e$, the dispersion relation (9.185) reduces to

$$n^2 = \frac{k^2 c^2}{\omega^2} \simeq \frac{\omega_{pe}^2}{\omega \, \Omega_e}. \tag{9.189}$$

As is well known, wave pulses propagate at the group velocity,

$$v_g = \frac{d\omega}{dk} = 2 \, c \, \frac{\sqrt{\omega \, \Omega_e}}{\omega_{pe}}. \tag{9.190}$$

Clearly, the low-frequency components of a pulse propagate more slowly than the high-frequency components. It follows that, by the time a pulse returns to ground level, it has been stretched out temporally, because its high-frequency components arrive slightly before its low-frequency components. This also accounts for the characteristic whistling-down effect observed at ground level.

The shape of whistler pulses, and the way in which the pulse frequency varies in time, can yield a considerable amount of information about the regions of the Earth's magnetosphere through which

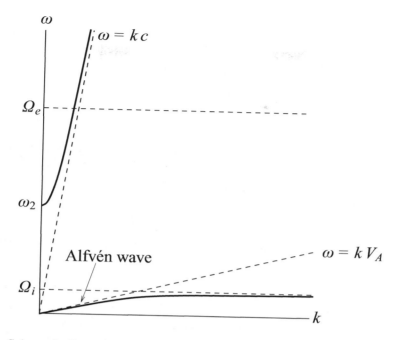

FIGURE 9.3 Schematic diagram showing the dispersion relation for a left-handed circularly polarized wave propagating parallel to the magnetic field in a magnetized plasma.

the pulses have passed. For this reason, many countries maintain observatories in polar regions—especially Antarctica—which monitor and collect whistler data.

For a left-handed circularly polarized wave, similar considerations to those described previously yield a dispersion curve of the form sketched in Figure 9.3. In this case, n^2 goes to infinity at the ion cyclotron frequency, Ω_i, corresponding to the so-called *ion cyclotron resonance* (at $L \to \infty$). At this resonance, the rotating electric field associated with a left-handed wave resonates with the gyration of the ions, allowing wave energy to be converted into perpendicular kinetic energy of the ions. There is a band of frequencies, lying above the ion cyclotron frequency, in which the left-handed wave does not propagate. At very high frequencies, a propagating mode exists, which is basically a standard left-handed circularly polarized electromagnetic wave, somewhat modified by the presence of the plasma. The cutoff frequency for this wave is

$$\omega_2 \simeq -\Omega_e/2 + \sqrt{\Omega_e^2/4 + \omega_{pe}^2}. \tag{9.191}$$

As before, the lower branch in Figure 9.3 describes a wave that can only propagate in the presence of an equilibrium magnetic field, whereas the upper branch describes a wave that can propagate in the absence of an equilibrium field. The continuation of the Alfvén wave to just below the ion cyclotron frequency is generally known as the *ion cyclotron wave*.

9.8 PERPENDICULAR EM WAVES IN MAGNETIZED PLASMAS

Consider electromagnetic wave propagation, at arbitrary frequencies, perpendicular to the equilibrium magnetic field in a magnetized plasma. When $\theta = \pi/2$, the eigenmode equation (9.168) simplifies to

$$\begin{pmatrix} S, & D, & 0 \\ D, & S-n^2, & 0 \\ 0, & 0, & P-n^2 \end{pmatrix} \begin{pmatrix} E_x \\ E_y \\ E_z \end{pmatrix} = \begin{pmatrix} 0 \\ 0 \\ 0 \end{pmatrix}. \qquad (9.192)$$

One obvious way of solving this equation is to have $P - n^2 = 0$, or

$$\omega^2 = \omega_{pe}^2 + k^2 c^2, \qquad (9.193)$$

with the eigenvector $(0, 0, \hat{E}_z)$. Because the wavevector now points in the x-direction, this is clearly a transverse wave polarized with its electric field parallel to the equilibrium magnetic field. Particle motions are along the magnetic field, so the mode dynamics are completely unaffected by this field. Thus, the wave is identical to the electromagnetic plasma wave found previously in an unmagnetized plasma. (See Section 9.3.) This wave is known as the *ordinary*, or *O-*, mode.

The other solution to Equation (9.192) is obtained by setting the 2×2 determinant involving the x- and y-components of the electric field to zero. With the help of the identity $S^2 - D^2 = RL$, the dispersion relation reduces to

$$n^2 = \frac{RL}{S}, \qquad (9.194)$$

with the associated eigenvector $E_x (1, -S/D, 0)$.

Let us, first of all, search for the cutoff frequencies, at which n^2 goes to zero. According to Equation (9.194), these frequencies are the roots of $R = 0$ and $L = 0$. In fact, we have already solved these equations; there are two cutoff frequencies, ω_1 and ω_2, that are specified by Equations (9.188) and (9.191), respectively.

Let us, next, search for the resonant frequencies, at which n^2 goes to infinity. According to Equation (9.194), the resonant frequencies are solutions of

$$S = 1 - \frac{\omega_{pe}^2}{\omega^2 - \Omega_e^2} - \frac{\omega_{pi}^2}{\omega^2 - \Omega_i^2} = 0. \qquad (9.195)$$

The roots of this equation can be obtained as follows. First, we note that if the first two terms in the middle are equated to zero then we obtain $\omega = \omega_{UH}$, where

$$\omega_{UH} = \sqrt{\omega_{pe}^2 + \Omega_e^2}. \qquad (9.196)$$

If this frequency is substituted into the third term in the middle then the result is far less than unity. We conclude that ω_{UH} is a good approximation of one of the roots of Equation (9.195). To obtain the second root, we make use of the fact that the product of the square of the roots is

$$\Omega_e^2 \Omega_i^2 + \omega_{pe}^2 \Omega_i^2 + \omega_{pi}^2 \Omega_e^2 \simeq \Omega_e^2 \Omega_i^2 + \omega_{pi}^2 \Omega_e^2. \qquad (9.197)$$

We, thus, obtain $\omega = \omega_{LH}$, where

$$\omega_{LH} = \sqrt{\frac{\Omega_e^2 \Omega_i^2 + \omega_{pi}^2 \Omega_e^2}{\omega_{pe}^2 + \Omega_e^2}}. \qquad (9.198)$$

The first resonant frequency, ω_{UH}, is greater than the electron cyclotron or plasma frequencies, and is called the *upper hybrid frequency*. The second resonant frequency, ω_{LH}, lies between the electron and ion cyclotron frequencies, and is called the *lower hybrid frequency*. Unfortunately, there is no simple explanation of the origins of the two hybrid resonances in terms of the motions of

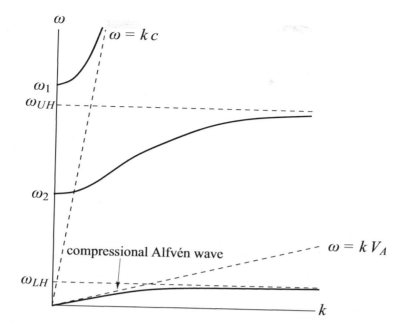

FIGURE 9.4   Schematic diagram showing the dispersion relation for an electromagnetic wave propagating perpendicular to the magnetic field in a magnetized plasma.

individual particles. At low frequencies, the mode in question reverts to the compressional-Alfvén wave discussed previously. Note that the shear-Alfvén wave does not propagate perpendicular to the magnetic field.

Using the previous information, and the easily demonstrated fact that

$$\omega_{LH} < \omega_2 < \omega_{UH} < \omega_1, \tag{9.199}$$

we deduce that the dispersion curve for the mode in question takes the form sketched in Figure 9.4. The lowest frequency branch corresponds to the compressional-Alfvén wave. The other two branches constitute the *extraordinary*, or *X*-, mode. The upper branch is basically a linearly polarized (in the *y*-direction) electromagnetic wave, somewhat modified by the presence of the plasma. This branch corresponds to a wave that propagates in the absence of an equilibrium magnetic field. The lowest branch corresponds to a wave that does not propagate in the absence of an equilibrium field. Finally, the middle branch corresponds to a wave that converts into an electrostatic plasma wave in the absence of an equilibrium magnetic field.

Wave propagation at oblique angles is generally more complicated than propagation parallel or perpendicular to the equilibrium magnetic field, but does not involve any new physical effects (Stix 1992; Swanson 2003).

9.9   ELECTROMAGNETIC WAVES IN CONDUCTORS

A so-called *Ohmic conductor* is a medium that satisfies Ohm's law, which can be written in the form

$$\mathbf{j} = \sigma\,\mathbf{E}, \tag{9.200}$$

where $\mathbf{j}$ is the current density (i.e., the current per unit area), $\mathbf{E}$ the electric field-strength, and σ a constant known as the *electrical conductivity* of the medium in question. The *z*-directed propagation of a plane electromagnetic wave, linearly polarized in the x-direction, through an Ohmic conductor of conductivity σ is governed by (see Appendix C)

$$\frac{\partial E_x}{\partial t} + \frac{\sigma}{\epsilon_0} E_x = -\frac{1}{\epsilon_0} \frac{\partial H_y}{\partial z}, \tag{9.201}$$

$$\frac{\partial H_y}{\partial t} = -\frac{1}{\mu_0} \frac{\partial E_x}{\partial z}. \tag{9.202}$$

For a so-called good conductor, which satisfies the inequality $\sigma \gg \epsilon_0 \omega$, the first term on the left-hand side of Equation (9.201) is negligible with respect to the second term, and the previous two equations reduce to

$$E_x \simeq -\frac{1}{\sigma} \frac{\partial H_y}{\partial z}, \tag{9.203}$$

$$\frac{\partial H_y}{\partial t} = -\frac{1}{\mu_0} \frac{\partial E_x}{\partial z}. \tag{9.204}$$

These equations can be solved to give

$$E_x(z, t) = E_0 \, e^{-z/d} \, \cos(\omega t - z/d), \tag{9.205}$$

$$H_y(z, t) = E_0 \, Z^{-1} \, e^{-z/d} \, \cos(\omega t - z/d - \pi/4), \tag{9.206}$$

where

$$d = \left(\frac{2}{\mu_0 \, \sigma \, \omega} \right)^{1/2}, \tag{9.207}$$

and

$$Z = \left(\frac{\mu_0 \, \omega}{\sigma} \right)^{1/2} = \left(\frac{\epsilon_0 \, \omega}{\sigma} \right)^{1/2} Z_0. \tag{9.208}$$

(See Exercise 9.7.) Equations (9.205) and (9.206) indicate that the amplitude of an electromagnetic wave propagating through a conductor decays exponentially on a characteristic lengthscale, d, that is known as the *skin-depth*. Consequently, an electromagnetic wave cannot penetrate more than a few skin-depths into a conducting medium. The skin-depth is smaller at higher frequencies. This implies that high-frequency waves penetrate a shorter distance into a conductor than low-frequency waves.

Consider a typical metallic conductor such as copper, whose electrical conductivity at room temperature is about $6 \times 10^7 \, (\Omega \, m)^{-1}$ (Wikipedia contributors 2018). Copper, therefore, acts as a good conductor for all electromagnetic waves of frequency below about 10^{18} Hz. The skin-depth in copper for such waves is thus

$$d = \sqrt{\frac{2}{\mu_0 \, \sigma \, \omega}} \simeq \frac{6}{\sqrt{f(\text{Hz})}} \text{ cm}. \tag{9.209}$$

It follows that the skin-depth is about 6 cm at 1 Hz, but only about 2 mm at 1 kHz. This gives rise to the so-called *skin effect* in copper wires, by which an oscillating electromagnetic signal of increasing frequency, transmitted along such a wire, is confined to an increasingly narrow layer (whose thickness is of order of the skin-depth) on the surface of the wire.

The conductivity of sea-water is only about $\sigma \simeq 5 \, (\Omega \, m)^{-1}$ (Wikipedia contributors 2018). However, this is still sufficiently high for sea-water to act as a good conductor for all radio frequency electromagnetic waves (i.e., $f = \omega/2\pi < 1$ GHz). The skin-depth at 1 MHz ($\lambda \sim 300$ m) is about 0.2 m, whereas that at 1 kHz ($\lambda \sim 300$ km) is still only about 7 m. This obviously poses quite severe restrictions for radio communication with submerged submarines. Either the submarines have to come quite close to the surface to communicate (which is dangerous), or the communication must be performed with extremely low-frequency (ELF) waves (i.e., $f < 100\,\text{Hz}$). Unfortunately, such waves have very large wavelengths ($\lambda > 3000\,\text{km}$), which means that they can only be efficiently generated by gigantic antennas.

According to Equation (9.206), the phase of the magnetic component of an electromagnetic wave propagating through a good conductor lags behind that of the electric component by $\pi/4$ radians. It follows that the mean energy flux into the conductor takes the form (see Appendix C)

$$\langle I_z \rangle = \langle E_x H_y \rangle = |E_x|^2 Z^{-1} \langle \cos(\omega t - z/d) \cos(\omega t - z/d - \pi/4)\rangle = \frac{|E_x|^2}{\sqrt{8}\,Z}, \tag{9.210}$$

where $|E_x| = E_0\,e^{-z/d}$ is the amplitude of the electric component of the wave. The fact that the mean energy flux is positive indicates that some of the wave energy is absorbed by the conductor. In fact, the absorbed energy corresponds to the energy lost due to Joule heating within the conductor. (See Exercise 9.15.)

According to Equation (9.208), the impedance of a good·conductor is far less than that of a vacuum (i.e., $Z \ll Z_0$). This implies that the ratio of the magnetic to the electric components of an electromagnetic wave propagating through a good conductor is far larger than that of a wave propagating through a vacuum. (This is because the conductor effectively "shorts out" the electric component of the wave.)

Suppose that the region $z < 0$ is a vacuum, and the region $z > 0$ is occupied by a good conductor of conductivity σ. Consider a linearly polarized plane wave, normally incident on the interface. Let the wave electric and magnetic fields in the vacuum region take the form of the incident and reflected waves specified in Equations (9.42) and (9.43). The wave electric and magnetic fields in the conductor are written

$$E_x(z, t) = E_t\,e^{-z/d}\cos(\omega t - z/d + \phi_t), \tag{9.211}$$

$$H_y(z, t) = E_t\,Z_0^{-1}\,\alpha^{-1}\,e^{-z/d}\cos(\omega t - z/d - \pi/4 + \phi_t), \tag{9.212}$$

where E_t is the amplitude of the evanescent wave that penetrates into the conductor, ϕ_t is the phase of this wave with respect to the incident wave, and

$$\alpha = \frac{Z}{Z_0} = \left(\frac{\epsilon_0 \omega}{\sigma}\right)^{1/2} \ll 1. \tag{9.213}$$

The appropriate matching conditions are the continuity of E_x and H_y at the vacuum/conductor interface ($z = 0$). (See Appendix C.) In other words,

$$E_i\cos(\omega t) + E_r\cos(\omega t + \phi_r) = E_t\cos(\omega t + \phi_t), \tag{9.214}$$

$$\alpha\,[E_i\cos(\omega t) - E_r\cos(\omega t + \phi_r)] = E_t\cos(\omega t - \pi/4 + \phi_t). \tag{9.215}$$

Equations (9.214) and (9.215), which must be satisfied at all times, can be solved, in the limit $\alpha \ll 1$, to give

$$E_r \simeq -\left(1 - \sqrt{2}\,\alpha\right) E_i, \tag{9.216}$$

$$\phi_r \simeq -\sqrt{2}\,\alpha, \tag{9.217}$$

$$E_t \simeq 2\,\alpha\,E_i, \tag{9.218}$$

$$\phi_t \simeq \frac{\pi}{4} - \frac{\alpha}{\sqrt{2}}. \tag{9.219}$$

(See Exercise 9.8.) Hence, the coefficient of reflection becomes

$$R \simeq \left(\frac{E_r}{E_i}\right)^2 \simeq 1 - 2\sqrt{2}\,\alpha = 1 - \left(\frac{8\,\epsilon_0\,\omega}{\sigma}\right)^{1/2}. \tag{9.220}$$

According to the previous analysis, a good conductor reflects a normally incident electromagnetic wave with a phase shift of almost π radians (i.e., $E_r \simeq -E_i$). The coefficient of reflection is just less than unity, indicating that, while most of the incident energy is reflected by the conductor, a small fraction of it is absorbed.

High-quality metallic mirrors are generally coated in silver, whose conductivity is 6.3×10^7 $(\Omega\,\mathrm{m})^{-1}$ (Wikipedia contributors 2018). It follows, from Equation (9.220), that at optical frequencies ($\omega = 4 \times 10^{15}\,\mathrm{rad.\,s^{-1}}$) the coefficient of reflection of a silvered mirror is $R \simeq 93.3\%$. This implies that about 7% of the light incident on the mirror is absorbed, rather than being reflected. This rather severe light loss can be problematic in instruments, such as astronomical telescopes, that are used to view faint objects.

9.10 WAVEGUIDES

As we saw in Section 6.6, transmission lines (e.g., ethernet cables) are used to carry high-frequency electromagnetic signals over distances that are long compared to the signal wavelength, $\lambda = c/f$, where c is the velocity of light and f the signal frequency (in hertz). Unfortunately, conventional transmission lines are subject to radiative losses (because the lines effectively act as antennas) that increase as the fourth power of the signal frequency (Fitzpatrick 2008). Above a certain critical frequency, which typically lies in the microwave band, the radiative losses become intolerably large. Under these circumstances, the transmission line must be replaced by a device known as a *waveguide*. A waveguide is basically a long hollow metal box within which electromagnetic signals propagate. Provided the walls of the box are much thicker than the skin-depth (see Section 9.9) in the wall material, the signal is essentially isolated from the outside world, and radiative losses are consequently negligible.

Consider an evacuated waveguide of rectangular cross-section that runs along the z-direction, and is enclosed by perfectly conducting (i.e., infinite conductivity) metal walls located at $x = 0$, $x = a$, $y = 0$, and $y = b$. Suppose that an electromagnetic wave propagates along the waveguide in the z-direction. For the sake of simplicity, let there be no y-variation of the wave electric or magnetic fields. The wave propagation inside the waveguide is governed by the two-dimensional wave equation [cf., Equation (7.9)]

$$\frac{\partial^2 \psi}{\partial t^2} = c^2 \left(\frac{\partial^2}{\partial x^2} + \frac{\partial^2}{\partial z^2}\right)\psi, \tag{9.221}$$

where $\psi(x, z, t)$ represents the electric component of the wave, which is assumed to be everywhere parallel to the y-axis, and c is the velocity of light in vacuum. The appropriate boundary conditions are

$$\psi(0, z, t) = 0, \tag{9.222}$$

$$\psi(a, z, t) = 0, \tag{9.223}$$

because the electric field inside a perfect conductor is zero (otherwise, an infinite current would flow), and, according to standard electromagnetic theory (see Appendix C), there cannot be a tangential discontinuity in the electric field at a conductor/vacuum boundary. (There can, however, be a normal discontinuity. This permits ψ to be non-zero at $y = 0$ and $y = b$.)

Let us search for a separable solution of Equation (9.221) of the form

$$\psi(x, z, t) = \psi_0 \sin(k_x x) \cos(\omega t - k z),$$
(9.224)

where k represents the z-component of the wavevector (rather than its magnitude), and is the effective wavenumber for propagation along the waveguide. The previous solution automatically satisfies the boundary condition (9.222). The second boundary condition (9.223) is satisfied provided

$$k_x = j \frac{\pi}{a},$$
(9.225)

where j is a positive integer. Suppose that j takes its smallest possible value 1. (j cannot be zero, because, in this case, $\psi = 0$ everywhere.) Substitution of expression (9.224) into the wave equation (9.221) yields the dispersion relation

$$\omega^2 = k^2 c^2 + \omega_0^2,$$
(9.226)

where

$$\omega_0 = \frac{\pi c}{a}.$$
(9.227)

This dispersion relation is analogous in form to the dispersion relation (9.27) for an electromagnetic wave propagating through an unmagnetized plasma, with the *cutoff frequency*, ω_0, playing the role of the (electron) plasma frequency, $\omega_{p e}$. The cutoff frequency is so-called because for $\omega < \omega_0$ the wavenumber is imaginary (i.e., $k^2 < 0$), which implies that the wave does not propagate along the waveguide, but, instead, decays exponentially with increasing z. On the other hand, for wave frequencies above the cutoff frequency the phase velocity,

$$v_p = \frac{\omega}{k} = \frac{c}{\sqrt{1 - \omega_0^2/\omega^2}},$$
(9.228)

is superluminal. This is not a problem, however, because the group velocity,

$$v_g = \frac{d\omega}{dk} = c \sqrt{1 - \omega_0^2/\omega^2},$$
(9.229)

which is the true signal velocity, remains sub-luminal. (Recall, from Section 9.3, that a high-frequency electromagnetic wave propagating through an unmagnetized plasma exhibits similar behavior.) Not surprisingly, the signal velocity goes to zero as $\omega \to \omega_0$, because the signal ceases to propagate at all when $\omega = \omega_0$.

It turns out that waveguides support many distinct modes of propagation. The type of mode discussed previously is termed a TE (for transverse electric-field) mode, because the electric field is transverse to the direction of propagation. (See Exercise 9.17.) There are many different sorts of TE mode, corresponding, for instance, to different choices of the mode number, j (Fitzpatrick 2008). However, the $j = 1$ mode has the lowest cutoff frequency. There are also TM (for transverse magnetic-field) modes (see Exercise 9.18), and TEM (for transverse electric- and magnetic-field) modes (ibid.). TM modes also only propagate when the wave frequency exceeds a cutoff frequency. On the other hand, TEM modes (which are the same type of mode as that supported by a conventional transmission line) propagate at all frequencies. However, TEM modes are only possible when the waveguide possesses an internal conductor running along its length (ibid.).

9.11 PULSE PROPAGATION IN TWO DIMENSIONS

In Section 6.10, we saw that a wave can propagate through an inhomogeneous medium, without significant reflection, provided that the properties of the medium vary on lengthscales that are much longer than the wavelength of the wave. In this section, we shall determine the path taken by a localized wave pulse as it travels though such a medium. For the sake of simplicity, we shall restrict our analysis to two dimensions (by neglecting any variation in the y-direction.)

The wavefunction of a two-dimensional wave propagating through a (slowly varying) inhomogeneous medium can be written in the general form

$$\psi(x, z, t) = A(x, z) \cos\left[\phi(x, z, t)\right]. \tag{9.230}$$

Here, $A(x, z)$ is a relatively slowly varying function that determines the wave amplitude, whereas $\phi(x, z, t)$ is a relatively rapidly varying function that determines the wave phase. We can expand $\phi(x, z, t)$ locally as a Taylor series (see Appendix B) to give

$$\phi \simeq \phi_0 + x \frac{\partial \phi}{\partial x} + z \frac{\partial \phi}{\partial z} + t \frac{\partial \phi}{\partial t}. \tag{9.231}$$

Comparison of the previous two equations with the standard expression for the wavefunction of a plane wave (see Section 7.2),

$$\psi(x, z, t) = A \cos(\omega t - k_x x - k_z z + \phi_0), \tag{9.232}$$

which is assumed to hold locally at each point in space, reveals that

$$\omega = \frac{\partial \phi}{\partial t}, \tag{9.233}$$

$$k_x = -\frac{\partial \phi}{\partial x}, \tag{9.234}$$

$$k_z = -\frac{\partial \phi}{\partial z}, \tag{9.235}$$

where ω is the wave angular frequency, and $\mathbf{k} = (k_x, 0, k_z)$ the wavevector. (The generalization to three dimensions is straightforward.) The previous three equations also imply that

$$\frac{\partial k_x}{\partial t} = -\frac{\partial^2 \phi}{\partial x \partial t} = -\frac{\partial \omega}{\partial x}, \tag{9.236}$$

$$\frac{\partial k_z}{\partial t} = -\frac{\partial^2 \phi}{\partial z \partial t} = -\frac{\partial \omega}{\partial z}. \tag{9.237}$$

As we saw in Section 9.2, a one-dimensional wave pulse propagating through a dispersive medium does so at the group velocity, $v_g = d\omega/dk$, where $\omega = \omega(k)$ is the dispersion relation. Thus, the equation of motion of the pulse is written

$$\frac{dx}{dt} = \frac{d\omega}{dk}. \tag{9.238}$$

In two dimensions, the dispersion relation (in a uniform medium) takes the form $\omega = \omega(k_x, k_z)$. I this case, the expression for the group velocity generalizes to (Landau and Lifshitz 1959)

$$\mathbf{v}_g = \left(\frac{\partial \omega}{\partial k_x}, 0, \frac{\partial \omega}{\partial k_z}\right). \tag{9.239}$$

(The generalization to three dimensions is straightforward.) Thus, in two dimensions, the equation of motion of a wave pulse are

$$\frac{\partial x}{\partial t} = \frac{\partial \omega}{\partial k_x}, \tag{9.240}$$

$$\frac{\partial z}{\partial t} = \frac{\partial \omega}{\partial k_z}. \tag{9.24}$$

It turns out that these equations also hold in a two-dimensional inhomogeneous medium; that is, a medium in which the dispersion relation is an explicit function of position, so that $\omega = \omega(x, z, k_x, k_z)$. (Actually, this is only true as long as the dispersion relation varies on lengthscales that are much longer than the wavelength.) Thus, in a two-dimensional inhomogeneous medium, once we know the local dispersion relation, $\omega = \omega(x, z, k_x, k_z)$, we can trace the path of a wave pulse using the following four equations [cf., Equations (9.236)–(9.237) and (9.240)–(9.241)]:

$$\frac{\partial x}{\partial t} = \frac{\partial \omega}{\partial k_x}, \tag{9.242}$$

$$\frac{\partial k_x}{\partial t} = -\frac{\partial \omega}{\partial x}, \tag{9.243}$$

$$\frac{\partial z}{\partial t} = \frac{\partial \omega}{\partial k_z}, \tag{9.244}$$

$$\frac{\partial k_z}{\partial t} = -\frac{\partial \omega}{\partial z}. \tag{9.245}$$

(Incidentally, x, z, k_x, and k_z are treated as independent variables in these equations.) It can be seen that the preceding equations are analogous to *Hamilton's equations* for two-dimensional motion in classical mechanics, with the wave vector playing the role of the momentum, and the function $\omega(x, z, k_x, k_z)$ playing the role of the Hamiltonian (Goldstein, Poole, and Safko 2002). (In fact, Hamilton's equations were first derived to determine the path of a light ray through an inhomogeneous medium, and only later applied to dynamical systems.)

Consider a radio wave pulse launched from the Earth's surface, and subsequently reflected by the ionosphere. Let x measure horizontal distance, and let z measure vertical height above the Earth's surface. Neglecting the Earth's weak magnetic field, the appropriate dispersion relation is (see Section 9.3)

$$\omega = \left(k_x^2 c^2 + k_z^2 c^2 + \omega_{pe}^2\right)^{1/2}, \tag{9.246}$$

where c is the velocity of light in vacuum, and ω_{pe} the (electron) plasma frequency. This frequency is proportional to the square root of the number density of free electrons in the atmosphere, which we would generally expect to be a function of height only. In other words, $\omega_{pe} = \omega_{pe}(z)$. Equations (9.242)–(9.244) yield

$$\frac{\partial x}{\partial t} = \frac{k_x c^2}{\omega}, \tag{9.247}$$

$$\frac{\partial z}{\partial t} = \frac{k_z c^2}{\omega}, \tag{9.248}$$

$$\frac{\partial k_x}{\partial t} = 0. \tag{9.249}$$

It is helpful to rewrite the dispersion relation as

$$n(z) = \frac{(k_x^2 + k_z^2)^{1/2} c}{\omega} = \left(1 - \frac{\omega_{pe}^2}{\omega^2}\right)^{1/2}, \tag{9.250}$$

where $n(z)$ is the refractive index.

Let us assume that $n = 1$ at $z = 0$, which is equivalent to the reasonable assumption that the atmosphere is non-ionized at ground level. It follows from Equation (9.249) that

$$k_x = k_x(z = 0) = \frac{\omega}{c} S, \tag{9.251}$$

where S is the sine of the angle of incidence of the pulse, with respect to the vertical axis, at ground level. The previous two equations can be combined to give

$$k_z = \pm \frac{\omega}{c} \sqrt{n^2 - S^2}. \tag{9.252}$$

According to Equation (9.248), the plus sign corresponds to the upward trajectory of the pulse, whereas the minus sign corresponds to the downward trajectory. Equations (9.247), (9.248), (9.251), and (9.252) give the following equations of motion of the pulse:

$$\frac{\partial x}{\partial t} = c\, S, \tag{9.253}$$

$$\frac{\partial z}{\partial t} = \pm c \sqrt{n^2 - S^2}. \tag{9.254}$$

It can be seen that the pulse attains its maximum altitude, $z = z_0$, when $n(z_0) = |S|$. The total distance traveled by the pulse (i.e., the distance from its launch point to the point where it intersects the Earth's surface again) is

$$x_0 = 2\,S \int_0^{z_0(S)} \frac{dz}{\sqrt{n^2(z) - S^2}}. \tag{9.255}$$

According to Equations (9.247), (9.248), and (9.252), the equations of motion of the pulse can also be written

$$\frac{\partial^2 x}{\partial t^2} = 0, \tag{9.256}$$

$$\frac{\partial^2 z}{\partial t^2} = \frac{\partial}{\partial z} \left(\frac{c^2\, n^2}{2} \right). \tag{9.257}$$

It follows that the trajectory of the pulse is the same as that of a particle moving in the gravitational potential $-c^2\, n^2/2$. In particular, if n^2 decreases linearly with increasing altitude (as would be the case if the number density of free electrons increased linearly with z) then the trajectory of the pulse is a parabola.

9.12 GRAVITY WAVES

Consider a stationary body of water, of uniform depth d, located on the surface of the Earth. Let us find the dispersion relation of a plane wave propagating across the water's surface. Suppose that the Cartesian coordinate x measures horizontal distance, while the coordinate z measures vertical height, with $z = 0$ corresponding to the unperturbed surface of the water. Let there be no variation in the y-direction. In other words, let the wavefronts of the wave all be parallel to the y-axis. Finally, let $v_x(x, z, t)$ and $v_z(x, z, t)$ be the perturbed horizontal and vertical velocity fields of the water due to the wave. It is assumed that there is no motion in the y-direction.

Water is essentially incompressible (i.e., its bulk modulus is very large) (Batchelor 2000). Thus, any (low speed) wave disturbance in water is constrained to preserve the volume of a co-moving volume element. Equivalently, the inflow rate of water into a stationary volume element must match the outflow rate. Consider a stationary cubic volume element lying between x and $x+dx$, y and $y+dy$, and z and $z + dz$. The element has two faces, of area $dy\, dz$, perpendicular to the x-axis, located at x and $x + dx$. Water flows into the element through the former face at the rate $v_x(x, z, t)\, dy\, dz$ (i.e., the product of the area of the face and the normal velocity), and out of the element though the latter face at the rate $v_x(x + dx, z, t)\, dy\, dz$. The element also has two faces perpendicular to the y-axis, but there is no flow through these faces, because $v_y = 0$. Finally, the element has two faces, of area

$dx\,dy$, perpendicular to the z-axis, located at z and $z + dz$. Water flows into the element through the former face at the rate $v_z(x, z, t)\,dx\,dy$, and out of the element though the latter face at the rate $v_z(x, z + dz, t)\,dx\,dy$. Thus, the net rate at which water flows into the element is $v_x(x, z, t)\,dy\,dz +$ $v_z(x, z, t)\,dx\,dy$. Likewise, the net rate at which water flows out of the element is $v_x(x+dx, z, t)\,dy\,dz +$ $v_z(x, z + dz, t)\,dx\,dy$. If the water is to remain incompressible then the inflow and outflow rates must match. In other words,

$$v_x(x, z, t)\,dy\,dz + v_z(x, z, t)\,dx\,dy = v_x(x + dx, z, t)\,dy\,dz + v_z(x, z + dz, t)\,dx\,dy, \tag{9.258}$$

or

$$\left[\frac{v_x(x + dx, z, t) - v_x(x, z, t)}{dx} + \frac{v_z(x, z + dz, t) - v_z(x, z, t)}{dz}\right] dx\,dy\,dz = 0. \tag{9.259}$$

Hence, the incompressibility constraint reduces to

$$\frac{\partial v_x}{\partial x} + \frac{\partial v_z}{\partial z} = 0, \tag{9.260}$$

which is consistent with Equation (7.194) in the incompressible limit $K \to \infty$.[1] Incidentally, it is reasonable to neglect the finite compressibility of water in this investigation because sound waves (which require finite compressibility) propagate much faster than surface waves, and, therefore, do not couple to them. (See the two footnotes in this section.)

Consider the equation of motion of a small volume element of water lying between x and $x+dx$, y and $y + dy$, and z and $z + dz$. The mass of this element is $\rho\,dx\,dy\,dz$, where $\rho = 1.0 \times 10^3\,\mathrm{m}^{-3}$ is the uniform mass density of water. Suppose that $p(x, z, t)$ is the pressure in the water, which is assumed to be isotropic (Batchelor 2000). The net horizontal force on the element is $p(x, z, t)\,dy\,dz -$ $p(x + dx, z, t)\,dy\,dz$ (because force is pressure times area, and the external pressure forces acting on the element are directed inward normal to its surface). Hence, the element's horizontal equation of motion is

$$\rho\,dx\,dy\,dz\,\frac{\partial v_x(x, z, t)}{\partial t} = -\left[\frac{p(x + dx, z, t) - p(x, z, t)}{dx}\right] dx\,dy\,dz, \tag{9.261}$$

which reduces to

$$\rho\,\frac{\partial v_x}{\partial t} = -\frac{\partial p}{\partial x}. \tag{9.262}$$

[See Equation (7.192).] The vertical equation of motion is similar, except that the element is subject to a downward acceleration, $g \simeq 9.8\,\mathrm{m\,s}^{-2}$, due to gravity. Hence, we obtain

$$\rho\,\frac{\partial v_z}{\partial t} = -\frac{\partial p}{\partial z} - \rho g. \tag{9.263}$$

[See Equation (7.193).]

We can write

$$p = p_0 - \rho g z + p_1, \tag{9.264}$$

where p_0 is atmospheric pressure (i.e., the air pressure just above the surface of the water), and p_1 is the pressure perturbation due to the wave. In the absence of the wave, the water pressure at a depth

---

[1] We can reproduce Equation (7.194) by realizing that $\partial v_x/\partial x + \partial v_z/\partial z = V^{-1}\,\partial V/\partial t$, where V is the volume of a small co-moving volume element. Combining this expression with the definition of bulk modulus, $K = \rho\,\partial p/\partial \rho = -V\,\partial p/\partial V$, we obtain $\partial p/\partial t = -K\,(\partial v_x/\partial x + \partial v_z/\partial z)$.

h below the surface is $p_0 + \rho g h$ (Batchelor 2000). Substitution into Equations (9.262) and (9.263) yields

$$\rho \frac{\partial v_x}{\partial t} = -\frac{\partial p_1}{\partial x}, \tag{9.265}$$

$$\rho \frac{\partial v_z}{\partial t} = -\frac{\partial p_1}{\partial z}. \tag{9.266}$$

It follows that

$$\rho \frac{\partial^2 v_x}{\partial z \, \partial t} - \rho \frac{\partial^2 v_z}{\partial x \, \partial t} = -\frac{\partial^2 p_1}{\partial z \, \partial x} + \frac{\partial^2 p_1}{\partial x \, \partial z} = 0, \tag{9.267}$$

which implies that

$$\rho \frac{\partial}{\partial t}\left(\frac{\partial v_x}{\partial z} - \frac{\partial v_z}{\partial x} \right) = 0, \tag{9.268}$$

or

$$\frac{\partial v_x}{\partial z} - \frac{\partial v_z}{\partial x} = 0. \tag{9.269}$$

(Actually, this quantity could be non-zero and constant in time, but this is not consistent with an oscillating wave-like solution.)

Equation (9.269) is automatically satisfied by writing the fluid velocity in terms of a velocity potential; that is,

$$v_x = \frac{\partial \phi}{\partial x}, \tag{9.270}$$

$$v_z = \frac{\partial \phi}{\partial z}. \tag{9.271}$$

(See Section 7.12.) Equation (9.260) then gives [2]

$$\frac{\partial^2 \phi}{\partial x^2} + \frac{\partial^2 \phi}{\partial z^2} = 0. \tag{9.272}$$

Finally, Equations (9.265) and (9.266) yield

$$p_1 = -\rho \frac{\partial \phi}{\partial t}. \tag{9.273}$$

As we have just demonstrated, surface waves in water are governed by Equation (9.272), which is known as *Laplace's equation*. We now need to derive the physical constraints that must be satisfied by the solution to this equation at the water's upper and lower boundaries. The water is bounded from below by a solid surface located at $z = -d$. Assuming that the water always remains in contact with this surface, the appropriate physical constraint at the lower boundary is $v_z(x, -d, t) = 0$ (i.e. there is no vertical motion of the water at the lower boundary), or

$$\frac{\partial \phi}{\partial z}\bigg|_{z=-d} = 0. \tag{9.274}$$

The physical constraint at the water's upper boundary is a little more complicated, because thi

---

[2] Taking the finite compressibility of water into account, this equation generalizes to $\partial^2 \phi / \partial t^2 = c^2 (\partial^2 \phi / \partial x^2 + \partial^2 \phi / \partial z^2$ where $c = \sqrt{K/\rho}$ is the velocity of sound. However, the left-hand side of the general equation is negligible for gravity wave whose propagation velocities are much less than c.

boundary is a free surface. Let $\zeta(x, t)$ represent the vertical displacement of the water's surface. It follows that

$$\frac{\partial \zeta}{\partial t} = v_z|_{z=0} = \frac{\partial \phi}{\partial z}\bigg|_{z=0}. \tag{9.275}$$

The physical constraint at the surface is that the water pressure is equal to atmospheric pressure, because there cannot be a pressure discontinuity across a free surface (in the absence of surface tension). Thus, it follows from Equation (9.264) that

$$p_0 = p_0 - \rho g \zeta(x, t) + p_1(x, 0, t). \tag{9.276}$$

Finally, differentiating with respect to t, and making use of Equations (9.273) and (9.275), we obtain

$$\frac{\partial \phi}{\partial z}\bigg|_{z=0} = -g^{-1} \frac{\partial^2 \phi}{\partial t^2}\bigg|_{z=0}. \tag{9.277}$$

Hence, the problem boils down to solving Laplace's equation, (9.272), subject to the physical constraints (9.274) and (9.277).

Let us search for a propagating wave-like solution of Equation (9.272) of the form

$$\phi(x, z, t) = F(z) \cos(\omega t - k x). \tag{9.278}$$

Substitution into Equation (9.272) yields

$$\frac{d^2 F}{dz^2} - k^2 F = 0, \tag{9.279}$$

whose independent solutions are $\exp(+k z)$ and $\exp(-k z)$. Hence, the most general wave-like solution to Laplace's equation is

$$\phi(x, z, t) = A e^{k z} \cos(\omega t - k x) + B e^{-k z} \cos(\omega t - k x), \tag{9.280}$$

where A and B are arbitrary constants. The boundary condition (9.274) is satisfied provided that $B = A \exp(-2 k d)$, giving

$$\phi(x, z, t) = A \left[e^{k z} + e^{-k(z + 2d)} \right] \cos(\omega t - k x), \tag{9.281}$$

The boundary condition (9.277) yields

$$A k \left(1 - e^{-2 k d}\right) \cos(\omega t - k x) = A \frac{\omega^2}{g} \left(1 + e^{-2 k d}\right) \cos(\omega t - k x), \tag{9.282}$$

which reduces to the dispersion relation

$$\omega^2 = g k \tanh(k d). \tag{9.283}$$

The type of wave just described is generally known as a *gravity wave*. We conclude that a gravity wave of arbitrary wavenumber k, propagating horizontally through water of depth d, has a phase velocity

$$v_p = \frac{\omega}{k} = (g d)^{1/2} \left[\frac{\tanh(k d)}{k d} \right]^{1/2}. \tag{9.284}$$

Moreover, the ratio of the group to the phase velocity is

$$\frac{v_g}{v_p} = \frac{k}{\omega} \frac{d\omega}{dk} = \frac{1}{2} \left[1 + \frac{2 k d}{\sinh(2 k d)} \right]. \tag{9.285}$$

Finally, the velocity fields associated with a gravity wave of surface amplitude a are

$$v_x(x,z,t) = a\omega\, \frac{\cosh[k\,(z+d)]}{\sinh(k\,d)}\, \sin(\omega t - k\,x),\qquad(9.286)$$

$$v_z(x,z,t) = a\omega\, \frac{\sinh[k\,(z+d)]}{\sinh(k\,d)}\, \cos(\omega t - k\,x).\qquad(9.287)$$

In *shallow water* (i.e., $kd \ll 1$), Equation (9.283) reduces to the linear dispersion relation

$$\omega = k\,\sqrt{g\,d}.\qquad(9.288)$$

Here, use has been made of the small argument expansion $\tanh x \simeq x$ for $|x| \ll 1$. (See Appendix B.) It follows that gravity waves in shallow water are non-dispersive in nature, and propagate at the phase velocity $\sqrt{g\,d}$. On the other hand, in *deep water* (i.e., $kd \gg 1$), Equation (9.283) reduces to the nonlinear dispersion relation

$$\omega = \sqrt{k\,g}.\qquad(9.289)$$

Here, use has been made of the large argument expansion $\tanh x \simeq 1$ for $x \gg 1$. We conclude that gravity waves in deep water are dispersive in nature. The phase velocity of the waves is $v_p = \omega/k = \sqrt{g/k}$, whereas the group velocity is $v_g = d\omega/dk = (1/2)\sqrt{g/k} = v_p/2$. In other words, the group velocity is half the phase velocity, and is largest for long wavelength (i.e., small k) waves.

The *mean kinetic energy per unit surface area* associated with a gravity wave is defined

$$K = \left\langle \int_{-d}^{\zeta} \frac{1}{2}\rho v^2\, dz \right\rangle,\qquad(9.290)$$

where

$$\zeta(x,t) = a\,\sin(\omega t - k\,x)\qquad(9.291)$$

is the vertical displacement at the surface, and

$$\langle \cdots \rangle = \int_0^{2\pi} (\cdots)\, \frac{d(k\,x)}{2\pi}\qquad(9.292)$$

represents an average over a wavelength. Given that $\langle \cos^2(\omega t - k\,x)\rangle = \langle \sin^2(\omega t - k\,x)\rangle = 1/2$, it follows from Equations (9.286) and (9.287) that, to second order in a,

$$K = \frac{1}{4}\rho a^2 \omega^2 \int_{-d}^{0} \frac{\cosh[2\,k\,(z+d)]}{\sinh^2(k\,d)}\, dz = \frac{1}{4}\rho g a^2\, \frac{\omega^2}{g k\,\tanh(k\,d)}.\qquad(9.293)$$

Making use of the general dispersion relation (9.283), we obtain

$$K = \frac{1}{4}\rho g a^2.\qquad(9.294)$$

The *mean potential energy perturbation per unit surface area* associated with a gravity wave is defined

$$U = \left\langle \int_{-d}^{\zeta} \rho g z\, dz \right\rangle + \frac{1}{2}\rho g d^2,\qquad(9.295)$$

which yields

$$U = \left\langle \frac{1}{2}\rho g\,(\zeta^2 - d^2)\right\rangle + \frac{1}{2}\rho g d^2 = \frac{1}{2}\rho g \langle \zeta^2\rangle,\qquad(9.296)$$

or

$$U = \frac{1}{4}\rho g a^2. \tag{9.297}$$

In other words, the mean potential energy per unit surface area of a gravity wave is equal to its mean kinetic energy per unit surface area.

Finally, the *mean total energy per unit surface area* associated with a gravity wave is

$$E = K + U = \frac{1}{2}\rho g a^2. \tag{9.298}$$

This energy depends on the wave amplitude at the surface, but is independent of the wavelength, or the water depth.

9.13 WAVE DRAG ON SHIPS

Under certain circumstances (see the following section), a ship traveling over a body of water leaves behind it a train of gravity waves whose wavefronts are transverse to the ship's direction of motion. Because these waves possess energy that is carried away from the ship, and eventually dissipated, this energy must have been produced at the ship's expense. The ship consequently experiences a drag force, D (Lamb 1932). Suppose that the ship is moving at the constant velocity V. We would expect the transverse waves making up the train to have a matching phase velocity, so that they maintain a constant phase relation with respect to the ship. To be more exact, we would generally expect the ship's bow to always correspond to a wave maximum (because of the pile up of water in front of the bow produced by the ship's forward motion). The condition $v_p = V$, combined with Equation (9.284), yields

$$\frac{\tanh(kd)}{kd} = \frac{V^2}{gd}. \tag{9.299}$$

Suppose, for the sake of argument, that the wave train is of uniform transverse width w. Consider a fixed line drawn downstream of the ship at right-angles to its path. The rate at which the length of the train is increasing ahead of this line is V. Therefore, the rate at which the energy of the train is increasing ahead of the line is $(1/2)\rho g a^2 w V$, where a is the typical amplitude of the transverse waves in the train. As is readily demonstrated [and is evident from Equation (9.35)], wave energy travels at the group velocity, rather than the phase velocity (Lighthill 1978). Thus, the energy flux per unit width of a propagating gravity wave is simply $E v_g$. Wave energy consequently crosses our fixed line in the direction of the ship's motion at the rate $(1/2)\rho g a^2 w v_g$. Finally, the ship does work against the drag force, which goes to increase the energy of the train in the region ahead of our line, at the rate DV. Energy conservation thus yields

$$\frac{1}{2}\rho g a^2 w V = \frac{1}{2}\rho g a^2 w v_g + DV. \tag{9.300}$$

However, because $V = v_p$, we obtain

$$D = \frac{1}{2}\rho g a^2 w \left(1 - \frac{v_g}{v_p}\right) = \frac{1}{4}\rho g a^2 w \left[1 - \frac{2kd}{\sinh(2kd)}\right], \tag{9.301}$$

where use has been made of Equation (9.285). Here, kd is determined implicitly in terms of the ship speed via Equation (9.299). However, this equation cannot be satisfied when the speed exceeds the critical value $(gd)^{1/2}$, because gravity waves cannot propagate at speeds in excess of this value. In this situation, no transverse wave train can keep up with the ship, and the drag associated with such waves consequently disappears. In fact, we can see, from the previous formulae, that when $V \rightarrow$

$(g\,d)^{1/2}$ then $k\,d \to 0$, and so $D \to 0$. Actually, the transverse wave amplitude, a, generally increases significantly as the ship speed approaches the critical value. Hence, the drag due to transverse waves actually peaks strongly at speeds just below the critical speed, before effectively falling to zero as this speed is exceeded. Consequently, it usually requires a great deal of propulsion power to force a ship to travel at speeds faster than $(g\,d)^{1/2}$.

In the deep water limit $k\,d \gg 1$, Equation (9.301) reduces to

$$D = \frac{1}{4}\rho g a^2 w. \tag{9.302}$$

At fixed wave amplitude, this expression is independent of the wavelength of the wave train, and, hence, independent of the ship's speed. This result is actually rather misleading. In fact, (at fixed wave amplitude) the drag acting on a ship traveling through deep water varies significantly with the ship's speed. We can account for this variation by incorporating the finite length of the ship into our analysis. A real ship moving through water generates a *bow wave* from its bow, and a *stern wave* from its stern. Moreover, the bow wave tends to have a positive vertical displacement, because water naturally piles up in front of the bow due to the forward motion of the ship, whereas the stern wave tends to have a negative vertical displacement, because water rushes into the void left by the stern. Very roughly speaking, suppose that the vertical displacement of the water surface caused by the ship is of the form

$$\zeta(x) \propto \cos\left(\pi\,\frac{x}{l}\right). \tag{9.303}$$

Here, l is the length of the ship. Moreover, the bow lies (instantaneously) at $x = 0$ [hence, $\zeta(0) > 0$], and the stern at $x = l$ [hence, $\zeta(l) < 0$]. For the sake of simplicity, the upward water displacement due to the bow is assumed to equal the downward displacement due to the stern. At fixed bow wave displacement, the amplitude of transverse gravity waves of wavenumber $k = g/V^2$ (chosen so that the phase velocity of the waves matches the ship's speed, V) produced by the ship is

$$a \propto \frac{1}{l}\int_0^l \cos\left(\pi\,\frac{x}{l}\right)\cos(k\,x)\,dx = \frac{1}{2}\left[\mathrm{sinc}(\pi - k\,l) + \mathrm{sinc}(\pi + k\,l)\right], \tag{9.304}$$

where $\mathrm{sinc}(x) \equiv \sin x/x$. In other words, the amplitude is proportional to the Fourier coefficient of the ship's vertical displacement pattern evaluated for a wavenumber that matches that of the wave train. Hence, (at fixed bow wave displacement) the drag produced by the transverse waves is

$$D \propto a^2 \propto \left[\mathrm{sinc}\left(\pi - F^{-2}\right) + \mathrm{sinc}\left(\pi + F^{-2}\right)\right]^2, \tag{9.305}$$

where the dimensionless parameter

$$F = \frac{V}{(g\,l)^{1/2}} \tag{9.306}$$

is known as the *Froude number*.

Figure 9.5 illustrates the variation of the wave drag with Froude number predicted by Equation (9.305). As we can see, if the Froude number is much less than unity, which implies that the wavelength of the wave train is much smaller than the length of the ship, then the drag is comparatively small. This is the case because the ship is extremely inefficient at driving short wavelength gravity waves. The drag increases as the Froude number increases, reaching a relatively sharp maximum when $\mathrm{Fr} = \mathrm{Fr}_c = 1/\sqrt{\pi} = 0.56$, and then falls rapidly. When $\mathrm{Fr} = \mathrm{Fr}_c$ the length of the ship is equal to half the wavelength of the wave train. In this situation, the bow and stern waves interfere constructively, leading to a particularly large amplitude wave train, and, hence, to a particularly large wave drag. The smaller peaks visible in the figure correspond to other situations in which the bow and stern waves interfere constructively. (For instance, when the length of the ship corresponds t

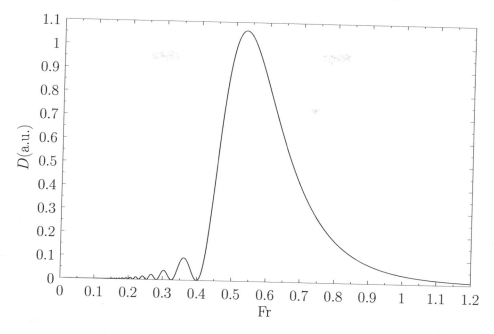

FIGURE 9.5 Variation of wave drag with Froude number for a ship traveling through deep water.

one and a half wavelengths of the wave train.) A heavy ship with a large displacement, and limited propulsion power, generally cannot overcome the peak in the wave drag that occurs when Fr = Fr_c. Such a ship is, therefore, limited to Froude numbers in the range $0 < Fr < Fr_c$, which implies a maximum speed of

$$V_c = 0.56\,(g\,l)^{1/2} = 1.75\,[l(\text{m})]^{1/2}\,\text{m/s} = 3.4\,[l(\text{m})]^{1/2}\,\text{kts}. \tag{9.307}$$

This characteristic speed is sometimes called the *hull speed*. It can be seen that the hull speed increases with the length of the ship. In other words, long ships have higher hull speeds than short ones.

9.14 SHIP WAKES

Let us now make a detailed investigation of the wake pattern generated behind a ship as it travels over a body of water, taking into account obliquely propagating gravity waves, in addition to transverse waves. For the sake of simplicity, the finite length of the ship is neglected in the following analysis. In other words, the ship is treated as a point source of gravity waves. Consider Figure 9.6. This shows a plane gravity wave generated on the surface of the water by a moving ship. The water surface corresponds to the x-y plane. The ship is traveling along the x-axis, in the negative x-direction, at the constant speed V. Suppose that the ship's bow is initially at point A', and has moved to point A after a time interval t. The only type of gravity wave that is continuously excited by the passage of the ship is one that maintains a constant phase relation with respect to its bow. In fact, as we have already mentioned, the bow should always correspond to a wave maximum. An oblique wavefront associated with such a wave is shown in the figure. Here, the wavefront $C'D'$, which initially passes through the bow at point A', has moved to CD after a time interval t, such that it again passes through the bow at point A. The wavefront propagates at the phase velocity, v_p. It follows that, in the right-angled triangle $AA'E$, the sides AA' and $A'E$ are of lengths $V\,t$ and $v_p\,t$,

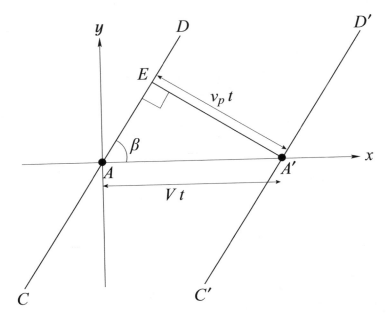

FIGURE 9.6   An oblique plane wave generated on the surface of the water by a moving ship.

respectively, so that

$$\sin\beta = \frac{v_p}{V}.$$ (9.308)

This, therefore, is the condition that must be satisfied in order for an obliquely propagating gravity wave to maintain a constant phase relation with respect to the ship.

In shallow water, all gravity waves propagate at the same phase velocity. That is,

$$v_p = (g\,d)^{1/2},$$ (9.309)

where d is the water depth. Hence, Equation (9.308) yields

$$\beta = \sin^{-1}\left[\frac{(g\,d)^{1/2}}{V}\right].$$ (9.310)

This equation can only be satisfied when

$$V > (g\,d)^{1/2}.$$ (9.311)

In other words, the ship must be traveling faster than the critical speed $(g\,d)^{1/2}$. Moreover, if this i the case then there is only one value of β that satisfies Equation (9.310). This implies the scenari illustrated in Figure 9.7. Here, the ship is instantaneously at A, and the wave maxima that it previ ously generated—which all propagate obliquely, subtending a fixed angle β with the x-axis—hav interfered constructively to produce a single strong wave maximum DAE. In fact, the wave maxim generated when the ship was at A' have travelled to B' and C', the wave maxima generated when th ship was at A'' have travelled to B'' and C'', et cetera. We conclude that a ship traveling over shallo water produces a V-shaped wake whose semi-angle, β, is determined by the ship's speed. Indeed, a is apparent from Equation (9.310), the faster the ship travels over the water, the smaller the angle becomes. Shallow water wakes are especially dangerous to other vessels, and particularly destruc tive of the coastline, because all of the wave energy produced by the ship is concentrated into

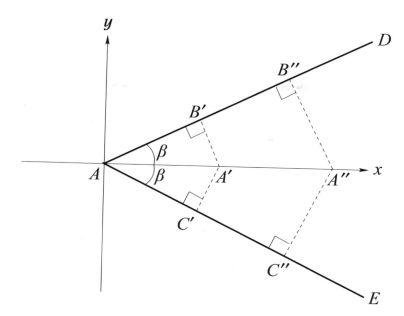

FIGURE 9.7   A shallow water wake.

single large wave maximum. The wake contains no transverse waves, because, as we have already mentioned, such waves cannot keep up with a ship traveling faster than the critical speed $(g\,d)^{1/2}$.

Let us now discuss the wake generated by a ship traveling over deep water. In this case, the phase velocity of gravity waves is $v_p = (g/k)^{1/2}$. Thus, Equation (9.308) yields

$$\sin\beta = \frac{v_p}{V} = \left(\frac{g}{k\,V^2}\right)^{1/2}.$$

(9.312)

It follows that in deep water any obliquely propagating gravity wave whose wavenumber exceeds the critical value

$$k_0 = \frac{g}{V^2}$$

(9.313)

can keep up with the ship, as long as its direction of propagation is such that Equation (9.312) is satisfied. In other words, the ship continuously excites gravity waves with a wide range of different wavenumbers and propagation directions. The wake is essentially the interference pattern generated by these waves. As described in Section 9.2, an interference maximum generated by the superposition of plane waves with a range of different wavenumbers propagates at the group velocity, v_g. Furthermore, as we have already seen, the group velocity of deep water gravity waves is half their phase velocity; that is, $v_g = v_p/2$.

Consider Figure 9.8. The curve APD corresponds to a particular interference maximum in the wake. Here, A is the ship's instantaneous position. Consider a point P on this curve. Let x and y be the coordinates of this point, relative to the ship. The interference maximum at P is part of the plane wavefront BC emitted some time t earlier, when the ship was at point A'. Let β be the angle subtended between this wavefront and the x-axis. Because interference maxima propagate at the group velocity, the distance $A'P$ is equal to $v_g\,t$. The distance AA' is equal to Vt. Simple trigonometry reveals that

$$x = Vt - v_g\,t\,\sin\beta,$$

(9.314)

$$y = v_g\,t\,\cos\beta.$$

(9.315)

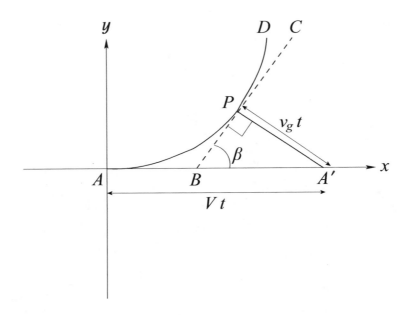

FIGURE 9.8   Formation of an interference maximum in a deep water wake.

Moreover,

$$\frac{dy}{dx} = \tan\beta, \tag{9.316}$$

because BC is the tangent to the curve APD—that is, the curve $y(x)$—at point P. It follows from Equation (9.312), and the fact that $v_g = v_p/2$, that

$$x = X\left(1 - \frac{1}{2}\sin^2\beta\right), \tag{9.317}$$

$$y = \frac{1}{2} X \sin\beta \cos\beta, \tag{9.318}$$

where $X(\beta) = V\,t$. The previous three equations can be combined to produce

$$\frac{dy}{dx} = \frac{dy/d\beta}{dx/d\beta} = \frac{(1/2)\,dX/d\beta\,\sin\beta\,\cos\beta + (1/2)\,X\,(\cos^2\beta - \sin^2\beta)}{dX/d\beta\,[1 - (1/2)\sin^2\beta] - X\,\sin\beta\,\cos\beta} = \frac{\sin\beta}{\cos\beta}, \tag{9.319}$$

which reduces to

$$\frac{dX}{d\beta} = \frac{X}{\tan\beta}. \tag{9.320}$$

This expression can be solved to give

$$X = X_0\,\sin\beta, \tag{9.321}$$

where X_0 is a constant. Hence, the locus of our interference maximum is determined parametrically by

$$x = X_0\,\sin\beta\left(1 - \frac{1}{2}\sin^2\beta\right), \tag{9.322}$$

$$y = \frac{1}{2} X_0\,\sin^2\beta\,\cos\beta. \tag{9.323}$$

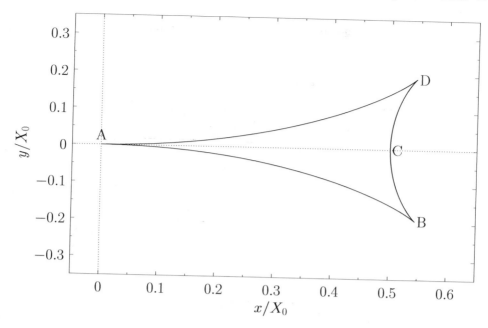

FIGURE 9.9 Locus of an interference maximum in a deep water wake.

Here, the angle β ranges from $-\pi/2$ to $+\pi/2$. The curve specified by the previous equations is plotted in Figure 9.9. As usual, A is the instantaneous position of the ship. It can be seen that the interference maximum essentially consists of the transverse maximum BCD, and the two radial maxima AB and AD. As is readily demonstrated, point C, which corresponds to $\beta = \pi/2$, lies at $x = X_0/2$, $y = 0$. Moreover, the two cusps, B and D, which correspond to $\beta = \pm \tan^{-1}\left(\sqrt{2}\right)$, lie at $x = (8/27)^{1/2} X_0$, $y = \pm(1/27)^{1/2} X_0$.

The complete interference pattern that constitutes the wake is constructed out of many different wave maximum curves of the form shown in Figure 9.9, corresponding to many different values of the parameter X_0. However, these X_0 values must be chosen such that the wavelength of the pattern along the x-axis corresponds to the wavelength $\lambda_0 = 2\pi/k_0 = 2\pi V^2/g$ of transverse (i.e., $\beta = \pi/2$) gravity waves whose phase velocity matches the speed of the ship. This implies that $X_0 = 2 j \lambda_0$, where j is a positive integer. A complete deep water wake pattern is shown in Figure 9.10. The pattern, which is made up of interlocking transverse and radial wave maxima, fills a wedge-shaped region—known as a *Kelvin wedge*—whose semi-angle takes the value $\tan^{-1}\left(1/\sqrt{8}\right) = 19.47°$. This angle is independent of the ship's speed. Finally, our initial assumption that the gravity waves that form the wake are all deep water waves is valid provided $k_0 d \gg 1$, which implies that

$$V \ll (g\,d)^{1/2}. \tag{9.324}$$

In other words, the ship must travel at a speed that is much less than the critical speed $(g\,d)^{1/2}$. This explains why the wake contains transverse wave maxima.

9.15 CAPILLARY WAVES

Water in contact with air actually possesses a finite *surface tension*, $T \simeq 7 \times 10^{-2}\,\mathrm{N\,m^{-1}}$ (Haynes and Lide 2011b), which allows there to be a small pressure discontinuity across a free surface that

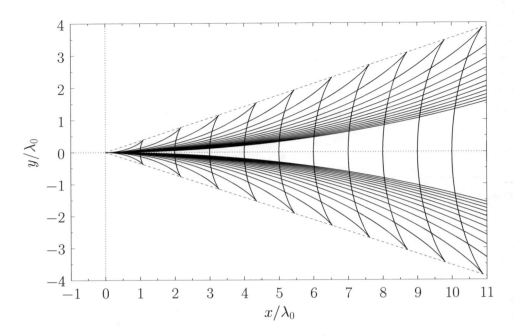

FIGURE 9.10   A deep water wake.

is curved. In fact,

$$[p]_{z=0_-}^{z=0_+} = T \frac{\partial^2 \zeta}{\partial x^2} \qquad (9.325)$$

(Batchelor 2000). Here, $(\partial^2 \zeta / \partial x^2)^{-1}$ is the *radius of curvature* of the surface. Thus, in the presence of surface tension, the boundary condition (9.276) takes the modified form

$$T \frac{\partial^2 \zeta}{\partial x^2} = \rho g \zeta - p_1|_{z=0}, \qquad (9.326)$$

which reduces to

$$\left. \frac{\partial \phi}{\partial z} \right|_{z=0} = \frac{T}{\rho g} \left. \frac{\partial^3 \phi}{\partial x^2 \partial z} \right|_{z=0} - \frac{1}{g} \left. \frac{\partial^2 \phi}{\partial t^2} \right|_{z=0}. \qquad (9.327)$$

This boundary condition can be combined with the solution (9.281), in the deep water limit $k\,d \gg 1$, to give the modified deep water dispersion relation (see Exercise 9.26)

$$\omega = \sqrt{g k + \frac{T}{\rho} k^3}. \qquad (9.328)$$

Hence, the phase velocity of the waves takes the form

$$v_p = \frac{\omega}{k} = \sqrt{\frac{g}{k} + \frac{T}{\rho} k}, \qquad (9.329)$$

and the ratio of the group velocity to the phase velocity can be shown to be

$$\frac{v_g}{v_p} = \frac{k}{\omega} \frac{d\omega}{dk} = \frac{1}{2} \left[\frac{1 + 3\,T\,k^2/(\rho\,g)}{1 + T\,k^2/(\rho\,g)} \right]. \qquad (9.330)$$

We conclude that the phase velocity of surface water waves attains a minimum value of $\sqrt{2}\,(g\,T/\rho)^{1/4} \sim 0.2\,\mathrm{m\,s^{-1}}$ when $k = k_0 \equiv (\rho\,g/T)^{1/2}$, which corresponds to $\lambda \sim 2\,\mathrm{cm}$. The group velocity equals the phase velocity at this wavelength. For long wavelength waves (i.e., $k \ll k_0$), gravity dominates surface tension, the phase velocity scales as $k^{-1/2}$, and the group velocity is half the phase velocity. As we have already mentioned, this type of wave is known as a gravity wave. On the other hand, for short wavelength waves (i.e., $k \gg k_0$), surface tension dominates gravity, the phase velocity scales as $k^{1/2}$, and the group velocity is $3/2$ times the phase velocity. This type of wave is known as a *capillary wave*. The fact that the phase velocity and the group velocity both attain minimum values when $\lambda \sim 2\,\mathrm{cm}$ means that when a wave disturbance containing a wide spectrum of wavelengths, such as might be generated by throwing a rock into the water, travels across the surface of a lake, and reaches the shore, the short and long wavelength components of the disturbance generally arrive before the components of intermediate wavelength.

EXERCISES

9.1 Demonstrate that the phase velocity of traveling waves on an infinitely long beaded string is

$$v_p = v_0\,\frac{\sin(k\,a/2)}{(k\,a/2)},$$

where $v_0 = \sqrt{T\,a/m}$, T is the tension in the string, a the spacing between the beads, m the mass of the beads, and k the wavenumber of the wave. What is the group velocity?

9.2 A uniform rope of mass per unit length ρ and length L hangs vertically. Determine the tension T in the rope as a function of height from the bottom of the rope. Show that the time required for a transverse wave pulse to travel from the bottom to the top of the rope is $2\sqrt{L/g}$.

9.3 Derive expressions (9.27) and (9.29) for propagating electromagnetic waves in a plasma from Equations (9.19) and (9.23)–(9.26).

9.4 Derive expressions (9.39) and (9.40) for evanescent electromagnetic waves in a plasma from Equations (9.24), (9.25), and (9.36)–(9.38).

9.5 Derive Equations (9.49)–(9.52) from Equations (9.47) and (9.48).

9.6 Derive Equations (9.60) and (9.61) from Equations (9.54)–(9.59).

9.7 Derive Equations (9.207) and (9.208) from Equations (9.203)–(9.206).

9.8 Derive Equations (9.216)–(9.219) from Equations (9.214) and (9.215), in the limit $\alpha \ll 1$.

9.9 A medium is such that the product of the phase and group velocities of electromagnetic waves is equal to c^2 at all wave frequencies, where c is the velocity of light in vacuum. Demonstrate that the dispersion relation for electromagnetic waves takes the form

$$\omega^2 = k^2\,c^2 \pm \omega_0^2,$$

where ω_0 is a constant.

9.10 The number density of free electrons in the ionosphere, n_e, as a function of vertical height, z, is measured by timing how long it takes a radio pulse launched vertically upward from the ground ($z = 0$) to return to ground level again, after reflection by the ionosphere, as a function of the pulse frequency, ω. It is conventional to define the *equivalent height*, $h(\omega)$, of the reflection layer as the height it would need to have above the ground if the pulse always traveled at the velocity of light in vacuum. Demonstrate that

$$h(\omega) = \int_0^{z_0} \frac{dz}{[1 - \omega_{pe}^2(z)/\omega^2]^{1/2}},$$

where $\omega_{pe}^2(z) = n_e(z)\,e^2/(\epsilon_0\,m_e)$, and $\omega_{pe}^2(z_0) = \omega^2$. Show that if $n_e \propto z^p$ then $h \propto \omega^{2/p}$.

9.11 Show that the general eigenmode equation, (9.168), yields the following dispersion relation for electromagnetic waves propagating through a magnetized plasma:

$$A\,n^4 - B\,n^2 + C- = 0,$$

where

$$A = S\,\sin^2\theta + P\,\cos^2\theta,$$
$$B = RL\,\sin^2\theta + PS\,(1 + \cos^2\theta),$$
$$C = PRL.$$

9.12 Show that the solution to the dispersion relation derived in the previous exercise can be written

$$n^2 = \frac{B \pm F}{2A},$$

where

$$F = (RL - PS)^2\,\sin^4\theta + 4\,P^2\,D^2\,\cos^2\theta.$$

Hence, deduce that n^2 is always real, and that n is either purely real, or purely imaginary. This implies that electromagnetic waves in (cold) magnetized plasmas either propagate without evanescence, or decay without spatial oscillation.

9.13 Show that the dispersion relation derived in Exercise 9.11 can also be written in the form

$$\tan^2\theta = -\frac{P\,(n^2 - R)\,(n^2 - L)}{(S\,n^2 - RL)\,(n^2 - P)}.$$

9.14 (a) Show that the cutoff frequencies for an electromagnetic wave propagating in a general direction through a magnetized plasma occur when

$$P = 0,$$

or

$$R = 0,$$

or

$$L = 0.$$

Hence, deduce that the cutoff frequencies are independent of the direction of wave propagation.

(b) Show that the resonant frequencies for an electromagnetic wave propagating in a general direction through a magnetized plasma occur when

$$\tan^2\theta = -\frac{P}{S}.$$

9.15 Consider an electromagnetic wave propagating in the positive z-direction through a conducting medium of conductivity σ. Suppose that the wave electric field is

$$E_x(z, t) = E_0\, e^{-z/d}\, \cos(\omega t - z/d),$$

where d is the skin-depth. Demonstrate that the mean electromagnetic energy flux across the plane $z = 0$ matches the mean rate at which electromagnetic energy is dissipated, per unit area, due to Joule heating in the region $z > 0$. [The rate of Joule heating per unit volume is σE_x^2 (Fitzpatrick 2008).]

9.16 The aluminum foil used in cooking has an electrical conductivity $\sigma = 3.5 \times 10^7\,(\Omega\,\mathrm{m})^{-1}$, and a typical thickness $\delta = 2 \times 10^{-4}$ m (Wikipedia contributors 2018). Show that such foil can be used to shield a region from electromagnetic waves of a given frequency, provided that the skin-depth of the waves in the foil is less than about a third of its thickness. Because skin-depth increases as frequency decreases, it follows that the foil can only shield waves whose frequency exceeds a critical value. Estimate this critical frequency (in hertz). What is the corresponding wavelength?

9.17 Consider a hollow, vacuum-filled, rectangular waveguide that runs parallel to the z-axis, and has perfectly conducting walls located at $x = 0$, a and $y = 0$, b. Maxwell's equations for a TE mode (which is characterized by $E_z = 0$) are (see Appendix C)

$$\frac{\partial E_x}{\partial t} = -\frac{1}{\epsilon_0}\left(\frac{\partial H_y}{\partial z} - \frac{\partial H_z}{\partial y}\right),$$

$$\frac{\partial E_y}{\partial t} = -\frac{1}{\epsilon_0}\left(\frac{\partial H_z}{\partial x} - \frac{\partial H_x}{\partial z}\right),$$

$$\frac{\partial H_x}{\partial t} = \frac{1}{\mu_0}\frac{\partial E_y}{\partial z},$$

$$\frac{\partial H_y}{\partial t} = -\frac{1}{\mu_0}\frac{\partial E_x}{\partial z},$$

$$\frac{\partial H_z}{\partial t} = \frac{1}{\mu_0}\left(\frac{\partial E_x}{\partial y} - \frac{\partial E_y}{\partial x}\right),$$

subject to the boundary conditions $E_y = 0$ at $x = 0$, a and $E_x = 0$ at $y = 0$, b. Show that the problem reduces to solving

$$\frac{\partial^2 H_z}{\partial t^2} = c^2\left(\frac{\partial^2}{\partial x^2} + \frac{\partial^2}{\partial y^2} + \frac{\partial^2}{\partial z^2}\right)H_z,$$

subject to the boundary conditions $\partial H_z/\partial x = 0$ at $x = 0$, a, and $\partial H_z/\partial y = 0$ at $y = 0$, b. Demonstrate that the various TE modes satisfy the dispersion relation

$$\omega^2 = k^2 c^2 + \omega_{m,n}^2,$$

where k is the z-component of the wavevector,

$$\omega_{m,n} = \pi c\left(\frac{m^2}{a^2} + \frac{n^2}{b^2}\right)^{1/2},$$

and m, n are non-negative integers, one of which must be non-zero.

9.18 Consider a hollow, vacuum-filled, rectangular waveguide that runs parallel to the z-axis, and has perfectly conducting walls located at $x = 0$, a and $y = 0$, b. Maxwell's equations for a TM mode (which is characterized by $H_z = 0$) are (see Appendix C)

$$\frac{\partial H_x}{\partial t} = \frac{1}{\mu_0}\left(\frac{\partial E_y}{\partial z} - \frac{\partial E_z}{\partial y}\right),$$

$$\frac{\partial H_y}{\partial t} = \frac{1}{\mu_0}\left(\frac{\partial E_z}{\partial x} - \frac{\partial E_x}{\partial z}\right),$$

$$\frac{\partial E_x}{\partial t} = -\frac{1}{\epsilon_0}\frac{\partial H_y}{\partial z},$$

$$\frac{\partial E_y}{\partial t} = \frac{1}{\epsilon_0}\frac{\partial H_x}{\partial z},$$

$$\frac{\partial E_z}{\partial t} = -\frac{1}{\epsilon_0}\left(\frac{\partial H_x}{\partial y} - \frac{\partial H_y}{\partial x}\right),$$

subject to the boundary conditions $E_y = 0$ at $x = 0$, a, $E_x = 0$ at $y = 0$, b, and $E_z = 0$ at $x = 0$, a and $y = 0$, b. Show that the problem reduces to solving

$$\frac{\partial^2 E_z}{\partial t^2} = c^2\left(\frac{\partial^2}{\partial x^2} + \frac{\partial^2}{\partial y^2} + \frac{\partial^2}{\partial z^2}\right)E_z,$$

subject to the boundary conditions $E_z = 0$ at $x = 0$, a and $y = 0$, b. Demonstrate that the various TM modes satisfy the dispersion relation

$$\omega^2 = k^2 c^2 + \omega_{m,n}^2,$$

where k is the z-component of the wavevector,

$$\omega_{m,n} = \pi c\left(\frac{m^2}{a^2} + \frac{n^2}{b^2}\right)^{1/2},$$

and m, n are positive integers.

9.19 Deduce that, for a hollow, vacuum-filled, rectangular waveguide, the mode with the lowest cutoff frequency is a TE mode.

9.20 Consider a vacuum-filled rectangular waveguide of internal dimensions 5×10 cm. What is the frequency (in MHz) of the lowest frequency TE mode that will propagate along the waveguide without attenuation? What are the phase and group velocities (expressed as multiples of c) of this mode when its frequency is $5/4$ times the cutoff frequency?

9.21 A wave pulse propagates in the x-z plane through an inhomogeneous medium with the linear dispersion relation

$$\omega = k v,$$

where

$$v(z) = v_0 - v_1 z.$$

Here, v_0 and v_1 are positive constants. Show that if $k_z = 0$ at $z = 0$ then the equations motion of the pulse can be written

$$\frac{dx}{ds} = \frac{v}{v_0},$$

$$\frac{dz}{ds} = \left(1 - \frac{v^2}{v_0^2}\right),$$

$$\frac{d^2x}{ds^2} = -\frac{v_1}{v_0}\frac{dz}{ds},$$

$$\frac{d^2z}{ds^2} = \frac{v_1}{v_0}\frac{dx}{ds},$$

where $s = vt$ denotes path-length. Hence, deduce that the pulse travels in the arc of a circle, of radius

$$R = \frac{v_0}{v_1},$$

whose center lies at $z = R$.

9.22 The speed of sound in the atmosphere decreases approximately linearly with increasing altitude (at relatively low altitude) due to an approximately linear decrease in the temperature of the atmosphere with height. Assuming that the sound speed varies with altitude, z, above the Earth's surface as

$$v(z) = v_0 - v_1 z,$$

where v_0 and v_1 are positive constants, show that sound generated by a source located a height $h \ll v_0/v_1$ above the ground is refracted upward by the atmosphere such that it never reaches ground level at points whose radial distances from the point lying directly beneath the source exceed the value

$$r = \left(\frac{2 v_0 h}{v_1}\right)^{1/2}.$$

This effect is known as *acoustic shadowing*.

9.23 A low amplitude sinusoidal gravity wave travels through shallow water of gradually decreasing depth d toward the shore. Assuming that the wave travels at right-angles to the shoreline, show that its wavelength and vertical amplitude vary as $\lambda \propto d^{1/2}$ and $a \propto d^{-1/4}$, respectively.

9.24 Demonstrate that a small amplitude gravity wave, of angular frequency ω and wavenumber k, traveling over the surface of a lake of uniform depth d causes an individual water volume element located at a depth h below the surface to execute a non-propagating elliptical orbit whose major and minor axes are horizontal and vertical, respectively. Show that the variation of the major and minor radii of the orbit with depth is $A \cosh[k(d-h)]$ and $A \sinh[k(d-h)]$, respectively, where A is a constant. Demonstrate that the volume elements are moving horizontally in the same direction as the wave at the top of their orbits, and in the opposite direction at the bottom. Show that a gravity wave traveling over the surface of a very deep lake causes water volume elements to execute non-propagating circular orbits whose radii decrease exponentially with depth.

9.25 Water fills a rectangular tank of length l and breadth b to a depth d. Show that the resonant frequencies of the water are

$$\omega_{m,n} = [g\, k_{m,n}\, \tanh(k_{m,n}\, d)]^{1/2}$$

where

$$k_{m,n} = \pi \left(\frac{m^2}{l^2} + \frac{n^2}{b^2} \right)^{1/2},$$

and n, m are non-negative integers that are not both zero. Neglect surface tension.

9.26 Derive the dispersion relation (9.328), and show that it generalizes to

$$\omega^2 = \left(gk + \frac{T}{\rho} k^3 \right) \tanh(kd)$$

in water of arbitrary depth.

9.27 Show that in water of uniform depth d the phase velocity of surface waves can only attain a stationary (i.e., maximum or minimum) value as a function of wavenumber, k, when

$$k = \left[\frac{\sinh(2kd) - 2kd}{\sinh(2kd) + 2kd} \right]^{1/2} k_c,$$

where $k_c = (\rho g/T)^{1/2}$. Hence, deduce that the phase velocity has just one stationary value (a minimum) for any depth greater than $3^{1/2} k_c^{-1} \simeq 4.8$ mm, but no stationary values for lesser depths.

9.28 Unlike gravity waves in deep water, whose group velocities are half their phase velocities, the group velocities of capillary waves are 3/2 times their phase velocities. Adapt the analysis of Section 9.14 to investigate the generation of capillary waves by a very small object traveling across the surface of the water at the constant speed V. Suppose that the unperturbed surface corresponds to the x-y plane. Let the object travel in the minus x-direction, such that it is instantaneously found at the origin. Find the present position of waves that were emitted with wavefronts traveling at an angle β to the object's direction of motion (see Figure 9.6), when it was located at $(X, 0)$. Show that along a given interference maximum the quantities X and β vary in such a manner that $X \sin^3 \beta$ takes a constant value, X_0 (say). Deduce that the interference maximum is given parametrically by the equations

$$x = \frac{X_0}{\sin^3 \beta} \left(1 - \frac{3}{2} \sin^2 \beta \right),$$

$$y = \frac{3}{2} X_0 \frac{\cos \beta}{\sin^2 \beta}.$$

Sketch the pattern of capillary waves generated by the object. [Modified from Lighthill (1978).]

CHAPTER 10

Wave Optics

10.1 INTRODUCTION

Visible light is a type of electromagnetic radiation whose wavelength lies in a relatively narrow band extending from about 400 to 700 nm. The area of physics that is devoted to the study of light is known as *optics*. This chapter is concerned with those optical phenomena that depend explicitly on the ultimate wave nature of light, and cannot be accounted for using the well-known laws of geometric optics. (See Section 7.8.) The branch of optics that deals with such phenomena is called *wave optics*. The two most important physical phenomena that are encountered in wave optics are *interference* and *diffraction*. Interference occurs when beams of light from multiple sources (but with similar frequencies), or multiple beams from the same source, intersect one another. Diffraction takes place, for instance, when a single beam of light passes through an opening in an opaque screen whose spatial extent is comparable to the wavelength of the light. Actually, interference and diffraction depend on the same underlying physics, and the distinction that is conventionally made between them is somewhat arbitrary.

10.2 TWO-SLIT INTERFERENCE

Consider a monochromatic plane light wave, propagating in the z-direction, through a transparent dielectric medium of refractive index unity (e.g., a vacuum). (Such a wave could be produced by a uniform line source, running parallel to the y-axis (say), that is located at $z = -\infty$.) Let the associated wavefunction take the form

$$\psi(z, t) = \psi_0 \cos(\omega t - kz - \phi). \tag{10.1}$$

Here, $\psi(z, t)$ represents the electric component of the wave, $\psi_0 > 0$ the wave amplitude, ϕ the phase angle, $k > 0$ the wavenumber, $\omega = kc$ the angular frequency, and c the velocity of light in vacuum. Let the wave be normally incident on an opaque screen that is coincident with the plane $z = 0$. See Figure 10.1. Suppose that there are two identical slits of width δ cut in the screen. Let the slits run parallel to the y-axis, and be located at $x = d/2$ and $x = -d/2$, where $d > \delta$ is the slit spacing. Suppose that the light that passes through the two slits travels to a cylindrical projection screen of radius R whose axis coincides with the line $x = z = 0$. In the following, it is assumed that there is no variation of wave quantities in the y-direction.

Provided the two slits are much narrower than the wavelength, $\lambda = 2\pi/k$, of the light (i.e., $\delta \ll \lambda$), we expect any radiation that passes through them to be strongly diffracted. (See Section 10.7.) Diffraction is a fundamental wave phenomenon that causes waves to bend around small (compared to the wavelength) obstacles, and spread out from narrow (compared to the wavelength) openings, while maintaining the same wavelength and frequency. The laws of geometric optics do not take diffraction into account, and are, therefore, restricted to situations in which light interacts

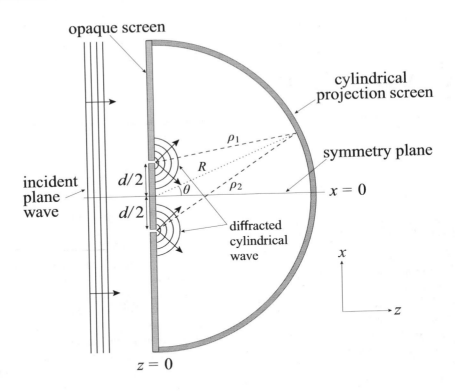

FIGURE 10.1 Two-slit interference at normal incidence.

with objects whose physical dimensions greatly exceed its wavelength. The assumption of strong diffraction suggests that each slit acts like a uniform line source that emits light isotropically in the forward direction (i.e., toward the region $z > 0$), but does not emit light in the backward direction (i.e., toward the region $z < 0$). [It is possible to demonstrate that this is, in fact, the case (with certain provisos; see Section 10.10), using electromagnetic theory (Jackson 1975), but such a demonstration lies beyond the scope of this course.] As discussed in Section 7.4, we would expect a uniform line source to emit a cylindrical wave. It follows that each slit emits a half-cylindrical light wave in the forward direction. See Figure 10.1. Moreover, these waves are emitted with equal amplitude and phase, because the incident plane wave that illuminates the slits has the same amplitude (i.e., ψ_0) and phase (i.e., $\omega t - \phi$) at both slits, and the slits are identical. Finally, we expect the cylindrical waves emitted by the two slits to interfere with one another (see Section 6.4) in such a manner as to generate a characteristic pattern on the projection screen. Let us determine the nature of this pattern.

Consider the wave amplitude at a point on the projection screen that lies an angular distance θ from the plane $x = 0$. See Figure 10.1. The wavefunction at this particular point is written

$$\psi(\theta, t) \propto \frac{\cos(\omega t - k\rho_1 - \phi)}{\rho_1^{1/2}} + O\left(\frac{1}{k\rho_1^{3/2}}\right) + \frac{\cos(\omega t - k\rho_2 - \phi)}{\rho_2^{1/2}} + O\left(\frac{1}{k\rho_2^{3/2}}\right), \tag{10.2}$$

assuming that $k\rho_1, k\rho_2 \gg 1$. In other words, the overall wavefunction in the region $z > 0$ is the superposition of cylindrical waves [see Equation (7.11)] of equal amplitude (i.e., $\rho^{-1/2}$) and phase (i.e., $\omega t - k\rho - \phi$) emanating from each slit. Here, $\rho = (x^2 + z^2)^{1/2}$. Moreover, ρ_1 and ρ_2 are the distances that the waves emitted by the first and second slits (located at $x = d/2$ and $x = -d/2$, respectively) have travelled by the time they reach the point on the projection screen in question.

Standard trigonometry (i.e., the law of cosines) reveals that

$$\rho_1 = R\left(1 - \frac{d}{R}\sin\theta + \frac{1}{4}\frac{d^2}{R^2}\right)^{1/2} = R\left[1 - \frac{1}{2}\frac{d}{R}\sin\theta + O\left(\frac{d^2}{R^2}\right)\right]. \tag{10.3}$$

Likewise,

$$\rho_2 = R\left[1 + \frac{1}{2}\frac{d}{R}\sin\theta + O\left(\frac{d^2}{R^2}\right)\right]. \tag{10.4}$$

Hence, Equation (10.2) yields

$$\psi(\theta, t) \propto \cos(\omega t - k\rho_1 - \phi) + \cos(\omega t - k\rho_2 - \phi) + O\left(\frac{1}{kR}\right) + O\left(\frac{d}{R}\right), \tag{10.5}$$

which, making use of the trigonometric identity $\cos x + \cos y \equiv 2\,\cos[(x+y)/2]\,\cos[(x-y)/2]$ (see Appendix B), gives

$$\psi(\theta, t) \propto \cos\left[\omega t - \frac{1}{2}k(\rho_1 + \rho_2) - \phi\right]\cos\left[\frac{1}{2}k(\rho_2 - \rho_1)\right] + O\left(\frac{1}{kR}\right) + O\left(\frac{d}{R}\right), \tag{10.6}$$

or

$$\psi(\theta, t) \propto \cos\left[\omega t - kR - \phi + O\left(\frac{kd^2}{R}\right)\right]\cos\left[\frac{1}{2}kd\sin\theta + O\left(\frac{kd^2}{R}\right)\right] + O\left(\frac{1}{kR}\right) + O\left(\frac{d}{R}\right). \tag{10.7}$$

Finally, assuming that

$$\frac{kd^2}{R}, \frac{1}{kR}, \frac{d}{R} \ll 1, \tag{10.8}$$

the previous expression reduces to

$$\psi(\theta, t) \propto \cos(\omega t - kR - \phi)\cos\left(\frac{1}{2}kd\sin\theta\right). \tag{10.9}$$

The orderings (10.8), which can also be written in the form

$$R \gg d, \lambda, \frac{d^2}{\lambda}, \tag{10.10}$$

are satisfied provided the projection screen is located sufficiently far away from the slits. Consequently, the type of interference described in this section is known as *far-field interference*. One characteristic feature of far-field interference is that the amplitudes of the cylindrical waves emitted by the two slits are approximately equal to one another when they reach a given point on the projection screen (i.e., $|\rho_1 - \rho_2|/\rho_1 \ll 1$), whereas the phases are, in general, significantly different (i.e., $k|\rho_1 - \rho_2| \gtrsim \pi$). In other words, the interference pattern generated on the projection screen is entirely a consequence of the phase difference between the cylindrical waves emitted by the two slits when they reach the screen. This phase difference is produced by the slight difference in the distance between the slits and a given point on the projection screen. (The phase difference becomes significant as soon as the path difference becomes comparable with the wavelength of the light.)

The mean energy flux, or intensity, of the light striking the projection screen at angular position θ is

$$I(\theta) \propto \langle\psi^2(\theta, t)\rangle \propto \langle\cos^2(\omega t - kR - \phi)\rangle \cos^2\left(\frac{1}{2}kd\sin\theta\right) \propto \cos^2\left(\frac{1}{2}kd\sin\theta\right), \tag{10.11}$$

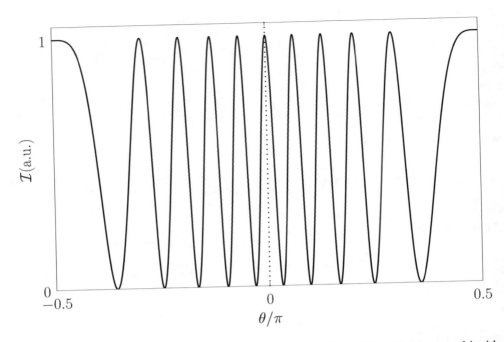

FIGURE 10.2 Two-slit far-field interference pattern calculated for $d/\lambda = 5$ with normal incidence and narrow slits.

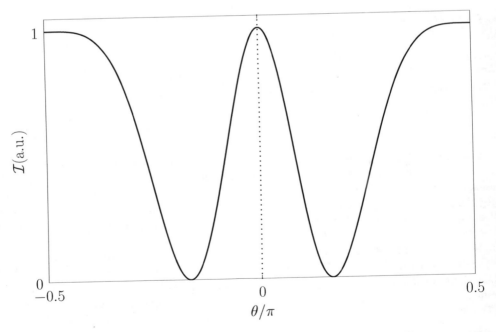

FIGURE 10.3 Two-slit far-field interference pattern calculated for $d/\lambda = 1$ with normal incidence and narrow slits.

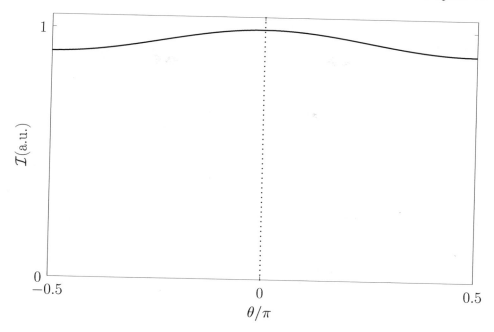

FIGURE 10.4   Two-slit far-field interference pattern calculated for $d/\lambda = 0.1$ with normal incidence and narrow slits.

where $\langle \cdots \rangle$ denotes an average over a wave period. [The previous expression follows from the standard result $\mathcal{I} \propto E^2/Z_0$, for an electromagnetic wave, where E is the electric component of the wave, and Z_0 the impedance of free space. (See Section 6.8.) Recall, also, that $\psi \propto E$.] Here, use has been made of the easily established result $\langle \cos^2(\omega t - kR - \phi) \rangle = 1/2$. The very high oscillation frequency of light waves (i.e., $f \sim 10^{14}$ Hz) ensures that experiments typically detect (e.g., by means of a photographic film, or a photo-multiplier tube) the intensity of light, rather than the rapidly oscillating amplitude of its electric component. For the case of two-slit, far-field interference, assuming normal incidence and narrow slits, the intensity of the characteristic interference pattern appearing on the projection screen is specified by

$$\mathcal{I}(\theta) \propto \cos^2\left(\pi \frac{d}{\lambda} \sin\theta\right). \tag{10.12}$$

Figure 10.2 shows the intensity of the typical two-slit, far-field interference pattern produced when the slit spacing, d, greatly exceeds the wavelength, λ, of the light. It can be seen that the pattern consists of multiple bright and dark fringes. A bright fringe is generated whenever the cylindrical waves emitted by the two slits interfere constructively at a given point on the projection screen. This occurs when the distances between the two slits and the point in question differ by an integer number of wavelengths; that is,

$$\rho_2 - \rho_1 = d \sin\theta = j\lambda, \tag{10.13}$$

where j is an integer. (This ensures that the effective phase difference of the two waves is zero.) Likewise, a dark fringe is generated whenever the cylindrical waves emitted by the two slits interfere destructively at a given point on the projection screen. This occurs when the distances between the two slits and the point in question differ by a half-integer number of wavelengths; that is,

$$\rho_2 - \rho_1 = d \sin\theta = (j + 1/2)\lambda. \tag{10.14}$$

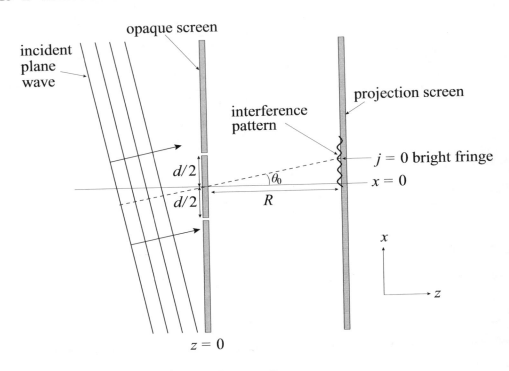

FIGURE 10.5 Two-slit interference at oblique incidence.

(This ensures that the effective phase difference between the two waves is π radians.) We conclude that the innermost (i.e., low j, small θ) bright fringes are approximately equally spaced, with a characteristic angular width $\Delta\theta \simeq \lambda/d$. This result, which follows from Equation (10.13), and the small-angle approximation $\sin\theta \simeq \theta$, can be used experimentally to determine the wavelength of a monochromatic light source from a two-slit interference apparatus. (See Exercise 10.5.)

Figure 10.3 shows the intensity of the interference pattern generated when the slit spacing is equal to the wavelength of the light. It can be seen that the width of the central (i.e., $j = 0$, $\theta = 0$) bright fringe has expanded to such an extent that the fringe occupies almost half of the projection screen, leaving room for just two dark fringes on either side of it.

Finally, Figure 10.4 shows the intensity of the interference pattern generated when the slit spacing is much less than the wavelength of the light. It can be seen that the width of the central bright fringe has expanded to such an extent that the band occupies the whole projection screen, and there are no dark fringes. Indeed, $I(\theta)$ becomes constant in the limit that $d/\lambda \ll 1$, in which case the interference pattern entirely disappears.

Figures 10.2–10.3 imply that the two-slit, far-field interference apparatus shown in Figure 10. only generates an interesting interference pattern when the slit spacing, d, is greater than the wave length, λ, of the light.

Suppose that the plane wave that illuminates the interference apparatus is not normally incider on the slits, but instead propagates at an angle θ_0 to the z-axis, as shown in Figure 10.5. In this case the incident wavefunction (10.1) becomes

$$\psi(x, z, t) = \psi_0 \cos(\omega t - kx \sin\theta_0 - kz \cos\theta_0 - \phi). \tag{10.1$}$$

Thus, the phase of the light incident on the first slit (located at $x = d/2$, $z = 0$) is ωt $(1/2) kd \sin\theta_0 - \phi$, whereas the phase of the light incident on the second slit (located at $x = -d/$

$z = 0$) is $\omega t + (1/2) kd \sin \theta_0 - \phi$. Assuming that the cylindrical waves emitted by each slit have the same phase (at the slits) as the plane wave that illuminates them, Equation (10.2) generalizes to

$$\psi(\theta, t) \propto \frac{\cos(\omega t - k\rho_1 - \phi_1)}{\rho_1^{1/2}} + \frac{\cos(\omega t - k\rho_2 - \phi_2)}{\rho_2^{1/2}}, \tag{10.16}$$

where $\phi_1 = (1/2) kd \sin \theta_0 + \phi$ and $\phi_2 = -(1/2) kd \sin \theta_0 + \phi$. Hence, making use of the far-field orderings (10.10), and a standard trigonometric identity, we obtain

$$\psi(\theta, t) \propto \cos\left[\omega t - kR - \frac{1}{2}(\phi_1 + \phi_2)\right] \cos\left[\frac{1}{2} kd \sin \theta - \frac{1}{2}(\phi_1 - \phi_2)\right]$$

$$\propto \cos(\omega t - kR - \phi) \cos\left[\frac{1}{2} kd (\sin \theta - \sin \theta_0)\right]. \tag{10.17}$$

For the sake of simplicity, let us consider the limit $d \gg \lambda$, in which the innermost (i.e., low j) interference fringes are located at small θ. (The projection screen is approximately planar in this limit, as indicated in Figure 10.5, because a sufficiently small section of a cylindrical surface looks like a plane.) Assuming that θ_0 is also small, the previous expression reduces to

$$\psi(\theta, t) \propto \cos(\omega - kR - \phi) \cos\left[\frac{1}{2} kd (\theta - \theta_0)\right], \tag{10.18}$$

and Equation (10.12) becomes

$$I(\theta) \propto \cos^2\left[\pi \frac{d}{\lambda} (\theta - \theta_0)\right]. \tag{10.19}$$

Thus, the bright fringes in the interference pattern are located at

$$\theta = \theta_0 + j \frac{\lambda}{d}, \tag{10.20}$$

where j is an integer. We conclude that if the slits in a two-slit interference apparatus, such as that shown in Figure 10.5, are illuminated by an obliquely incident plane wave then the consequent phase difference between the cylindrical waves emitted by each slit produces an angular shift in the interference pattern appearing on the projection screen. To be more exact, the angular shift is equal to the angle of incidence, θ_0, of the plane wave, so that the central ($j = 0$) bright fringe in the interference pattern is located at $\theta = \theta_0$. See Figure 10.5. This is equivalent to saying that the position of the central bright fringe can be determined via the rules of geometric optics. (This conclusion holds even when θ_0 is not small.)

10.3 COHERENCE

A practical monochromatic light source consists of a collection of similar atoms that are continually excited by collisions, and then spontaneously decay back to their electronic ground states, in the process emitting photons of characteristic angular frequency $\omega = \Delta\mathcal{E}/\hbar$, where $\Delta\mathcal{E}$ is the difference in energy between the excited state and the ground state, and $\hbar = 1.055 \times 10^{-34}\,\mathrm{J\,s^{-1}}$ is Planck's constant divided by 2π (Hecht and Zajac 1974). An excited electronic state of an atom has a characteristic lifetime, τ, which can be calculated from quantum mechanics, and is typically 10^{-8} s (ibid.). It follows that when an atom in an excited state decays back to its ground state it emits a burst of electromagnetic radiation of duration τ and angular frequency ω. However, according to the bandwidth theorem (see Section 8.5), a sinusoidal wave of finite duration τ has the finite bandwidth

$$\Delta\omega \sim \frac{2\pi}{\tau}. \tag{10.21}$$

In other words, if the emitted wave is Fourier transformed in time then it will be found to consist of a linear superposition of sinusoidal waves of infinite duration whose frequencies lie in the approximate range $\omega - \Delta\omega/2$ to $\omega + \Delta\omega/2$. We conclude that there is no such thing as a truly monochromatic light source. In reality, all such sources have small, but finite, bandwidths that are inversely proportional to the lifetimes, τ, of the associated excited atomic states.

How do we take the finite bandwidth of a practical "monochromatic" light source into account in our analysis? In fact, all we need to do is to assume that the phase angle, ϕ, appearing in Equations (10.1) and (10.15), is only constant on timescales much less that the lifetime, τ, of the associated excited atomic state, and is subject to abrupt random changes on timescales much greater than τ. We can understand this phenomenon as being due to the fact that the radiation emitted by a single atom has a fixed phase angle, ϕ, but only lasts a finite time period, τ, combined with the fact that there is generally no correlation between the phase angles of the radiation emitted by different atoms. Alternatively, we can account for the variation in the phase angle in terms of the finite bandwidth of the light source. To be more exact, because the light emitted by the source consists of a superposition of sinusoidal waves of frequencies extending over the range $\omega - \Delta\omega/2$ to $\omega + \Delta\omega/2$, even if all the component waves start off in phase, the phases will be completely scrambled after a time period $2\pi/\Delta\omega = \tau$ has elapsed. In effect, what we are saying is that a practical monochromatic light source is temporally coherent on timescales much less than its characteristic *coherence time*, τ (which, for visible light, is typically of order 10^{-8} seconds), and temporally incoherent on timescales much greater than τ. Incidentally, two waves are said to be *coherent* if their phase difference is constant in time, and *incoherent* if their phase difference varies significantly in time. In this case, the two waves in question are the same wave observed at two different times.

What effect does the temporal incoherence of a practical monochromatic light source on timescales greater than $\tau \sim 10^{-8}$ seconds have on the two-slit interference patterns discussed in the previous section? Consider the case of oblique incidence. According to Equation (10.16), the phase angles, $\phi_1 = (1/2) k d \sin \theta_0 + \phi$, and $\phi_2 = -(1/2) k d \sin \theta_0 + \phi$, of the cylindrical waves emitted by each slit are subject to abrupt random changes on timescales much greater than τ, because the phase angle, ϕ, of the plane wave that illuminates the two slits is subject to identical changes. Nevertheless, the relative phase angle, $\phi_1 - \phi_2 = k d \sin \theta_0$, between the two cylindrical waves remains constant. Moreover, according to Equation (10.17), the interference pattern appearing on the projection screen is produced by the phase difference $(1/2) k d \sin \theta - (1/2) (\phi_1 - \phi_2)$ between the two cylindrical waves at a given point on the screen, and this phase difference only depends on the relative phase angle. Indeed, the intensity of the interference pattern is $I(\theta) \propto \cos^2[(1/2) k d \sin \theta - (1/2) (\phi_1 - \phi_2)]$. Hence, the fact that the relative phase angle, $\phi_1 - \phi_2$, between the two cylindrical waves emitted by the slits remains constant on timescales much longer than the characteristic coherence time, τ, of the light source implies that the interference pattern generated in a conventional two-slit interference apparatus is unaffected by the temporal incoherence of the source. Strictly speaking, however, the preceding conclusion is only accurate when the spatial extent of the light source is negligible. Let us now broaden our discussion to take spatially extended light sources into account.

Up until now, we have assumed that our two-slit interference apparatus is illuminated by a single plane wave, such as might be generated by a line source located at infinity. Let us now consider a more realistic situation in which the light source is located a finite distance from the slits, and also has a finite spatial extent. Figure 10.6 shows the simplest possible case. Here, the slits are illuminated by two identical line sources, A and B, that are a distance D apart, and a perpendicular distance L from the opaque screen containing the slits. Assuming that $L \gg D, d$, the light incident on the slits from source A is effectively a plane wave whose direction of propagation subtends an angle $\theta_0/2 \simeq D/(2 L)$ with the z-axis. Likewise, the light incident on the slits from source B is a plane wave whose direction of propagation subtends an angle $-\theta_0/2$ with the z-axis. Moreover, the net interference pattern (i.e., wavefunction) appearing on the projection screen is the linear

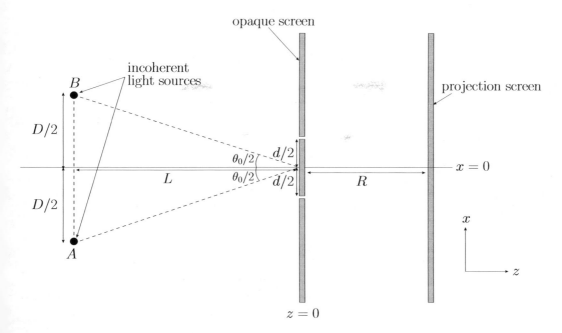

FIGURE 10.6 Two-slit interference with two line sources.

superposition of the patterns generated by each source taken individually (because light propagation is ultimately governed by a linear wave equation with superposable solutions; see Section 7.3.). Let us determine whether these patterns reinforce, or interfere with, one another.

The light emitted by source A has a phase angle, ϕ_A, that is constant on timescales much less than the characteristic coherence time of the source, τ, but is subject to abrupt random changes on timescale much longer than τ. Likewise, the light emitted by source B has a phase angle, ϕ_B, that is constant on timescales much less than τ, and varies significantly on timescales much greater than τ. In general, there is no correlation between ϕ_A and ϕ_B. In other words, our composite light source, consisting of the two line sources A and B, is both temporally and spatially incoherent on timescales much longer than τ.

Again working in the limit $d \gg \lambda$, with $\theta, \theta_0 \ll 1$, Equation (10.18) yields the following expression for the wavefunction at the projection screen:

$$\psi(\theta, t) \propto \cos(\omega t - kR - \phi_A) \cos\left[\frac{1}{2} k d (\theta - \theta_0/2)\right]$$

$$+ \cos(\omega t - kR - \phi_B) \cos\left[\frac{1}{2} k d (\theta + \theta_0/2)\right]. \tag{10.22}$$

Hence, the intensity of the interference pattern is

$$I(\theta) \propto \langle \psi^2(\theta, t) \rangle \propto \langle \cos^2(\omega t - kR - \phi_A) \rangle \cos^2 \left[\frac{1}{2} k d (\theta - \theta_0/2) \right]$$

$$+ 2 \langle \cos(\omega t - kR - \phi_A) \cos(\omega t - kR - \phi_B) \rangle$$

$$\times \cos \left[\frac{1}{2} k d (\theta - \theta_0/2) \right] \cos \left[\frac{1}{2} k d (\theta + \theta_0/2) \right]$$

$$+ \langle \cos^2(\omega t - kR - \phi_B) \rangle \cos^2 \left[\frac{1}{2} k d (\theta + \theta_0/2) \right]. \qquad (10.23)$$

However, $\langle \cos^2(\omega t - kR - \phi_A) \rangle = \langle \cos^2(\omega t - kR - \phi_B) \rangle = 1/2$, and $\langle \cos(\omega t - kR - \phi_A) \cos(\omega t - kR - \phi_B) \rangle = 0$, because the phase angles ϕ_A and ϕ_B are uncorrelated. Hence, the previous expression reduces to

$$I(\theta) \propto \cos^2 \left[\frac{1}{2} k d (\theta - \theta_0/2) \right] + \cos^2 \left[\frac{1}{2} k d (\theta + \theta_0/2) \right]$$

$$= 1 + \cos \left(2\pi \frac{d}{\lambda} \theta \right) \cos \left(\pi \frac{d}{\lambda} \theta_0 \right), \qquad (10.24)$$

where use has been made of the trigonometric identities $\cos^2 \theta \equiv (1 + \cos 2\theta)/2$, and $\cos x + \cos y \equiv 2 \cos[(x + y)/2] \cos[(x - y)/2]$. (See Appendix B.) If $\theta_0 = \lambda/(2d)$ then $\cos[\pi (d/\lambda) \theta_0] = 0$ and $I(\theta) \propto 1$. In this case, the bright fringes of the interference pattern generated by source A exactly overlay the dark fringes of the pattern generated by source B, and vice versa, and the net interference pattern is completely washed out. On the other hand, if $\theta_0 \ll \lambda/d$ then $\cos[\pi (d/\lambda) \theta_0] \simeq 1$ and $I(\theta) \propto 1 + \cos[2\pi (d/\lambda) \theta] = 2 \cos^2[\pi (d/\lambda) \theta]$. In this case, the two interference patterns reinforce one another, and the net interference pattern is the same as that generated by a light source of negligible spatial extent.

Suppose that our light source consists of a regularly spaced array of very many identical incoherent line sources, filling the region between the sources A and B in Figure 10.6. In other words, suppose that our light source is a uniform incoherent source of angular extent θ_0. As is readily demonstrated, the associated interference pattern is obtained by averaging expression (10.24) over all θ_0 values in the range 0 to θ_0; that is, by operating on this expression with $\theta_0^{-1} \int_0^{\theta_0} \cdots d\theta_0$. In this manner, we obtain

$$I(\theta) \propto 1 + \cos \left(2\pi \frac{d}{\lambda} \theta \right) \mathrm{sinc} \left(\pi \frac{d}{\lambda} \theta_0 \right), \qquad (10.25)$$

where $\mathrm{sinc}(x) \equiv \sin x/x$. We can conveniently parameterize the *visibility* of the interference pattern, appearing on the projection screen, in terms of the quantity

$$V = \frac{I_{max} - I_{min}}{I_{max} + I_{min}}, \qquad (10.26)$$

where the maximum and minimum values of the intensity are taken with respect to variation in θ (rather than θ_0). Thus, $V = 1$ corresponds to a sharply defined pattern, and $V = 0$ to a pattern that is completely washed out. It follows from Equation (10.25) that

$$V = \left| \mathrm{sinc} \left(\pi \frac{d}{\lambda} \theta_0 \right) \right|. \qquad (10.27)$$

The predicted visibility, V, of a two-slit interference pattern generated by an extended incoherent

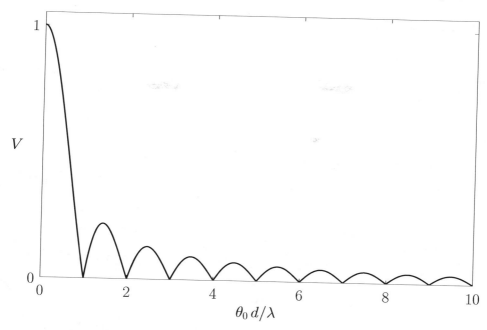

FIGURE 10.7   Visibility of a two-slit far-field interference pattern generated by an extended incoherent light source.

light source is plotted as a function of the angular extent, θ_0, of the source in Figure 10.7. It can be
seen that the pattern is highly visible (i.e., $V \sim 1$) when $\theta_0 \ll \lambda/d$, but becomes washed out (i.e.,
$V \sim 0$) when $\theta_0 \gtrsim \lambda/d$.

   We conclude that a spatially extended incoherent light source only generates a visible interference pattern in a conventional two-slit interference apparatus when the angular extent of the source
is sufficiently small that

$$\theta_0 \ll \frac{\lambda}{d}. \tag{10.28}$$

Equivalently, if the source is of linear extent D, and located a distance L from the slits, then the
source only generates a visible interference pattern when it is sufficiently far away from the slits
that

$$L \gg \frac{dD}{\lambda}. \tag{10.29}$$

This follows because $\theta_0 \simeq D/L$.

   The whole of the preceding discussion is premised on the assumption that an extended light
source is both temporally and spatially incoherent on timescales much longer than a typical atomic
coherence time, which is about 10^{-8} seconds. This is generally the case. However, there is one type
of light source—namely, a *laser*—for which this is not necessarily the case. In a laser (in single-
mode operation), excited atoms are stimulated in such a manner that they emit radiation that is both
temporally and spatially coherent on timescales much longer than the relevant atomic coherence
time.

   Let us consider the two-slit far-field interference pattern generated by an extended coherent
light source of angular extent θ_0. In this case, as is readily demonstrated (see Exercise 10.2), Equa-

tion (10.25) is replaced by

$$\mathcal{I}(\theta) \propto \cos^2\left(\pi \frac{d}{\lambda}\theta\right) \text{sinc}^2\left(\frac{\pi}{2}\frac{d}{\lambda}\theta_0\right). \tag{10.30}$$

It follows, from Equation (10.26), that the visibility of the interference pattern is unity; that is, the pattern is sharply defined, irrespective of the angular extent of the light source. (However, the overall brightness of the pattern is considerably reduced when $\theta_0 \gtrsim \lambda/d$.) It follows that lasers generally produce much clearer interference patterns than conventional incoherent light sources.

10.4 MULTI-SLIT INTERFERENCE

Suppose that the interference apparatus pictured in Figure 10.1 is modified such that N identical slits of width $\delta \ll \lambda$, running parallel to the y-axis, are cut in the opaque screen that occupies the plane $z = 0$. Let the slits be located at $x = x_n$, for $n = 1, N$. For the sake of simplicity, the arrangement of slits is assumed to be symmetric with respect to the plane $x = 0$. In other words, if there is a slit at $x = x_n$ then there is also a slit at $x = -x_n$. The distance between the nth slit and a point on the projection screen that is an angular distance θ from the plane $x = 0$ is [cf., Equation (10.3)]

$$\rho_n = R\left[1 - \frac{x_n}{R}\sin\theta + O\left(\frac{x_n^2}{R^2}\right)\right]. \tag{10.31}$$

Thus, making use of the far-field orderings (10.10), where d now represents the typical spacing between neighboring slits, and assuming normally incident collimated light, Equation (10.5) generalizes to

$$\psi(\theta, t) \propto \sum_{n=1,N} \cos(\omega t - kR - \phi + k x_n \sin\theta), \tag{10.32}$$

which can also be written

$$\psi(\theta, t) \propto \cos(\omega t - kR - \phi)\sum_{n=1,N}\cos(k x_n \sin\theta)\sin(\omega t - kR - \phi)\sum_{n=1,N}\sin(k x_n \sin\theta), \tag{10.33}$$

or

$$\psi(\theta, t) \propto \cos(\omega t - kR - \phi)\sum_{n=1,N}\cos(k x_n \sin\theta). \tag{10.34}$$

Here, we have made use of the fact that arrangement of slits is symmetric with respect to the plane $x = 0$ [which implies that $\sum_{n=1,N}\sin(k x_n \sin\theta) = 0$]. We have also employed the trigonometric identity $\cos(x - y) \equiv \cos x \cos y + \sin x \sin y$. (See Appendix B.) It follows that the intensity of the interference pattern appearing on the projection screen is specified by

$$\mathcal{I}(\theta) \propto \langle\psi^2(\theta, t)\rangle \propto \left[\sum_{n=1,N}\cos\left(2\pi\frac{x_n}{\lambda}\sin\theta\right)\right]^2, \tag{10.35}$$

because $\langle\cos^2(\omega t - kR - \phi)\rangle = 1/2$. The previous expression is a generalization of Equation (10.12)
Suppose that the slits are evenly spaced a distance d apart, so that

$$x_n = [n - (N + 1)/2]\,d \tag{10.36}$$

for $n = 1, N$. It follows that

$$\mathcal{I}(\theta) \propto \left[\sum_{n=1,N}\cos\left(2\pi[n - (N+1)/2]\frac{d}{\lambda}\sin\theta\right)\right]^2, \tag{10.37}$$

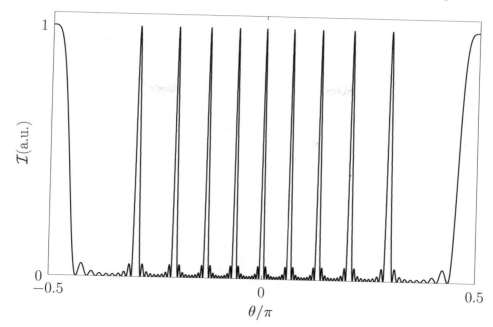

FIGURE 10.8 Multi-slit far-field interference pattern calculated for $N = 10$ and $d/\lambda = 5$ with normal incidence and narrow slits.

which can be summed to give

$$I(\theta) \propto \frac{\sin^2[\pi N (d/\lambda) \sin \theta]}{\sin^2[\pi (d/\lambda) \sin \theta)]}. \tag{10.38}$$

(See Exercise 10.1.)

The multi-slit interference function, (10.38), exhibits strong maxima in situations in which its numerator and denominator are simultaneously zero; that is, when

$$\sin \theta = j \frac{\lambda}{d}, \tag{10.39}$$

where j is an integer. In this situation, application of l'Hopital's rule yields $I = N^2$. The heights of these so-called *principal maxima* in the interference function are very large, being proportional to N^2, because there is constructive interference of the light from all N slits. This occurs because the distances between neighboring slits and the point on the projection screen at which a given maximum is located differ by an integer number of wavelengths; that is, $\rho_n - \rho_{n-1} = d \sin \theta = j\lambda$. All of the principal maxima have the same height.

The multi-slit interference function (10.38) is zero when its numerator is zero, but its denominator non-zero; that is, when

$$\sin \theta = \frac{l}{N} \frac{\lambda}{d}, \tag{10.40}$$

where l is an integer that is not an integer multiple of N. It follows that there are $N-1$ zeros between neighboring principal maxima. It can also be demonstrated that there are $N - 2$ *secondary maxima* between neighboring principal maxima. However, these maxima are much lower in height, by a factor of order N^2, than the primary maxima.

Figure 10.8 shows the typical far-field interference pattern produced by a system of ten identical, equally spaced, parallel slits, assuming normal incidence and narrow slits, when the slit spacing,

d, greatly exceeds the wavelength, λ, of the light (which, as we saw in Section 10.2, is the most interesting case). It can be seen that the pattern consists of a series of bright fringes of equal intensity, separated by much wider (relatively) dark fringes. The bright fringes correspond to the principal maxima discussed previously. As is the case for two-slit interference, the innermost (i.e., low j, small θ) principal maxima are approximately equally spaced, with a characteristic angular spacing $\Delta\theta \simeq \lambda/d$. [This result follows from Equation (10.39), and the small-angle approximation $\sin\theta \simeq \theta$.] However, the typical angular width of a principal maximum (i.e., the angular distance between the maximum and the closest zeroes on either side of it) is $\delta\theta \simeq (1/N)(\lambda/d)$. [This result follows from Equation (10.40), and the small-angle approximation.] The ratio of the angular width of a principal maximum to the angular spacing between successive maxima is thus

$$\frac{\delta\theta}{\Delta\theta} \simeq \frac{1}{N}. \tag{10.41}$$

Hence, we conclude that, as the number of slits increases, the bright fringes in a multi-slit interference pattern become progressively sharper.

The most common practical application of multi-slit interference is the *transmission diffraction grating*. Such a device consists of N identical, equally spaced, parallel scratches on one side of a thin uniform transparent glass, or plastic, film. When the film is illuminated, the scratches strongly scatter the incident light, and effectively constitute N identical, equally spaced, parallel line sources. Hence, the grating generates the type of N-slit interference pattern discussed previously, with one major difference. Namely, the central ($j = 0$) principal maximum has contributions not only from the scratches, but also from all the transparent material between the scratches. Thus, the central principal maximum is considerably brighter than the other ($j \neq 0$) principal maxima.

Diffraction gratings are often employed in *spectroscopes*, which are instruments used to decompose light that is made up of a mixture of different wavelengths into its constituent wavelengths. As a simple example, suppose that a spectroscope contains an N-line diffraction grating that is illuminated, at normal incidence, by a mixture of light of wavelength λ, and light of wavelength $\lambda + \Delta\lambda$ where $\Delta\lambda \ll \lambda$. As always, the overall interference pattern (i.e., the overall wavefunction at the projection screen) produced by the grating is a linear superposition of the pattern generated by the light of wavelength λ, and the pattern generated by the light of wavelength $\lambda + \Delta\lambda$. Consider the jth-order principal maximum associated with the wavelength λ interference pattern, which is located at θ_j where $\sin\theta_j = j(\lambda/d)$. [See Equation (10.39).] Here, d is the spacing between neighboring lines on the diffraction grating, which is assumed to be greater than λ. (Incidentally, the width of the line is assumed to be much less than λ.) The maximum in question has a finite angular width. We can determine this width by locating the zeros in the interference pattern on either side of the maximum. Let the zeros be located at $\theta_j \pm \delta\theta_j$. The maximum itself corresponds to $\pi N (d/\lambda)\sin\theta_j = \pi N j$. Hence, the zeros correspond to $\pi N (d/\lambda)\sin(\theta_j \pm \delta\theta_j) = \pi (N j \pm 1)$ (i.e., they correspond to the first zeros of the function $\sin[\pi N (d/\lambda)\sin\theta]$ on either side of the zero at θ_j.) [See Equation (10.38)] Taylor expanding to first order in $\delta\theta_j$, we obtain

$$\delta\theta_j = \frac{\tan\theta_j}{N j}. \tag{10.42}$$

Hence, the maximum in question effectively extends from $\theta_j - \delta\theta_j$ to $\theta_j + \delta\theta_j$. Consider the jth-order principal maximum associated with the wavelength $\lambda + \Delta\lambda$ interference pattern, which is located at $\theta_j + \Delta\theta_j$, where $\sin(\theta_j + \Delta\theta_j) = j(\lambda + \Delta\lambda)/d$. [See Equation (10.39).] Taylor expanding to first order in $\Delta\theta_j$, we obtain

$$\Delta\theta_j = \tan\theta_j \frac{\Delta\lambda}{\lambda}. \tag{10.43}$$

In order for the spectroscope to resolve the incident light into its two constituent wavelengths,

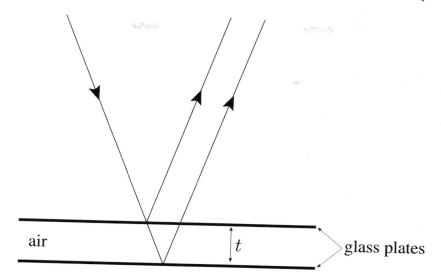

FIGURE 10.9   Interference of light due to a thin film of air trapped between two pieces of glass.

the jth spectral order, the angular spacing, $\Delta\theta_j$, between the jth-order maxima associated with these two wavelengths must be greater than the angular widths, $\delta\theta_j$, of the maxima themselves. If this is the case then the overall jth-order maximum will consist of two closely spaced maxima, or "spectral lines" (centered at θ_j and $\theta_j + \Delta\theta_j$). On the other hand, if this is not the case then the two maxima will merge to form a single maximum, and it will consequently not be possible to tell that the incident light consists of a mixture of two different wavelengths. Thus, the condition for the spectroscope to be able to resolve the spectral lines at the jth spectral order is $\Delta\theta_j > \delta\theta_j$, or

$$\frac{\Delta\lambda}{\lambda} > \frac{1}{N\,j}. \tag{10.44}$$

We conclude that the resolving power of a diffraction grating spectroscope increases as the number of illuminated lines (i.e., N) increases, and also as the spectral order (i.e., j) increases. Incidentally, there is no resolving power at the lowest (i.e., $j = 0$) spectral order, because the corresponding principal maximum is located at $\theta = 0$ irrespective of the wavelength of the incident light. Moreover, there is a limit to how large j can become (i.e., a given diffraction grating, illuminated by light of a given wavelength, has a finite number of principal maxima). This follows because $\sin\theta_j$ cannot exceed unity, so, according to Equation (10.39), j cannot exceed d/λ.

10.5   THIN-FILM INTERFERENCE

In everyday life, the most common manifestation of interference occurs when light impinges on a thin film of some transparent material. For instance, the brilliant colors seen in soap bubbles, in oil films floating on puddles of water, and in the feathers of a peacock's tail, are due to this type of interference.

Suppose that a very thin film of air is trapped between two pieces of glass, as shown in Figure 10.9. If monochromatic light is normally (or almost normally) incident on the film then some of the light is reflected from the interface between the bottom of the upper plate and the air, and some is reflected from the interface between the air and the top of the lower plate. These two reflected light rays interfere either destructively or constructively with one another.

Let t be the thickness of the air film. The difference in distance traveled between the two light rays shown in the figure is $\delta = 2\,t$. This difference introduces a phase difference $\Delta\phi = k\,\delta = 2\pi\,\delta/\lambda$ between the rays. Naively, we might expect that constructive interference would occur when $\Delta\phi = m\,2\pi$, where m is an integer, and destructive interference would occur when $\Delta\phi = (m + 1/2)\,2\pi$. However, this is not the case, because an additional phase difference is introduced between the rays on reflection. The first ray is reflected at an interface between an optically dense medium (glass) and a less dense medium (air). There is no phase change on reflection from such an interface. (See Section 6.8.) The second ray is reflected at an interface between an optically less dense medium (air) and a dense medium (glass). There is a π radian phase change on reflection from such an interface. (See Section 6.8.) Thus, an additional π radian phase change is introduced between the two rays, which is equivalent to an additional path difference of $\lambda/2$. When this additional phase change is taken into account, the condition for constructive interference becomes

$$2\,t = (m + 1/2)\,\lambda, \tag{10.45}$$

where m is an integer. Similarly, the condition for destructive interference becomes

$$2\,t = m\,\lambda. \tag{10.46}$$

For white light, the previous criteria yield constructive interference for some wavelengths, and destructive interference for others. Thus, the light reflected back from the film exhibits those colors for which constructive interference occurs.

If the thin film consists of water, oil, or some other transparent material, of refractive index n then the results are basically the same as those for an air film, except that the wavelength of the light in the film is reduced from λ (the vacuum wavelength) to λ/n. It follows that the modified criteria for constructive and destructive interference are

$$2\,n\,t = (m + 1/2)\,\lambda, \tag{10.47}$$

and

$$2\,n\,t = m\,\lambda, \tag{10.48}$$

respectively.

10.6   ONE-DIMENSIONAL FOURIER OPTICS

We have already considered the interference of monochromatic light produced when a plane wave is incident on an opaque screen, coincident with the plane $z = 0$, in which a number of narrow (i.e., $\delta \ll \lambda$, where δ is the slit width) slits, running parallel to the y-axis, have been cut. Let us now generalize our analysis to take slits of finite width (i.e., $\delta \gtrsim \lambda$) into account. In order to achieve this goal, it is convenient to define the so-called *aperture function*, $F(x)$, of the screen. This function takes the value zero if the screen is opaque at position x, and some constant positive value if it is transparent, and is normalized such that $\int_{-\infty}^{\infty} F(x)\,dx = 1$. Thus, for the case of a screen with N identical slits of negligible width, located at $x = x_n$, for $n = 1, N$, the appropriate aperture function is

$$F(x) = \frac{1}{N} \sum_{n=1,N} \delta(x - x_n), \tag{10.49}$$

where $\delta(x)$ is a Dirac delta function.

The wavefunction at the projection screen, generated by the previously mentioned arrangement of slits, when the opaque screen is illuminated by a plane wave of phase angle ϕ, wavenumber k, and angular frequency ω, whose direction of propagation subtends an angle θ_0 with the z-axis, is (see the analysis in Sections 10.2 and 10.4)

$$\psi(\theta, t) \propto \cos(\omega t - kR - \phi) \sum_{n=1,N} \cos[k \, x_n \, (\sin \theta - \sin \theta_0)]. \tag{10.50}$$

Here, for the sake of simplicity, we have assumed that the arrangement of slits is symmetric with respect to the plane $x = 0$, so that $\sum_{n=1,N} \sin(\alpha \, x_n) = 0$ for any α. Using the well-known properties of the delta function [see Equation (8.26)], Equation (10.50) can also be written

$$\psi(\theta, t) \propto \cos(\omega t - kR - \phi) \, \bar{F}(\theta), \tag{10.51}$$

where

$$\bar{F}(\theta) = \int_{-\infty}^{\infty} F(x) \cos[k \, (\sin \theta - \sin \theta_0) \, x] \, dx. \tag{10.52}$$

In the following, we shall assume that Equation (10.51) is a general result, and is valid even when the slits in the opaque screen are of finite width (i.e., $\delta \gtrsim \lambda$). This assumption is equivalent to the assumption that each unblocked section of the screen emits a cylindrical wave in the forward direction that is in phase with the plane wave which illuminates it from behind. The latter assumption is known as *Huygen's principle*. [Huygen's principle can be justified, with certain provisos (see Section 10.10), using electromagnetic theory (Jackson 1975), but such a proof lies beyond the scope of this course.] The interference/diffraction function, $\bar{F}(\theta)$, is the Fourier transform of the aperture function, $F(x)$. This is an extremely powerful result. It implies that we can calculate the far-field interference/diffraction pattern associated with any arrangement of parallel slits, of arbitrary width, by Fourier transforming the associated aperture function. Once we have determined the interference/diffraction function, $\bar{F}(\theta)$, the intensity of the interference/diffraction pattern appearing on the projection screen is readily obtained from

$$\mathcal{I}(\theta) \propto \left[\bar{F}(\theta) \right]^2. \tag{10.53}$$

10.7 SINGLE-SLIT DIFFRACTION

Suppose that the opaque screen contains a single slit of finite width. In fact, let the slit in question be of width δ, and extend from $x = -\delta/2$ to $x = \delta/2$. The associated aperture function is

$$F(x) = \begin{cases} 1/\delta & |x| \leq \delta/2 \\ 0 & |x| > \delta/2 \end{cases}. \tag{10.54}$$

It follows from Equation (10.52) that

$$\bar{F}(\theta) = \frac{1}{\delta} \int_{-\delta/2}^{\delta/2} \cos[k \, (\sin \theta - \sin \theta_0) \, x] \, dx = \text{sinc}\left[\pi \frac{\delta}{\lambda} (\sin \theta - \sin \theta_0) \right], \tag{10.55}$$

where $\text{sinc}(x) \equiv \sin(x)/x$. Finally, assuming, for the sake of simplicity, that $\theta, \theta_0 \ll 1$, which is most likely to be the case when $\delta \gg \lambda$, the diffraction pattern appearing on the projection screen is specified by

$$\mathcal{I}(\theta) \propto \text{sinc}^2\left[\pi \frac{\delta}{\lambda} (\theta - \theta_0) \right]. \tag{10.56}$$

According to L'Hopital's rule, $\text{sinc}(0) = \lim_{x \to 0} \sin x/x = \lim_{x \to 0} \cos x/1 = 1$. Furthermore, it is easily demonstrated that the zeros of the $\text{sinc}(x)$ function occur at $x = j\pi$, where j is a non-zero integer.

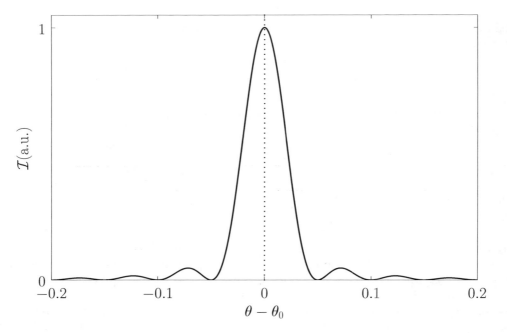

FIGURE 10.10 Single-slit far-field diffraction pattern calculated for $\delta/\lambda = 20$.

Figure 10.10 shows a typical single-slit diffraction pattern calculated for a case in which the slit width greatly exceeds the wavelength of the light. The pattern consists of a dominant central maximum, flanked by subsidiary maxima of fairly negligible amplitude. The situation is shown schematically in Figure 10.11. When the incident plane wave, whose direction of propagation subtends an angle θ_0 with the z-axis, passes through the slit it is effectively transformed into a divergent beam of light (the beam corresponds to the central peak in Figure 10.10) that is centered on $\theta = \theta_0$. The angle of divergence of the beam, which is obtained from the first zero of the single-slit diffraction function (10.56), is

$$\delta\theta = \frac{\lambda}{\delta}; \qquad (10.57)$$

that is, the beam effectively extends from $\theta_0 - \delta\theta$ to $\theta_0 + \delta\theta$. Thus, if the slit width, δ, is very much greater than the wavelength, λ, of the light then the beam divergence is negligible, and the beam is, thus, governed by the laws of geometric optics (according to which there is no beam divergence). On the other hand, if the slit width is comparable with the wavelength of the light then the beam divergence is significant, and the behavior of the beam is, consequently, very different to that predicted by the laws of geometric optics.

The diffraction of light is an important physical phenomenon because it sets a limit on the angular resolution of optical instruments. For instance, consider a telescope whose objective lens is of diameter D. When a plane wave from a distant light source of negligible angular extent (e.g., a star) enters the lens it is diffracted, and forms a divergent beam of angular width λ/D. Thus, instead of being a point, the resulting image of the star is a disk of finite angular width λ/D. (See Section 10.9.) Suppose that two stars are an angular distance $\Delta\theta$ apart in the sky. As we have just seen, when viewed through the telescope, each star appears as a disk of angular extent $\delta\theta = \lambda/D$. If $\Delta\theta > \delta\theta$ then the two stars appear as separate disks. On the other hand, if $\Delta\theta < \delta\theta$ then the two disks merge to form a single disk, and it becomes impossible to tell that there are, in fact, two stars. It follows that the minimum angular resolution of a telescope whose objective lens is of diameter D

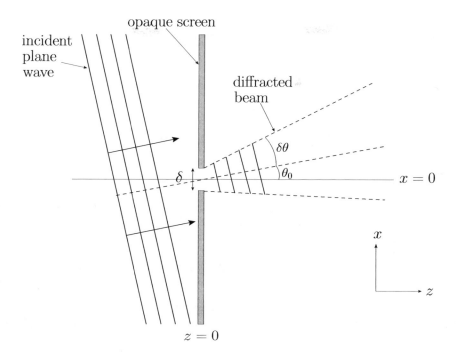

FIGURE 10.11 Single-slit diffraction at oblique incidence.

is

$$\delta\theta \simeq \frac{\lambda}{D}. \tag{10.58}$$

This result is called the *Rayleigh criterion*. (See Section 10.9 for a more accurate version of this criterion.) It can be seen that the angular resolution of the telescope increases (i.e., $\delta\theta$ decreases) as the diameter of its objective lens increases.

10.8 MULTI-SLIT DIFFRACTION

Suppose that the opaque screen in our interference/diffraction apparatus contains N identical, equally spaced, parallel slits of finite width. Let the slit spacing be d, and the slit width δ, where $\delta < d$. It follows that the aperture function for the screen is written

$$F(x) = \frac{1}{N} \sum_{n=1,N} F_2(x - x_n), \tag{10.59}$$

where

$$x_n = [n - (N + 1)/2]\, d, \tag{10.60}$$

and

$$F_2(x) = \begin{cases} 1/\delta & |x| \leq \delta/2 \\ 0 & |x| > \delta/2 \end{cases}. \tag{10.61}$$

We recognize $F_2(x)$ as the aperture function for a single slit, of finite width δ, that is centered on $x = 0$. [See Equation (10.54).]

Assuming normal incidence (i.e., $\theta_0 = 0$), the interference/diffraction function, which is the Fourier transform of the aperture function, takes the form [see Equation (10.52)]

$$\bar{F}(\theta) = \int_{-\infty}^{\infty} F(x) \cos(k \sin \theta x) \, dx. \tag{10.62}$$

Hence,

$$\bar{F}(\theta) = \frac{1}{N} \sum_{n=1,N} \int_{-\infty}^{\infty} F_2(x - x_n) \cos(k \sin \theta x) \, dx$$

$$= \frac{1}{N} \sum_{n=1,N} \left[\cos(k \sin \theta x_n) \int_{-\infty}^{\infty} F_2(x') \cos(k \sin \theta x') \, dx' \right.$$

$$\left. - \sin(k \sin \theta x_n) \int_{-\infty}^{\infty} F_2(x') \sin(k \sin \theta x') \, dx' \right]$$

$$= \left[\frac{1}{N} \sum_{n=1,N} \cos(k \sin \theta x_n) \right] \int_{-\infty}^{\infty} F_2(x') \cos(k \sin \theta x') \, dx', \tag{10.63}$$

where $x' = x - x_n$. Here, we have made use of the result $\int_{-\infty}^{\infty} F_2(x') \sin(\alpha x') \, dx' = 0$, for any α, which follows because $F_2(x')$ is even in x', whereas $\sin(\alpha x')$ is odd. We have also employed the trigonometric identity $\cos(x - y) \equiv \cos x \cos y - \sin x \sin y$. (See Appendix B.) The previous expression reduces to

$$\bar{F}(\theta) = \bar{F}_1(\theta) \bar{F}_2(\theta). \tag{10.64}$$

Here [cf., Equation (10.38)],

$$\bar{F}_1(\theta) = \int_{-\infty}^{\infty} F_1(x) \cos(k \sin \theta x) \, dx = \frac{1}{N} \sum_{n=1,N} \cos(k \sin \theta x_n) = \frac{1}{N} \frac{\sin[\pi N (d/\lambda) \sin \theta]}{\sin[\pi (d/\lambda) \sin \theta]}, \tag{10.65}$$

is the interference/diffraction function for N identical parallel slits of negligible width that are equally spaced a distance d apart, and

$$F_1(x) = \frac{1}{N} \sum_{n=1,N} \delta(x - x_n), \tag{10.66}$$

is the corresponding aperture function. Furthermore [cf., Equation (10.55)],

$$\bar{F}_2(\theta) = \int_{-\infty}^{\infty} F_2(x) \cos(k \sin \theta x) \, dx = \mathrm{sinc}\left[\pi \frac{\delta}{\lambda} \sin \theta \right], \tag{10.67}$$

is the interference/diffraction function for a single slit of width δ.

We conclude, from the preceding analysis, that the interference/diffraction function for N identical, equally spaced, parallel slits of finite width is the product of the interference/diffraction function for N identical, equally spaced, parallel slits of negligible width, $\bar{F}_1(\theta)$, and the interference/diffraction function for a single slit of finite width, $\bar{F}_2(\theta)$. We have already encountered both of these functions. The former function (see Figure 10.8, which shows $[\bar{F}_1(\theta)]^2$) consists of a series of sharp maxima of equal amplitude located at [see Equation (10.39)]

$$\theta_j = \sin^{-1}\left(j \frac{\lambda}{d} \right), \tag{10.68}$$

where j is an integer. The latter function (see Figure 10.10, which shows $[\bar{F}_2(\theta - \theta_0)]^2$) is of order

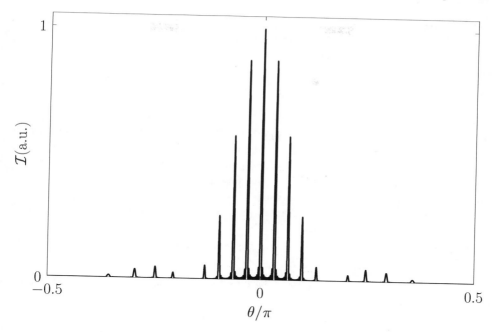

FIGURE 10.12   Multi-slit far-field interference pattern calculated for $N = 10$, $d/\lambda = 10$, and $\delta/\lambda = 2$, assuming normal incidence.

unity for $|\theta| \lesssim \sin^{-1}(\lambda/\delta)$, and much less than unity for $|\theta| \gtrsim \sin^{-1}(\lambda/\delta)$. It follows that the interference/diffraction pattern associated with N identical, equally spaced, parallel slits of finite width, which is given by

$$\mathcal{I}(\theta) \propto \left[\bar{F}_1(\theta)\,\bar{F}_2(\theta)\right]^2 \propto \left[\bar{F}_1(\theta)\right]^2 \left[\bar{F}_2(\theta)\right]^2, \qquad (10.69)$$

is similar to that for N identical, equally spaced, parallel slits of negligible width, $[\bar{F}_1(\theta)]^2$, except that the intensities of the various maxima in the pattern are modulated by $[\bar{F}_2(\theta)]^2$. Hence, those maxima lying in the angular range $|\theta| < \sin^{-1}(\lambda/\delta)$ are of similar intensity, whereas those lying in the range $|\theta| > \sin^{-1}(\lambda/\delta)$ are of negligible intensity. This is illustrated in Figure 10.12, which shows the multi-slit interference/diffraction pattern calculated for $N = 10$, $d/\lambda = 10$, and $\delta/\lambda = 2$. As expected, the maxima lying in the angular range $|\theta| < \sin^{-1}(0.5) = \pi/6$ have relatively large intensities, whereas those lying in the range $|\theta| > \pi/6$ have negligibly small intensities.

10.9   TWO-DIMENSIONAL FOURIER OPTICS

Consider a monochromatic plane light wave, propagating in the z-direction, which is normally incident on an opaque screen that occupies the plane $z = 0$. Suppose that there is an irregularly shaped aperture cut in the screen, and that the light that passes through this aperture travels to a flat projection screen occupying the plane $z = R$. We wish to determine the interference/diffraction pattern that appears on the projection screen, as a function of the size and shape of the aperture. We can achieve this goal by employing a modified version of Huygen's principle. In Section 10.6, we divided the aperture into infinitesimal parallel strips of equal height, and argued that each strip emits a cylindrical wave in the forward direction that has the same amplitude and phase as the light that illuminates it from behind. For the case of an irregularly shaped aperture, we can similarly divide the aperture into infinitesimal squares of equal area, and argue that each square emits a spherical wave in the forward direction that has the same amplitude and phase as the light that illuminates it from behind. [Recall that line sources and point sources emit cylindrical and spherical waves, respectively. (See Sections 7.4 and 7.5.)] Using analogous arguments to those employed in Section 10.2, we can show that provided

$$R \gg d, \frac{d^2}{\lambda}, \qquad (10.70)$$

where d is the characteristic size of the aperture, and λ is the wavelength of the light that illuminates it from behind, we are in the far-field limit. In this limit, the interference/diffraction pattern appearing on the projection screen is entirely due to the phase differences between the spherical waves that travel to a given point on the screen from different parts of the aperture. These phase differences are produced by the slightly different distances traveled by these waves.

Consider a spherical wave emitted from the point $(x, y, 0)$ in the aperture that travels to the point (x', y', R) on the projection screen. The distance traveled by this wave is

$$r = \left[(x - x')^2 + (y - y')^2 + R^2\right]^{1/2} = R\left[1 + \frac{1}{2}(\theta_x^2 + \theta_y^2) - \theta_x \frac{x}{R} - \theta_y \frac{y}{R} + O\left(\frac{d^2}{R^2}\right)\right], \qquad (10.71)$$

where $\theta_x = x'/R$ and $\theta_y = y'/R$ are the angular coordinates of the point on the projection screen. Here, we are assuming that

$$\frac{d}{R} \ll |\theta_x|, |\theta_y| \ll 1. \qquad (10.72)$$

The wavefunction at the point θ_x, θ_y on the projection screen is a linear superposition of the spherical waves that travel to it from all parts of the aperture. It follows that

$$\psi(\theta_x, \theta_y, t) \propto \int_{-\infty}^{\infty} \int_{-\infty}^{\infty} F(x, y) \cos(\omega t - k r - \phi) \, dx \, dy. \qquad (10.73)$$

Here, $F(x, y)$ is an aperture function that takes the value $1/A$ in the unblocked parts of the opaque screen, where A is the aperture area, and the value zero in the blocked parts. We have neglected the r^{-1} variation of the wave amplitudes (see Section 7.5) because we are working in the far-field limit. It follows, from Equations (10.71) and (10.73), that

$$\psi(\theta_x, \theta_y, t) \propto \int_{-\infty}^{\infty} \int_{-\infty}^{\infty} F(x, y) \cos\left[\omega t - k R\left(1 + \theta_x^2/2 + \theta_y^2/2\right) - \phi + k \theta_x x + k \theta_y y\right] dx \, dy$$

$$\propto \cos(\omega t - k R - \phi) \int_{-\infty}^{\infty} \int_{-\infty}^{\infty} F(x, y) \cos(k \theta_x x + k \theta_y y) \, dx \, dy$$

$$- \sin(\omega t - k R - \phi) \int_{-\infty}^{\infty} \int_{-\infty}^{\infty} F(x, y) \sin(k \theta_x x + k \theta_y y) \, dx \, dy$$

$$\propto \cos(\omega t - k R - \phi) \int_{-\infty}^{\infty} \int_{-\infty}^{\infty} F(x, y) \cos(k \theta_x x + k \theta_y y) \, dx \, dy. \qquad (10.74)$$

Here, we have made use of the fact that $|\theta_x|, |\theta_y| \ll 1$, and have also assumed that the aperture function is an even function of both x and y (i.e., the aperture is symmetric about $x = 0$ and $y = 0$). The intensity of the light at the point θ_x, θ_y on the projection screen is thus

$$I(\theta_x, \theta_y) \propto \langle \psi^2(\theta_x, \theta_y, t) \rangle \propto \left[\int_{-\infty}^{\infty} \int_{-\infty}^{\infty} F(x, y) \cos(k \theta_x x + k \theta_y y) \, dx \, dy\right]^2. \qquad (10.75)$$

Consider a rectangular aperture that occupies the region $-a/2 < x < a/2$ and $-b/2 < y < b/2$ in

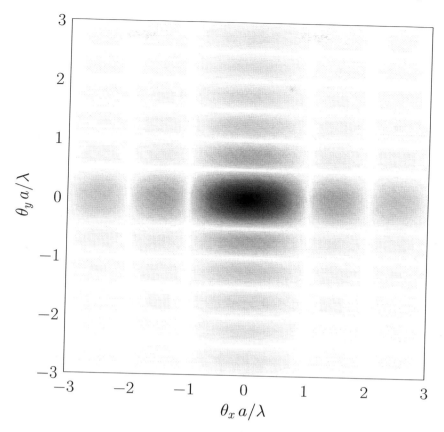

FIGURE 10.13 Far-field interference/diffraction pattern produced by a rectangular aperture for which $b = 2a$. Dark regions indicate high light intensity.

the x-y plane. The intensity of the interference/diffraction pattern on the projection screen produced by such an aperture is

$$I(\theta_x, \theta_y) \propto \left[\int_{-b/2}^{b/2} \int_{-a/2}^{a/2} \cos(k\,\theta_x\,x + k\,\theta_y\,y)\,\frac{dx}{a}\,\frac{dy}{b} \right]^2, \qquad (10.76)$$

which yields

$$I(\theta_x, \theta_y) \propto \operatorname{sinc}^2\!\left(\pi\,\frac{a}{\lambda}\,\theta_x\right) \operatorname{sinc}^2\!\left(\pi\,\frac{b}{\lambda}\,\theta_y\right). \qquad (10.77)$$

This pattern is shown in Figure 10.13. It consists of a strong central maximum, that extends over the region $|\theta_x| < \lambda/a$ and $|\theta_y| < \lambda/b$, surrounded by much weaker secondary maxima arranged on a rectangular grid. Incidentally, our assumption that $|\theta_x|, |\theta_y| \ll 1$ is only self-consistent provided $\lambda \ll a, b$ (i.e., provided the wavelength of the light is much less than the dimensions of the aperture).

Consider a circular aperture that occupies the region $(x^2 + y^2)^{1/2} < a$ in the x-y plane. Let $x = \rho \cos\phi$, $y = \rho \sin\phi$, $\theta_x = \theta \cos\phi'$, and $\theta_y = \theta \sin\phi'$. The intensity of the interference/diffraction pattern on the projection screen produced by such an aperture is

$$I(\theta, \phi') \propto \left(\int_0^a \int_0^{2\pi} \cos[k\rho\,\theta\,\cos(\phi - \phi')]\,\frac{\rho\,d\phi\,d\rho}{\pi\,a^2} \right)^2, \qquad (10.78)$$

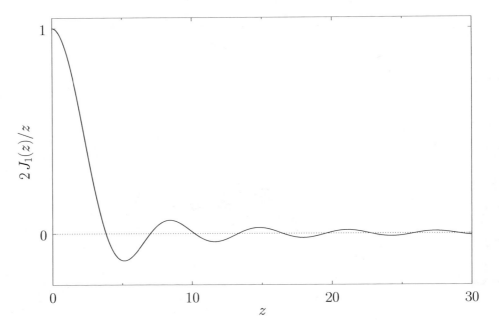

FIGURE 10.14   The function $2\,J_1(z)/z$.

or

$$I(\theta) \propto \left(\int_0^a \int_0^{2\pi} \cos(k\rho\,\theta\,\cos v)\,dv\, \frac{\rho\,d\rho}{\pi\,a^2} \right)^2, \tag{10.79}$$

where $v = \phi - \phi'$. It is readily demonstrated that (Abramowitz and Stegun 1965)

$$\frac{1}{2\pi} \int_0^{2\pi} \cos(z\,\cos v)\,dv = \frac{1}{\pi} \int_0^{\pi} \cos(z\,\sin v)\,dv = J_0(z), \tag{10.80}$$

where $J_0(z)$ is the Bessel function of degree zero introduced in Section 7.6. It follows that

$$I(\theta) \propto \left[\frac{2}{\pi\,a^2} \int_0^a J_0(k\,\theta\rho)\,\rho\,d\rho \right]^2 = \left[\frac{2}{(k\,\theta\,a)^2} \int_0^{k\theta a} J_0(z')\,z'\,dz' \right]^2. \tag{10.81}$$

It can be shown that (Gradshteyn and Ryzhik 1980)

$$\int_0^z J_0(z')\,z'\,dz' = z\,J_1(z), \tag{10.82}$$

where (Abramowitz and Stegun 1965)

$$J_1(z) = \frac{1}{\pi} \int_0^{\pi} \cos(z\,\sin v - v)\,dv \tag{10.83}$$

is a Bessel function of degree one. Hence,

$$I(\theta) \propto \left[\frac{2\,J_1(k\,a\,\theta)}{k\,a\,\theta} \right]^2. \tag{10.84}$$

Figure 10.14 shows the function $2\,J_1(z)/z$, whereas Table 10.1 lists the first few values of z at which

j	z_j	j	z_j
1	3.83171	6	19.61586
2	7.01559	7	22.76008
3	10.17347	8	25.90367
4	13.32369	9	29.04683
5	16.47063	10	32.18968

TABLE 10.1   First few zeros of the function $2\,J_1(z)/z$. Source: Abramowitz and Stegun 1965.

this function is zero. Finally, Figure 10.15 shows the interference/diffraction pattern associated with a circular aperture. The pattern consists of a central disk, known as an *Airy disk*, surrounded by much fainter concentric rings. The angular radius of the Airy disk is $\delta\theta = z_1/(k\,a)$ where $z_1 = 3.83171$ is the first zero of the function $2\,J_1(z)/z$. It follows that

$$\delta\theta = \frac{3.83171}{\pi}\frac{\lambda}{D} = 1.22\,\frac{\lambda}{D}, \tag{10.85}$$

where $D = 2\,a$ is the diameter of the aperture.

As we have already mentioned (see Section 10.7), when a point light source, such as a star, is observed in a telescope, what is actually seen is the diffraction patten of the source produced by the objective aperture. If the aperture is circular then then the telescope images the star as an Airy disk, surrounded by much fainter rings. Likewise, the telescope images two neighboring stars as two Airy disks. The Rayleigh criterion (see Section 10.7) for resolving the two stars (i.e., for being able to tell that there are two stars, rather than one) is that the angular distance, $\Delta\theta$, between the stars be greater than the angular radii of their Airy disks. This yields

$$\Delta\theta > 1.22\,\frac{\lambda}{D}, \tag{10.86}$$

where D is the diameter of the telescope's objective aperture. Thus, the minimum angular resolution of the telescope is

$$\delta\theta = 1.22\,\frac{\lambda}{D}, \tag{10.87}$$

which is a slightly more accurate version of the criterion given in Section 10.7.

10.10   HUYGENS–FRESNEL PRINCIPLE

Let us again consider a monochromatic plane light wave, propagating in the z-direction, which is normally incident on an opaque screen that occupies the plane $z = 0$. Suppose that there is an irregularly shaped aperture cut in the screen, and that the light that passes through this aperture travels to a flat projection screen occupying the plane $z = R$. Let $\psi_0\,\cos(\omega t - \phi)$ be the electric component of the wave illuminating the opaque screen from behind (i.e., from $z < 0$). We can determine the diffraction pattern that appears on the projection screen by using a more accurate version of Huygen's principle known as the *Hugyens–Fresnel principle*. According to this principle, each unblocked element of the opaque screen emits a secondary wave of the form (Jackson 1975)

$$\psi(r,\theta,t) = \frac{\psi_0\,K(\theta)\,\cos(\omega t - k\,r - \phi + \pi/2)}{\lambda\,r}\,dS, \tag{10.88}$$

where r is the distance that the wave has propagated, θ the angle subtended between the direction of propagation and the direction of incidence (i.e., the z-axis), dS the area of the element, and

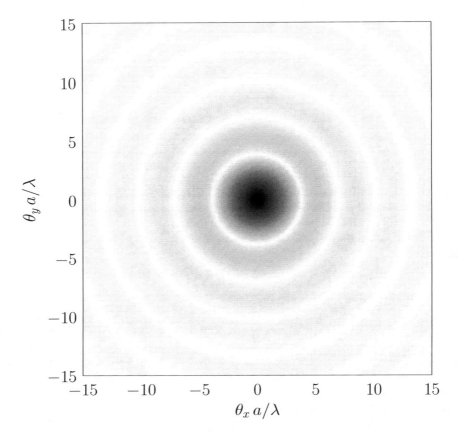

FIGURE 10.15 Far-field interference/diffraction pattern produced by a circular aperture of radius a. Dark regions indicate high light intensity.

$\lambda = 2\pi/k$ the wavelength. Here, $K(\theta)$ is known as the *obliquity factor*. Previously (in Section 10.9), we effectively assumed that $K(\theta) = 1$ for $0 \le \theta \le \pi/2$, and $K(\theta) = 0$ for $\pi/2 < \theta \le \pi$. In other words, we assumed that the secondary wave is a half-spherical wave that is emitted isotropically in all forward directions, and has zero amplitude in all backward directions. In fact, the true obliquity factor is (Jackson 1975)

$$K(\theta) = \frac{1}{2}(1 + \cos\theta),\qquad(10.89)$$

which implies that, although the secondary wave propagates predominately in the forward direction, there is some backward propagation. According to Equation (10.88), there is a $-\pi/2$ phase difference between the secondary wave emitted by an element of the aperture, and the light that illuminates the element from behind. We previously (in Section 10.9) assumed that there was no phase difference. However, it is easily demonstrated that none of our previous results would be modified had we taken this phase difference into account.

Let us define a modified aperture function, $f(x, y)$, which is such that $f = 1$ if the point $(x, y, 0)$ on the opaque screen falls within the aperture, and $f = 0$ otherwise. It follows from Equation (10.88) that the wave amplitude at the point (x', y', R) on the projection screen, which is the resultant of all of the secondary waves that are emitted by the aperture and travel to this particular point, is given by

$$\psi(x',y',t) = \frac{\psi_0}{\lambda} \int_{-\infty}^{\infty}\int_{-\infty}^{\infty} f(x,y)\, K(\theta)\, \frac{\cos(\omega t - kr - \phi + \pi/2)}{r}\, dx\, dy. \quad (10.90)$$

Now,

$$r = \left[(x-x')^2 + (y-y')^2 + R^2\right]^{1/2}$$

$$= R + \frac{(x-x')^2}{2R} + \frac{(y-y')^2}{2R} + O\left[\frac{(x-x')^4}{R^3}\right] + O\left[\frac{(y-y')^4}{R^3}\right], \quad (10.91)$$

and

$$\cos\theta \equiv \frac{R}{r} = 1 - \frac{(x-x')^2}{2R^2} - \frac{(y-y')^2}{2R^2} + O\left[\frac{(x-x')^4}{R^4}\right] + O\left[\frac{(y-y')^4}{R^4}\right]. \quad (10.92)$$

Let us assume that

$$\frac{d}{R} \ll 1, \quad (10.93)$$

where d is the typical aperture dimension (or, to be more exact, the typical value of $|x - x'|$ and $|y - y'|$). In this limit, Equation (10.90) reduces to

$$\psi(x',y',t) \simeq -\frac{\psi_0}{\lambda R} \int_{-\infty}^{\infty}\int_{-\infty}^{\infty} f(x,y)\, \sin\left[\omega t - kR - \phi - \frac{k(x-x')^2}{2R} - \frac{k(y-y')^2}{2R}\right] dx\, dy, \quad (10.94)$$

where use has been made of the trigonometric identity $\cos(x + \pi/2) = -\sin(x)$. The neglect of terms involving $(x-x')^4$ and $(y-y')^4$ inside the argument of the sine function in the previous equation is justified provided that

$$\frac{d^4}{\lambda R^3} \ll 1. \quad (10.95)$$

Note that this is a far less stringent criterion than the far-field criterion (see Section 10.2)

$$\frac{d^2}{\lambda R} \ll 1. \quad (10.96)$$

Hence, Equation (10.94) is valid not only in the far-field limit (specified by the previous inequality), but also in the so-called *near-field limit* specified by the inequality

$$1 \lesssim \frac{d^2}{\lambda R} \ll \left(\frac{R}{d}\right)^2. \quad (10.97)$$

Expression (10.94) is more general than our previous far-field expression, (10.74), because we have retained terms in the argument of the sine function that are quadratic in x and y, whereas these terms were previously neglected. Incidentally, far-field diffraction is often referred to as *Fraunhofer diffraction*, whereas near-field diffraction is termed *Fresnel diffraction*. Note, finally, that because the ordering $d/R \ll 1$ implies that $|\theta| \ll 1$, our previous incorrect assumption for the form of the obliquity factor makes no difference, because our previous form and the correct form are both characterized by $K(0) = 1$.

Let

$$u = \left(\frac{2}{\lambda R}\right)^{1/2} x, \tag{10.98}$$

$$u' = \left(\frac{2}{\lambda R}\right)^{1/2} x', \tag{10.99}$$

$$v = \left(\frac{2}{\lambda R}\right)^{1/2} y, \tag{10.100}$$

$$v' = \left(\frac{2}{\lambda R}\right)^{1/2} y'. \tag{10.101}$$

Thus, u and v are dimensionless coordinates that locate a point within the aperture, whereas u' and v' are corresponding coordinates that locate a point on the projection screen. Equation (10.94) transforms to give

$$\psi(u', v', t) \simeq -\psi_0 \sin(\omega t - kR - \phi) f_c(u', v') + \psi_0 \cos(\omega t - kR - \phi) f_s(u', v'), \tag{10.102}$$

where

$$f_c(u', v') = \frac{1}{2} \int_{-\infty}^{\infty} \int_{-\infty}^{\infty} f(u, v) \cos\left[\frac{\pi}{2}(u - u')^2 + \frac{\pi}{2}(v - v')^2\right] du\, dv, \tag{10.103}$$

$$f_s(u', v') = \frac{1}{2} \int_{-\infty}^{\infty} \int_{-\infty}^{\infty} f(u, v) \sin\left[\frac{\pi}{2}(u - u')^2 + \frac{\pi}{2}(v - v')^2\right] du\, dv. \tag{10.104}$$

Suppose that the aperture is completely transparent, so that $f(u, v) = 1$ for all u and v. In this case, the two-dimensional integrals in the previous two equations become separable. Making use of some standard trigonometric identities (see Section B.3), we obtain

$$f_c(u', v') = \frac{1}{2}\left[\int_{-\infty}^{\infty} \cos\left(\frac{\pi}{2} t^2\right) dt\right]^2 - \frac{1}{2}\left[\int_{-\infty}^{\infty} \sin\left(\frac{\pi}{2} t^2\right) dt\right]^2, \tag{10.105}$$

$$f_s(u', v') = \left[\int_{-\infty}^{\infty} \cos\left(\frac{\pi}{2} t^2\right) dt\right]\left[\int_{-\infty}^{\infty} \sin\left(\frac{\pi}{2} t^2\right) dt\right]. \tag{10.106}$$

However, (Abramowitz and Stegun 1965)

$$\int_{-\infty}^{\infty} \cos\left(\frac{\pi}{2} t^2\right) dt = \int_{-\infty}^{\infty} \sin\left(\frac{\pi}{2} t^2\right) dt = 1. \tag{10.107}$$

Hence, we deduce that $f_c(u', v') = 0$ and $f_s(u', v') = 1$, which, from Equation (10.102), implies that

$$\psi(u', v', t) = \psi_0 \cos(\omega t - kR - \phi). \tag{10.108}$$

Of course, this is the correct answer; if the aperture is completely transparent (i.e., if $f = 1$ everywhere) then the incident wave, $\psi_0 \cos(\omega t - \phi)$, illuminating the aperture from behind, propagates in the z-direction without changing amplitude, and acquires a phase shift kR by the time it reaches the projection screen. The previous result is the ultimate justification for the Hugyens–Fresnel formula, (10.88).

The intensity at which a given point on the projection screen is illuminated is (see Section 10.2)

$$\mathcal{I}(u', v') \propto \langle \psi^2(u', v', t)\rangle, \tag{10.109}$$

where $\langle\cdots\rangle$ denotes an average over a wave period. It follows from Equation (10.102) that

$$\frac{I(u',v')}{I_0} = f_c^2(u',v') + f_s^2(u',v'),\qquad(10.110)$$

where I_0 is the illumination intensity when the aperture is completely transparent. Here, use has been made of the standard results $\langle\cos^2(\omega t - kR - \phi)\rangle = \langle\sin^2(\omega t - kR - \phi)\rangle = 1/2$ and $\langle\cos(\omega t - kR - \phi)\sin(\omega t - kR - \phi)\rangle = 0$.

10.11 BABINET'S PRINCIPLE

Suppose that a given aperture is characterized by the aperture function $f(u,v)$. Consider a so-called *complementary aperture* whose aperture function, $\bar{f}(u,v)$, satisfies

$$f(u,v) + \bar{f}(u,v) = 1\qquad(10.111)$$

for all u and v. In other words, the complementary aperture is transparent everywhere where the original aperture is opaque, and vice versa. Let

$$\bar{f}_c(u',v') = \frac{1}{2}\int_{-\infty}^{\infty}\bar{f}(u,v)\cos\left[\frac{\pi}{2}(u-u')^2 + \frac{\pi}{2}(v-v')^2\right]du\,dv,\qquad(10.112)$$

$$\bar{f}_s(u',v') = \frac{1}{2}\int_{-\infty}^{\infty}\bar{f}(u,v)\sin\left[\frac{\pi}{2}(u-u')^2 + \frac{\pi}{2}(v-v')^2\right]du\,dv.\qquad(10.113)$$

It is clear from the analysis in the previous section that

$$f_c(u',v') + \bar{f}_c(u',v') = 0,\qquad(10.114)$$

$$f_s(u',v') + \bar{f}_s(u',v') = 1.\qquad(10.115)$$

Let $\bar{\psi}(u',v',t)$ denote the wave amplitude at the projection screen when the complementary aperture is illuminated. By analogy with Equation (10.102),

$$\bar{\psi}(u',v',t) \simeq -\psi_0\sin(\omega t - kR - \phi)\,\bar{f}_c(u',v') + \psi_0\cos(\omega t - kR - \phi)\,\bar{f}_s(u',v').\qquad(10.116)$$

Thus, it follows from Equation (10.102), and the previous three equations, that

$$\psi(u',v',t) + \bar{\psi}(u',v',t) = \psi_0\cos(\omega t - kR - \phi).\qquad(10.117)$$

In other words, the sum of the wave amplitude at the projection screen seen when the original aperture is illuminated and the wave amplitude seen when the complementary aperture is illuminated is equal to the wave amplitude seen when the aperture is completely transparent. This result is known as *Babinet's principle*.

10.12 DIFFRACTION FROM RECTANGULAR APERTURE

Suppose that the aperture is rectangular. In other words,

$$f(u,v) = \begin{cases} 1 & u_1 < u < u_2 \text{ and } v_1 < v < v_2 \\ 0 & \text{otherwise} \end{cases},\qquad(10.118)$$

where $u_1 < u_2$ and $v_1 < v_2$. In this case, the two-dimensional integrals in Equations (10.103) and (10.104) are separable. Using some standard trigonometric identities (see Section B.3), we obtain

$$f_c(u',v') = \frac{1}{2}C(u',u_1,u_2)\,C(v',v_1,v_2) - \frac{1}{2}S(u',u_1,u_2)\,S(v',v_1,v_2),\qquad(10.119)$$

$$f_s(u',v') = \frac{1}{2}S(u',u_1,u_2)\,C(v',v_1,v_2) + \frac{1}{2}C(u',u_1,u_2)\,S(v',v_1,v_2),\qquad(10.120)$$

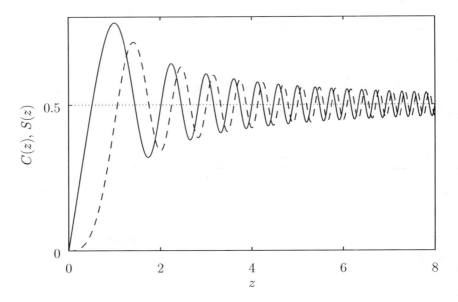

FIGURE 10.16   The Fresnel integrals. The solid curve shows $C(z)$, whereas the dashed curve shows $S(z)$.

where

$$C(z', z_1, z_2) = C(z_2 - z') + C(z' - z_1), \tag{10.121}$$

$$S(z', z_1, z_2) = S(z_2 - z') + S(z' - z_1). \tag{10.122}$$

Here, the functions

$$C(z) = \int_0^z \cos\left(\frac{\pi}{2} t^2\right) dt, \tag{10.123}$$

$$S(z) = \int_0^z \sin\left(\frac{\pi}{2} t^2\right) dt \tag{10.124}$$

are known as *Fresnel integrals* (Abramowitz and Stegun 1965). Note that $C(-z) = -C(z)$, $S(-z) = -S(z)$, and $C(\infty) = S(\infty) = 1/2$. The Frensel integrals are plotted in Figure 10.16.

Suppose that the aperture is of infinite extent in the v-direction, which implies that $v_1 = -\infty$ and $v_2 = \infty$. In this case, the aperture becomes a rectangular slit running parallel to the v-axis, and extending from $u = u_1$ to $u = u_2$. It is easily seen that $C(v', -\infty, \infty) = S(v', -\infty, \infty) = 1$. Hence, Equations (10.119) and (10.120) yield

$$f_c(u') = \frac{1}{2} C(u', u_1, u_2) - \frac{1}{2} S(u', u_1, u_2) \tag{10.125}$$

$$f_s(u') = \frac{1}{2} S(u', u_1, u_2) + \frac{1}{2} C(u', u_1, u_2). \tag{10.126}$$

Furthermore, Equation (10.110) gives

$$\frac{\mathcal{I}(u')}{\mathcal{I}_0} = f_c^2(u') + f_s^2(u'). \tag{10.127}$$

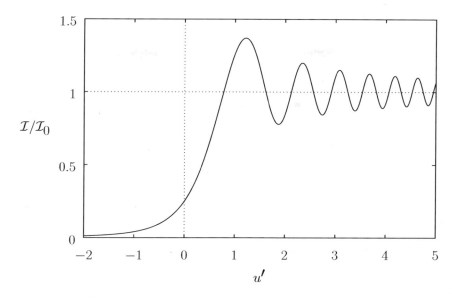

FIGURE 10.17   Near-field diffraction pattern of a semi-infinite opaque plane occupying the region $u < 0$.

10.13   DIFFRACTION FROM STRAIGHT EDGE

Consider the diffraction pattern of a semi-infinite opaque plane bounded by a sharp straight edge. Suppose that plane occupies the region $u < 0$, which implies that the edge corresponds to $u = 0$. In other words, suppose that $u_1 = 0$ and $u_2 = \infty$. It is easily demonstrated that

$$C(u', 0, \infty) = \frac{1}{2} + C(u'),$$
(10.128)

$$S(u', 0, \infty) = \frac{1}{2} + S(u').$$
(10.129)

Thus, Equations (10.125) and (10.126) yield

$$f_c(u') = \frac{1}{2}\left[C(u') - S(u')\right],$$
(10.130)

$$f_s(u') = \frac{1}{2}\left[1 + C(u') + S(u')\right],$$
(10.131)

whereas Equation (10.127) gives

$$\frac{I(u')}{I_0} = \frac{1}{2}\left[C(u') + \frac{1}{2}\right]^2 + \frac{1}{2}\left[S(u') + \frac{1}{2}\right]^2.$$
(10.132)

The intensity of the diffraction pattern of a semi-infinite opaque plane is shown in Figure 10.17. Note that this is an intrinsically near-field diffraction pattern [because the aperture is of infinite extent; see Equation (10.96)]. In the directly illuminated region, $u' > 0$, the intensity oscillates with diminishing amplitude, as the distance from the edge increases, and asymptotically approaches the value I_0, as would be expected on the basis of geometric optics. In the shadow region, $u' < 0$, the intensity decreases monotonically towards zero as the distance from the edge increases. Note that the maximum value of the intensity is not at the edge of the geometric shadow (i.e., $u' = 0$), but some distance away from it, in the directly illuminated region. Finally, at the edge of the shadow, $I = I_0/4$. This is to be expected because half the wavefront is obstructed, the amplitude of the wave at the projection screen is thus halved, and the intensity consequently drops to one quarter of the unobstructed intensity.

10.14 DIFFRACTION FROM RECTANGULAR SLIT

Consider the diffraction pattern of a rectangular slit that runs parallel to the v-axis and extends from $u = -\Delta u/2$ to $u = \Delta u/2$. In other words, let $u_1 = -\Delta u/2$ and $u_2 = \Delta u/2$. It is easily demonstrated, from Equations (10.121), (10.122), (10.125), and (10.126), that

$$f_c(u') = \frac{1}{2}\left[C(\Delta u/2 - u') + C(u' + \Delta u/2)\right] - \frac{1}{2}\left[S(\Delta u/2 - u') + S(u' + \Delta u/2)\right], \qquad (10.133)$$

$$f_s(u') = \frac{1}{2}\left[C(\Delta u/2 - u') + C(u' + \Delta u/2)\right] + \frac{1}{2}\left[S(\Delta u/2 - u') + S(u' + \Delta u/2)\right]. \qquad (10.134)$$

It follows from Equation (10.127) that

$$\frac{I(u')}{I_0} = \frac{1}{2}\left[C(\Delta u/2 - u') + C(u' + \Delta u/2)\right]^2 + \frac{1}{2}\left[S(\Delta u/2 - u') + S(u' + \Delta u/2)\right]^2. \qquad (10.135)$$

Note that the far-field limit corresponds to $\Delta u \ll 1$ [see Equations (10.96) and (10.98)], whereas the near-field limit corresponds to $\Delta u \gtrsim 1$ [see Equation (10.97)].

Figure 10.18 shows the diffraction pattern of a rectangular slit. It can be seen that in the far-field limit, $\Delta u \ll 1$, the diffraction pattern is the same as that calculated in Section 10.7. However, in the near-field limit, $\Delta u > 1$, the diffraction pattern is significantly modified. For instance, in the far-field limit, the diffraction pattern is much wider than the slit, whereas, in the near-field limit, the diffraction pattern is similar in size to the slit. In fact, in the extreme near-field limit, $\Delta u \gg 1$, the diffraction pattern is fairly similar in form to the geometric image of the slit, apart from the presence of fringes within the image.

10.15 DIFFRACTION FROM STRAIGHT WIRE

Consider the diffraction pattern of a straight wire that runs parallel to the v-axis and extends from $u = -\Delta u/2$ to $u = \Delta u/2$. Because the wire is the complementary aperture of the rectangular slit discussed in the previous section, we can use Babinet's principle (see Section 10.11) to deduce that

$$f_c(u') = -\frac{1}{2}\left[C(\Delta u/2 - u') + C(u' + \Delta u/2)\right] + \frac{1}{2}\left[S(\Delta u/2 - u') + S(u' + \Delta u/2)\right], \qquad (10.136)$$

$$f_s(u') = 1 - \frac{1}{2}\left[C(\Delta u/2 - u') + C(u' + \Delta u/2)\right] - \frac{1}{2}\left[S(\Delta u/2 - u') + S(u' + \Delta u/2)\right]. \qquad (10.137)$$

Figure 10.19 shows the diffraction pattern of a straight wire. In the far-field limit, $\Delta u \ll 1$, the diffraction pattern consists of a bright spot, centered on the wire, surrounded by a set of interference fringes. The fringes can be thought of as the interference pattern of the light passing immediately to either side of the wire. In fact, if we think of the wire as two slits whose spacing is the diameter of the wire then straightforward interference theory (see Section 10.2) suggests that the fringe width should be

$$\delta u = \frac{2}{\Delta u}. \qquad (10.138)$$

It is clear from the figure that this is indeed the case. Hence, it is possible to determine the diameter of a thin wire from its diffraction pattern. In the extreme near-field limit, $\Delta u \gg 1$, the diffraction pattern of the wire is essentially a geometric shadow, bounded on either side by a straight-edge diffraction pattern. (See Section 10.13.)

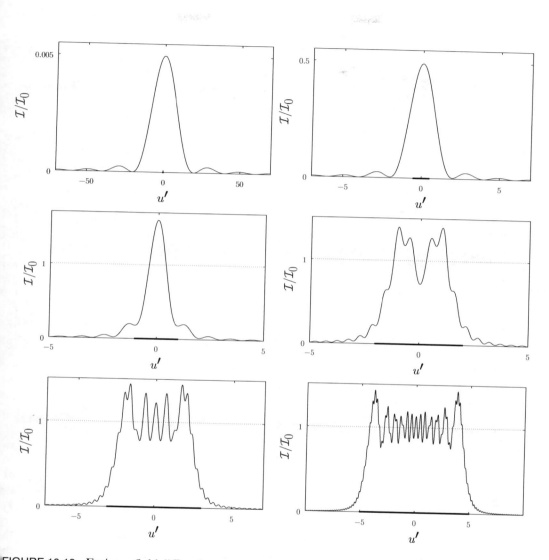

FIGURE 10.18 Far/near-field diffraction pattern of a rectangular slit. The top-left, top-right, middle-left, middle-right, bottom-left, and bottom-right panels correspond to Δu = 0.1, 1.0, 2.0, 4.0, 6.0, and 10.0, respectively. The thick black line indicates the physical extent of the slit.

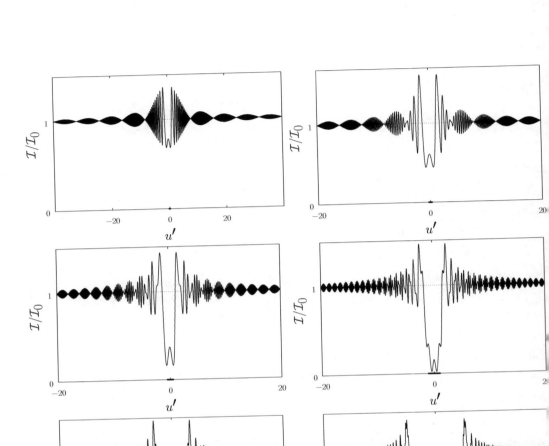

FIGURE 10.19 Far/near-field diffraction pattern of a straight wire. The top-left, top-right, middle-left, middle-right, bottom-left, and bottom-right panels correspond to $\Delta u = 0.25, 0.5, 1.0, 2.0, 4.0$ and 8.0, respectively. The thick black line indicates the physical extent of the wire.

10.16 DIFFRACTION FROM CIRCULAR APERTURE

Consider the diffraction pattern of a circular aperture of radius a whose center lies at $x = y = 0$. (See Section 10.10.) We expect the pattern to be rotationally symmetric about the z-axis. In other words, we expect the intensity of the illumination on the projection screen to be only a function of the radial coordinate $r' = (x'^2 + y'^2)^{1/2}$. It is helpful to redefine the dimensionless parameters u and v as follows:

$$u = \frac{2\pi}{\lambda} \frac{a}{R} a, \tag{10.139}$$

$$v = \frac{2\pi}{\lambda} \frac{a}{R} r'. \tag{10.140}$$

Thus, u now parameterizes the aperture radius, whereas v is a normalized radial coordinate on the projection screen. Note, from Equation (10.96), that the far-field limit corresponds to $u \lesssim 1$, whereas the near-field limit corresponds to $u \gg 1$. Furthermore, a point on the projection screen lies in the geometric (i.e., as predicted by geometric optics) lit part of the screen if $v < u$, and vice versa. Finally, the aperture function takes the form

$$f(v) = \begin{cases} 1 & v < u \\ 0 & \text{otherwise} \end{cases}. \tag{10.141}$$

When expressed in terms of the new variables, Equations (10.103) and (10.104) transform to give

$$f_c(u, v) = u \int_0^1 \oint \cos\left(\frac{v^2}{2u} + \frac{u z^2}{2} - v z \cos\theta\right) z \, dz \frac{d\theta}{2\pi}, \tag{10.142}$$

$$f_s(u, v) = u \int_0^1 \oint \sin\left(\frac{v^2}{2u} + \frac{u z^2}{2} - v z \cos\theta\right) z \, dz \frac{d\theta}{2\pi}, \tag{10.143}$$

where $z = (x^2 + y^2)^{1/2}/a$. Now, (Abramowitz and Stegun 1965)

$$\oint \cos(v z \cos\theta) \frac{d\theta}{2\pi} = J_0(v z), \tag{10.144}$$

$$\oint \sin(v z \cos\theta) \frac{d\theta}{2\pi} = 0, \tag{10.145}$$

where (ibid.)

$$J_n(z) = \frac{1}{\pi} \int_0^\pi \cos(z \sin\theta - n\theta) \, d\theta \tag{10.146}$$

denotes a Bessel function of degree n. Hence, making use of some trigonometric identities (see Appendix B), Equations (10.142) and (10.143) reduce to

$$f_c(u, v) = \cos\left(\frac{v^2}{2u}\right) C(u, v) - \sin\left(\frac{v^2}{2v}\right) S(u, v), \tag{10.147}$$

$$f_s(u, v) = \sin\left(\frac{v^2}{2u}\right) C(u, v) + \cos\left(\frac{v^2}{2v}\right) S(u, v), \tag{10.148}$$

where

$$C(u, v) = u \int_0^1 \cos\left(\frac{u z^2}{2}\right) J_0(v z) z \, dz, \tag{10.149}$$

$$S(u, v) = u \int_0^1 \sin\left(\frac{u z^2}{2}\right) J_0(v z) z \, dz. \tag{10.150}$$

It is helpful, at this stage, to introduce the so-called *Lommel functions* (of two arguments) (Watson 1962)

$$U_n(u, v) = \sum_{s=0,\infty} (-1)^s \left(\frac{u}{v}\right)^{n+2s} J_{n+2s}(v),$$ (10.151)

$$V_n(u, v) = \sum_{s=0,\infty} (-1)^s \left(\frac{v}{u}\right)^{n+2s} J_{n+2s}(v).$$ (10.152)

In the geometric lit region, $v < u$, the integrals $C(u, v)$ and $S(u, v)$ are conveniently expanded in terms of the convergent V_n Lommel functions (Born and Wolf 1980)

$$C(u, v) = \sin\left(\frac{v^2}{2u}\right) + \sin\left(\frac{u}{2}\right) V_0(u, v) - \cos\left(\frac{u}{2}\right) V_1(u, v),$$ (10.153)

$$S(u, v) = \cos\left(\frac{v^2}{2u}\right) - \cos\left(\frac{u}{2}\right) V_0(u, v) - \sin\left(\frac{u}{2}\right) V_1(u, v).$$ (10.154)

(See Exercise 10.22.) Likewise, in the geometric shadow region, $v > u$, the integrals can be expanded in term of the convergent U_n Lommel functions (Born and Wolf 1980)

$$C(u, v) = \cos\left(\frac{u}{2}\right) U_1(u, v) + \sin\left(\frac{u}{2}\right) U_2(u, v),$$ (10.155)

$$S(u, v) = \sin\left(\frac{u}{2}\right) U_1(u, v) - \cos\left(\frac{u}{2}\right) U_2(u, v).$$ (10.156)

(See Exercise 10.22.) It follows (with the aid of some trigonometric identities) that

$$f_c(u, v) = \sin\left(\frac{u^2 + v^2}{2u}\right) V_0(u, v) - \cos\left(\frac{u^2 + v^2}{2u}\right) V_1(u, v),$$ (10.157)

$$f_s(u, v) = 1 - \cos\left(\frac{u^2 + v^2}{2u}\right) V_0(u, v) - \sin\left(\frac{u^2 + v^2}{2u}\right) V_1(u, v)$$ (10.158)

when $v < u$, and

$$f_c(u, v) = \cos\left(\frac{u^2 + v^2}{2u}\right) U_1(u, v) + \sin\left(\frac{u^2 + v^2}{2u}\right) U_2(u, v),$$ (10.159)

$$f_s(u, v) = \sin\left(\frac{u^2 + v^2}{2u}\right) U_1(u, v) - \cos\left(\frac{u^2 + v^2}{2u}\right) U_2(u, v)$$ (10.160)

when $v > u$.

Finally, making use of Equation (10.110), the previous four equations imply that the illumination intensity on the projection screen can be written

$$\frac{I(v)}{I_0} = \left[V_0(u, v) - \cos\left(\frac{u^2 + v^2}{2u}\right)\right]^2 + \left[V_1(u, v) - \sin\left(\frac{u^2 + v^2}{2u}\right)\right]^2$$ (10.16)

when $v < u$, and

$$\frac{I(v)}{I_0} = \left[U_1^2(u, v) + U_2^2(u, v)\right]$$ (10.16)

when $v > u$. Here, I_0 is the intensity of the light illuminating the aperture from behind.

Figure 10.20 shows a typical far-field (i.e., $u \lesssim 1$) and near-field (i.e., $u \gg 1$) diffraction patte

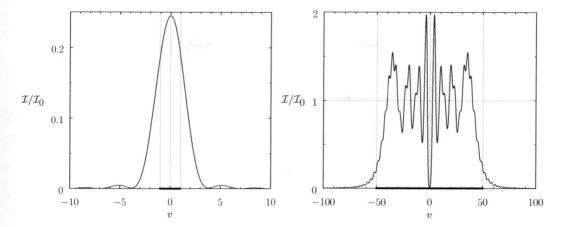

FIGURE 10.20 Far/near-field diffraction pattern of a circular aperture. The left and right panels correspond to $u = 1$ and $u = 50$, respectively. The thick black line indicates the physical extent of the aperture.

of a circular aperture, as determined from the previous analysis. It can be seen that the far-field diffraction pattern is similar in form to that predicted by the simplified Fourier analysis of Section 10.9. On the other hand, the near-field diffraction pattern is quite different. In fact, the near-field diffraction pattern is fairly similar in form to the geometric image of the aperture, apart from the presence of fringes within the image.

10.17   DIFFRACTION FROM CIRCULAR DISK

Finally, consider the diffraction pattern of a circular disk of radius a whose center lies at $x = y = 0$. Because the disk is complementary to the circular aperture discussed in the previous section, we can use Babinet's principle (see Section 10.11) to deduce that

$$\frac{I(v)}{I_0} = V_0^2(u, v) + V_1^2(u, v) \tag{10.163}$$

in the geometric shadow region, $v < u$, and

$$\frac{I(v)}{I_0} = \left[U_1(u, v) - \sin\left(\frac{u^2 + v^2}{2u}\right)\right]^2 + \left[U_2(u, v) - \cos\left(\frac{u^2 + v^2}{2u}\right)\right]^2 \tag{10.164}$$

in the geometric lit region, $v > u$.

Figure 10.20 shows a typical far-field (i.e., $u \ll 1$) and near-field (i.e., $u \gg 1$) diffraction pattern of a circular disk, as determined by the previous two formulae.

It can be seen that the far-field diffraction pattern is an axially symmetric version of the far-field diffraction pattern of a thin wire, which was discussed in Section 10.15. To be more exact, the far-field diffraction pattern consists of a central bright disk surrounded by concentric diffraction fringes. As is clear from Exercise 10.24, the centers of the fringes correspond (approximately) to the maxima of the function $J_1[2\pi a\, r'/(\lambda R)]$. Hence, it is possible to determine the diameter of a circular disk (or a sphere) from its diffraction pattern.

In the near-field limit, $u \gg 1$, the diffraction pattern of the disk is essentially a geometric shadow, bounded by a circular straight-edge diffraction pattern. (See Section 10.13.) Note, however, that

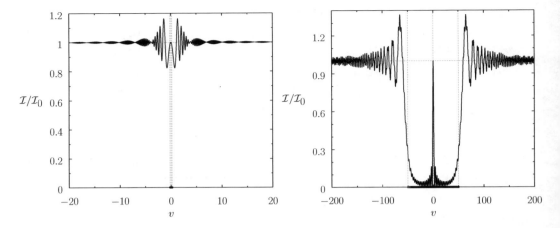

FIGURE 10.21 Far/near-field diffraction pattern of a circular disk. The left and right panels correspond to $u = 0.2$ and $u = 50$, respectively. The thick black line indicates the physical extent of the disk.

there is a bright spot at the center of the shadow. In fact, it can be demonstrated (see Exercise 10.24) that the central intensity of the diffraction pattern is the same as that of the light illuminating the disk [i.e., $I(0) = I_0$], irrespective of the disk's radius (i.e., for all values of u). This is one of the most famous results in optics.

At the beginning of the 19th century, most scientists favored Isaac Newton's corpuscular theory of light; among these was the theoretician Siméon Denis Poisson (Wikipedia contributors 2018). In 1818, the French Academy of Sciences launched a competition to explain the properties of light, and Poisson was one of the members of the judging committee. The civil engineer Augustin-Jean Fresnel entered this competition by submitting a new wave theory of light. Poisson studied Fresnel's theory in detail, and, being a supporter of the particle theory of light, looked for a way to prove it wrong. Poisson thought that he had found a flaw when he argued that one consequence of Fresnel's theory was that there would exist an on-axis bright spot in the shadow of a circular obstacle, where there should be complete darkness according to the particle theory of light. Because this spot is not easily observed in everyday situations, Poisson interpreted it as an absurd result that disproved Fresnel's theory. However, the head of the committee, Dominique-François-Jean Arago (who incidentally later became Prime Minister of France) decided to perform the experiment in more detail. He succeeded in observing the predicted spot, which convinced most scientists of the wave nature of light, and gave Fresnel victory in the competition. Ever since, the spot in question has been known, somewhat ironically, as *Poisson's spot*.

EXERCISES

10.1   (a) Consider the geometric series

$$S = \sum_{n=0,N-1} z^n,$$

where z is a complex number. Demonstrate that

$$S = \frac{1 - z^N}{1 - z}.$$

(b) Suppose that $z = e^{i\theta}$, where θ is real. Employing the well-known identity

$$\sin\theta \equiv \frac{1}{2i}\left(e^{i\theta} - e^{-i\theta}\right),$$

show that

$$S = e^{i(N-1)\theta/2}\frac{\sin(N\theta/2)}{\sin(\theta/2)}.$$

(c) Finally, making use of Euler's theorem,

$$e^{in\theta} \equiv \cos(n\theta) + i\,\sin(n\theta),$$

demonstrate that

$$C = \sum_{n=1,N}\cos(\alpha\,x_n),$$

where

$$x_n = [n - (N+1)/2]\,d,$$

evaluates to

$$C = \frac{\sin(N\alpha\,d/2)}{\sin(\alpha\,d/2)}.$$

10.2 Derive Equation (10.30).

10.3 An interference experiment employs two narrow parallel slits of separation 0.25 mm, and monochromatic light of wavelength $\lambda = 500$ nm. Estimate the minimum distance that the projection screen must be placed behind the slits in order to obtain a far-field interference pattern.

10.4 A double-slit of slit separation 0.5 mm is illuminated at normal incidence by a parallel beam from a helium-neon laser that emits monochromatic light of wavelength 632.8 nm. A projection screen is located 5 m behind the slit. What is the separation of the central interference fringes on the screen? [From Crawford 1968.]

10.5 Consider a double-slit interference experiment in which the slit spacing is 0.1 mm, and the projection screen is located 50 cm behind the slits. Assuming monochromatic illumination at normal incidence, if the observed separation between neighboring interference maxima at the center of the projection screen is 2.5 mm, what is the wavelength of the light illuminating the slits?

10.6 What is the mean length of the classical wavetrain (wave packet) corresponding to the light emitted by an atom whose excited state has a mean lifetime $\tau \sim 10^{-8}$ s? In an ordinary gas-discharge source, the excited atomic states do not decay freely, but instead have an effective lifetime $\tau \sim 10^{-9}$ s, due to collisions and Doppler effects. What is the length of the corresponding classical wavetrain? [From Crawford 1968.]

10.7 If a "monochromatic" incoherent "line" source of visible light is not really a line, but has a finite width of 1 mm, estimate the minimum distance it can be placed in front of a double-slit, of slit separation 0.5 mm, if the light from the slit is to generate a clear interference pattern.

10.8 The visible emission spectrum of a sodium atom is dominated by a yellow line which actually consists of two closely spaced lines of wavelength 589.0 nm and 589.6 nm. Demonstrate that a diffraction grating must have at least 328 lines in order to resolve this doublet at the third spectral order.

10.9 Consider a diffraction grating having 5000 lines per centimeter. Find the angular locations of the principal maxima when the grating is illuminated at normal incidence by (a) red light of wavelength 700 nm, and (b) violet light of wavelength 400 nm.

10.10 A soap bubble 250 nm thick is illuminated by white light. The index of refraction of the soap film is 1.36. Which colors are not seen in the reflected light? Which colors appear bright in the reflected light? What color does the soap film appear at normal incidence?

10.11 Suppose that a monochromatic laser of wavelength 632.8 nm emits a diffraction-limited beam of initial diameter 2 mm. Estimate how large a light spot the beam would produce on the surface of the Moon (which is a mean distance 3.76×10^5 km from the surface of the Earth). Neglect any effects of the Earth's atmosphere. [From Hecht and Zajac 1974.]

10.12 Estimate how far away an automobile is when you can only just barely resolve the two headlights with your eyes. [From Crawford 1968.]

10.13 Venus has a diameter of about 8000 miles. When it is prominently visible in the sky, in the early morning or late evening, it is about as far away as the Sun; that is, about 93 million miles away. Venus commonly appears larger than a point to the unaided eye. Are we seeing the true size of Venus? [From Crawford 1968.]

10.14 The world's largest steerable radio telescope, at the National Radio Astronomy Observatory, Green Bank, West Virginia, consists of a parabolic disk that is 300 ft in diameter. Estimate the angular resolution (in minutes of an arc) of the telescope when it is observing the well-known 21-cm radiation of hydrogen. [From Crawford 1968.]

10.15 Estimate how large the lens of a camera carried by an artificial satellite orbiting the Earth at an altitude of 150 miles would have to be in order to resolve features on the Earth's surface a foot in diameter.

10.16 Demonstrate that the secondary maxima in the far-field interference pattern generated by three identical, equally spaced, parallel slits of negligible width are nine times less intense than the principal maxima.

10.17 Consider a double-slit interference/diffraction experiment in which the slit spacing is d and the slit width δ. Show that the intensity of the far-field interference pattern, assuming normal incidence by monochromatic light of wavelength λ, is

$$I(\theta) \propto \cos^2\left(\pi \frac{d}{\lambda} \sin\theta\right)\mathrm{sinc}^2\left(\pi \frac{\delta}{\lambda} \sin\theta\right).$$

Plot the intensity pattern for $d/\lambda = 8$ and $\delta/\lambda = 2$.

10.18 What is the theoretical maximum angular resolving power (in arc seconds) of a conventional reflecting telescope with a 12 in main mirror?

10.19 (a) Demonstrate that, in the far-field limit, Equation (10.110) reduces to

$$\frac{I(u',v')}{I_0} = F_c^2(u',v') + F_s^2(u',v'),$$

where

$$F_c(u',v') = \frac{1}{2}\int_{-\infty}^{\infty}\int_{-\infty}^{\infty} f(u,v)\cos\left[\pi\left(u\,u' + v\,v'\right)\right]du\,dv,$$

$$F_s(u',v') = \frac{1}{2}\int_{-\infty}^{\infty}\int_{-\infty}^{\infty} f(u,v)\sin\left[\pi\left(u\,u' + v\,v'\right)\right]du\,dv.$$

(b) Hence, deduce that

$$\frac{\mathcal{I}(0,0)}{\mathcal{I}_0} = \left(\frac{A}{\lambda R}\right)^2,$$

where A is the area of the aperture, R the distance of the projection screen behind the aperture, and λ the wavelength.

10.20 Consider the far-field diffraction pattern of a circular aperture of radius a, normally illuminated by monochromatic light of wavelength λ. Let θ denote the angular distance from the optic axis (i.e., the line that passes through the center of the aperture, and is perpendicular to the plane of the aperture and the projection screen) on the projection screen. Demonstrate that the mean energy flux that falls within the region $0 < \theta < \theta_0$ is

$$\mathcal{I}_0\,\pi\,a^2\left[1 - J_0^2(k\,a\,\theta_0) - J_1^2(k\,a\,\theta_0)\right],$$

where $\mathcal{I}_0$ is the intensity of the incident wave, and $k = 2\pi/\lambda$. [Hint: $J_1'(z) = J_0(z) - J_1(z)/z$ and $J_0'(z) = -J_1(z)$, where $'$ denotes a derivative with respect to argument.] Hence, deduce that the total energy flux that illuminates the projection screen is the same as that predicted by geometric optics. Finally, show that the Airy disk contains approximately 84% of the total energy flux.

10.21 The Fresnel integrals have the asymptotic forms (Abramowitz and Stegun 1965)

$$C(z) \simeq \frac{1}{2} + \frac{1}{\pi z}\,\sin\left(\frac{\pi z^2}{2}\right),$$

$$S(z) \simeq \frac{1}{2} - \frac{1}{\pi z}\,\cos\left(\frac{\pi z^2}{2}\right)$$

in the large-argument limit, $z \gg 1$. Use these forms to demonstrate that the intensity of a diffraction pattern of a semi-infinite opaque plane bounded by a sharp straight edge, which is specified in Equation (10.132), reduces to

$$\frac{\mathcal{I}(u')}{\mathcal{I}_0} \simeq \frac{1}{2\pi^2\,|u'|^2}$$

in the extreme shadow region (i.e., $u' < 0$ and $|u'| \gg 1$). This implies that the intensity in the shadow region attenuates as the inverse-square of the distance from the straight edge.

10.22 (a) The J_n Bessel functions satisfy the recursion relation (Abramowitz and Stegun 1965)

$$\frac{d}{dz}\left[z^{n+1}\,J_{n+1}(z)\right] = z^{n+1}\,J_n(z),$$

where n is a non-negative integer. By repeatedly integrating by parts, demonstrate that the functions

$$C(u, v) = u\int_0^1 \cos\left(\frac{u\,z^2}{2}\right)J_0(v\,z)\,z\,dz,$$

$$S(u, v) = u\int_0^1 \sin\left(\frac{u\,z^2}{2}\right)J_0(v\,z)\,z\,dz$$

can be expanded in the forms

$$C(u, v) = \cos\left(\frac{u}{2}\right) U_1(u, v) + \sin\left(\frac{u}{2}\right) U_2(u, v),$$

$$S(u, v) = \sin\left(\frac{u}{2}\right) U_1(u, v) - \cos\left(\frac{u}{2}\right) U_2(u, v),$$

where

$$U_n(u, v) = \sum_{s=0,\infty} (-1)^s \left(\frac{u}{v}\right)^{n+2s} J_{n+2s}(v).$$

is a Lommel function.

(b) The J_n Bessel functions also satisfy the recursion relation (Abramowitz and Stegun 1965)

$$\frac{d}{dz}[z^{-n} J_n(z)] = -z^{-n} J_{n+1}(z),$$

where n is a non-negative integer. By repeatedly integrating by parts, demonstrate that

$$C(u, v) = \sin\left(\frac{v^2}{2u}\right) + \sin\left(\frac{u}{2}\right) V_0(u, v) - \cos\left(\frac{u}{2}\right) V_1(u, v),$$

$$S(u, v) = \cos\left(\frac{v^2}{2u}\right) - \cos\left(\frac{u}{2}\right) V_0(u, v) - \sin\left(\frac{u}{2}\right) V_1(u, v),$$

where

$$V_n(u, v) = \sum_{s=0,\infty} (-1)^s \left(\frac{v}{u}\right)^{n+2s} J_{n+2s}(v).$$

is a Lommel function. Note that $\lim_{z\to 0} J_n(z)/z^n = 1/(2^n n!)$ (Abramowitz and Stegun 1965).

10.23 Consider the near-field diffraction pattern a circular aperture of radius a, normally illumi‐ nated by monochromatic light of wavelength λ.

(a) Show that the illumination intensity at the point where the optic axis (see Exer‐ cise 10.20) meets the projection screen is

$$I = 4 I_0 \sin^2\left(\frac{\pi}{2} \frac{a^2}{\lambda R}\right),$$

where I_0 is the intensity of the incident wave, and R is the distance of the projectio‐ screen behind the aperture. Hence, deduce that if the so-called *Fresnel number*,

$$N = \frac{a^2}{\lambda R},$$

takes an even integer value then $I = 0$, and if the Fresnel number takes an odd-integer value then $I = 4 I_0$.

(b) Show that the illumination intensity a radial distance r' from the optic axis asymptot‐ to

$$I = I_0 \left[\frac{a}{r'} J_1\left(2\pi \frac{a r'}{\lambda R}\right)\right]^2$$

in the limit $r' \gg a$.

10.24 Consider the near-field diffraction pattern a circular disk of radius a, normally illuminated by monochromatic light of wavelength λ.

(a) Show that the illumination intensity at the point where the optic axis (see Exercise 10.20) meets the projection screen is

$$I = I_0,$$

where I_0 is the intensity of the incident wave.

(b) Show that the illumination intensity a radial distance r' from the optic axis asymptotes to

$$I = I_0\left[1 - \frac{2a}{r'}J_1\left(2\pi\frac{ar'}{\lambda R}\right)\sin\left(\pi\frac{r'^2}{\lambda R}\right)\right]$$

in the limit $r' \gg a$. Here, R is the distance of the projection screen behind the aperture.

CHAPTER 11

Wave Mechanics

11.1   INTRODUCTION

According to classical physics (i.e., physics prior to the 20th century), particles and waves are distinct classes of physical entities that possess markedly different properties. For instance, particles are discrete, which means that they cannot be arbitrarily divided. In other words, it makes sense to talk about one electron, or two electrons, but not about a third of an electron. Waves, on the other hand, are continuous, which means that they can be arbitrarily divided. In other words, given a wave whose amplitude has a certain value, it makes sense to talk about a similar wave whose amplitude is one third, or any other fraction whatsoever, of this value. Particles are also highly localized in space. For example, atomic nuclei have very small radii of order 10^{-15} m, whereas electrons act like point particles (i.e., they have no discernible spatial extent). Waves, on the other hand, are non-localized in space. In fact, a wave is defined as a disturbance that is periodic in space, with some finite periodicity length (i.e., wavelength). Hence, it is fairly meaningless to talk about a disturbance being a wave unless it extends over a region of space that is at least a few wavelengths in size.

The classical scenario, just described, in which particles and waves are distinct phenomena, had to be significantly modified in the early decades of the 20th century (Gasiorowicz 1996). During this time period, physicists discovered, much to their surprise, that, under certain circumstances, waves act as particles, and particles act as waves. This bizarre behavior is known as *wave-particle duality*. For instance, the *photoelectric effect* (see Section 11.2) shows that electromagnetic waves sometimes act like swarms of massless particles called *photons*. Moreover, the phenomenon of *electron diffraction* by atomic lattices (see Section 11.3) implies that electrons sometimes possess wave-like properties. Wave-particle duality usually only manifests itself on atomic and sub-atomic length-scales [i.e., on lengthscales less than, or of order, 10^{-10} m; see Section 11.3.] The classical picture remains valid on significantly longer lengthscales. Thus, on macroscopic lengthscales, waves only act like waves, particles only act like particles, and there is no wave-particle duality. However, on atomic lengthscales, classical mechanics, which governs the macroscopic behavior of massive particles, and classical electrodynamics, which governs the macroscopic behavior of electromagnetic fields—neither of which take wave-particle duality into account—must be replaced by new theories (Dirac 1982). The theories in question are called *quantum mechanics* and *quantum electrodynamics*, respectively. In this chapter, we shall discuss a simple version of quantum mechanics in which the microscopic dynamics of massive particles (i.e., particles with finite mass) is described entirely in terms of wavefunctions. This particular version of quantum mechanics is known as *wave mechanics*.

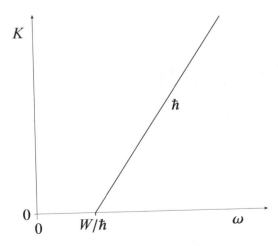

FIGURE 11.1   Variation of the kinetic energy, K, of photoelectrons with the wave angular frequency, ω.

11.2   PHOTOELECTRIC EFFECT

The so-called *photoelectric effect*, by which a polished metal surface emits electrons when illumi-nated by visible or ultra-violet light, was discovered by Heinrich Hertz in 1887. The following facts regarding this effect can be established via careful observation (Gasiorowicz 1996). First, a given surface only emits electrons when the frequency of the light with which it is illuminated exceeds a certain threshold value that is a property of the metal. Second, the current of photoelectrons, when it exists, is proportional to the intensity of the light falling on the surface. Third, the energy of the pho-toelectrons is independent of the light intensity, but varies linearly with the light frequency. These facts are inexplicable within the framework of classical physics.

   In 1905, Albert Einstein proposed a radical new theory of light in order to account for the pho-toelectric effect. According to this theory, light of fixed angular frequency ω consists of a collection of indivisible discrete packages, called *quanta*,[1] whose energy is

$$E = \hbar \omega. \tag{11.1}$$

Here, $\hbar = 1.055 \times 10^{-34}$ J s is a new constant of nature, known as *Planck's constant*. (Strictly speaking, it is Planck's constant divided by 2π.) Incidentally, $\hbar$ is called Planck's constant, rather than Einstein's constant, because Max Planck first introduced the concept of the quantization of light, in 1900, when trying to account for the electromagnetic spectrum of a black body (i.e., a perfect emitter and absorber of electromagnetic radiation) (Gasiorowicz 1996).

   Suppose that the electrons at the surface of a piece of metal lie in a potential well of depth W. In other words, the electrons have to acquire an energy W in order to be emitted from the surface. Here, W is generally called the *workfunction* of the surface, and is a property of the metal. Suppose that an electron absorbs a single quantum of light, otherwise known as a *photon*. Its energy therefore increases by $\hbar \omega$. If $\hbar \omega$ is greater than W then the electron is emitted from the surface with the residual kinetic energy

$$K = \hbar \omega - W. \tag{11.}$$

Otherwise, the electron remains trapped in the potential well, and is not emitted. Here, we are assum-ing that the probability of an electron absorbing two or more photons is negligibly small compared

---

[1] Plural of *quantum*: Latin neuter of *quantus*: how much?

to the probability of it absorbing a single photon (as is, indeed, the case for relatively low intensity illumination). Incidentally, we can determine Planck's constant, as well as the workfunction of the metal, by plotting the kinetic energy of the emitted photoelectrons as a function of the wave frequency, as shown in Figure 11.1. This plot is a straight line whose slope is $\hbar$, and whose intercept with the ω axis is $W/\hbar$. Finally, the number of emitted electrons increases with the intensity of the light because, the more intense the light, the larger the flux of photons onto the surface. Thus, Einstein's quantum theory of light is capable of accounting for all three of the previously mentioned observational facts regarding the photoelectric effect. In the following, we shall assume that the central component of Einstein's theory—namely, Equation (11.1)—is a general result that applies to all particles, not just photons.

11.3 ELECTRON DIFFRACTION

In 1927, George Paget Thomson discovered that if a beam of electrons is made to pass through a thin metal film then the regular atomic array in the metal acts as a sort of diffraction grating, so that when a photographic film, placed behind the metal, is developed an interference pattern is discernible. This implies that electrons have wave-like properties. Moreover, the electron wavelength, λ, or, alternatively, the wavenumber, $k = 2\pi/\lambda$, can be deduced from the spacing of the maxima in the interference pattern. (See Chapter 10.) Thomson found that the momentum, p, of an electron is related to its wavenumber, k, according to the following simple relation:

$$p = \hbar k \tag{11.3}$$

(Gasiorowicz 1996). The associated wavelength, $\lambda = 2\pi/k$, is known as the *de Broglie wavelength*, because this relation was first hypothesized by Louis de Broglie in 1926. In the following, we shall assume that Equation (11.3) is a general result that applies to all particles, not just electrons.

It turns out that wave-particle duality only manifests itself on lengthscales less than, or of order, the de Broglie wavelength (Dirac 1982). Under normal circumstances, this wavelength is fairly small. For instance, the de Broglie wavelength of an electron is

$$\lambda_e = 1.2 \times 10^{-9} \, [E(\text{eV})]^{-1/2} \, \text{m}, \tag{11.4}$$

where the electron energy is conveniently measured in units of electron-volts (eV). (An electron accelerated from rest through a potential difference of 1000 V acquires an energy of 1000 eV, and so on. Electrons in atoms typically have energies in the range 10 to 100 eV.) Moreover, the de Broglie wavelength of a proton is

$$\lambda_p = 2.9 \times 10^{-11} \, [E(\text{eV})]^{-1/2} \, \text{m}. \tag{11.5}$$

11.4 REPRESENTATION OF WAVES VIA COMPLEX NUMBERS

In mathematics, the symbol i is conventionally used to represent the square-root of minus one; that is, the solution of $i^2 = -1$ (Riley 1974). A real number, x (say), can take any value in a continuum of values lying between $-\infty$ and $+\infty$. On the other hand, an *imaginary number* takes the general form $i\,y$, where y is a real number. It follows that the square of a real number is a positive real number, whereas the square of an imaginary number is a negative real number. In addition, a general *complex number* is written

$$z = x + i\,y, \tag{11.6}$$

where x and y are real numbers. In fact, x is termed the real part of z, and y the imaginary part of z. This is written mathematically as $x = \text{Re}(z)$ and $y = \text{Im}(z)$. Finally, the *complex conjugate* of z is defined $z^* = x - i\,y$.

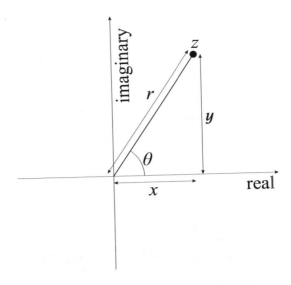

FIGURE 11.2 Representation of a complex number as a point in a plane.

Just as we can visualize a real number as a point on an infinite straight line, we can visualize a complex number as a point in an infinite plane. The coordinates of the point in question are the real and imaginary parts of the number; that is, $z \equiv (x, y)$. This idea is illustrated in Figure 11.2. The distance, $r = (x^2 + y^2)^{1/2}$, of the representative point from the origin is termed the *modulus* of the corresponding complex number, z. This is written mathematically as $|z| = (x^2 + y^2)^{1/2}$. Incidentally, it follows that $z z^* = x^2 + y^2 = |z|^2$. The angle, $\theta = \tan^{-1}(y/x)$, that the straight line joining the representative point to the origin subtends with the real axis is termed the *argument* of the corresponding complex number, z. This is written mathematically as $\arg(z) = \tan^{-1}(y/x)$. It follows from standard trigonometry that $x = r \cos\theta$, and $y = r \sin\theta$. Hence, $z = r \cos\theta + i r \sin\theta$.

Complex numbers are often used to represent waves and wavefunctions. All such representations ultimately depend on a fundamental mathematical identity, known as *Euler's theorem* (see Exercise 11.2), which takes the form

$$e^{i\phi} \equiv \cos\phi + i \sin\phi, \tag{11.?}$$

where ϕ is a real number (Riley 1974). Incidentally, given that $z = r \cos\theta + i r \sin\theta = r[\cos\theta + i \sin\theta]$, where z is a general complex number, $r = |z|$ its modulus, and $\theta = \arg(z)$ its argument, follows from Euler's theorem that any complex number, z, can be written

$$z = r e^{i\theta}, \tag{11.?}$$

where $r = |z|$ and $\theta = \arg(z)$ are real numbers.

A one-dimensional wavefunction takes the general form

$$\psi(x, t) = A \cos(\omega t - k x - \phi), \tag{11.?}$$

where $A > 0$ is the wave amplitude, ϕ the phase angle, k the wavenumber, and ω the angular frequency. Consider the complex wavefunction

$$\psi(x, t) = \psi_0 e^{-i(\omega t - k x)}, \tag{11.?}$$

where ψ_0 is a complex constant. We can write

$$\psi_0 = A\,e^{i\phi}, \tag{11.11}$$

where A is the modulus, and ϕ the argument, of ψ_0. Hence, we deduce that

$$\text{Re}\left[\psi_0\,e^{-i(\omega t-kx)}\right] = \text{Re}\left[A\,e^{i\phi}\,e^{-i(\omega t-kx)}\right] = \text{Re}\left[A\,e^{-i(\omega t-kx-\phi)}\right] = A\,\text{Re}\left[e^{-i(\omega t-kx-\phi)}\right]. \tag{11.12}$$

Thus, it follows from Euler's theorem, and Equation (11.9), that

$$\text{Re}\left[\psi_0\,e^{-i(\omega t-kx)}\right] = A\,\cos(\omega t - kx - \phi) = \psi(x,t). \tag{11.13}$$

In other words, a general one-dimensional real wavefunction, (11.9), can be represented as the real part of a complex wavefunction of the form (11.10). For ease of notation, the "take the real part" aspect of the previous expression is usually omitted, and our general one-dimension wavefunction is simply written

$$\psi(x,t) = \psi_0\,e^{-i(\omega t-kx)}. \tag{11.14}$$

The main advantage of the complex representation, (11.14), over the more straightforward real representation, (11.9), is that the former enables us to combine the amplitude, A, and the phase angle, ϕ, of the wavefunction into a single complex amplitude, ψ_0.

11.5 SCHRÖDINGER'S EQUATION

The basic premise of wave mechanics is that a massive particle of energy E and linear momentum p, moving in the x-direction (say), can be represented by a one-dimensional *complex wavefunction* of the form

$$\psi(x,t) = \psi_0\,e^{-i(\omega t-kx)}, \tag{11.15}$$

where the complex amplitude, ψ_0, is arbitrary, while the wavenumber, k, and the angular frequency, ω, are related to the particle momentum, p, and energy, E, via the fundamental relations (11.3) and (11.1), respectively. The previous one-dimensional wavefunction is the solution of a one-dimensional wave equation that determines how the wavefunction evolves in time. As described below, we can guess the form of this wave equation by drawing an analogy with classical physics.

A classical particle of mass m, moving in a one-dimensional potential $U(x)$, satisfies the energy conservation equation

$$E = K + U, \tag{11.16}$$

where

$$K = \frac{p^2}{2m} \tag{11.17}$$

is the particle's kinetic energy (Fitzpatrick 2012). Hence,

$$E\psi = (K+U)\psi \tag{11.18}$$

is a valid, but not obviously useful, wave equation.

However, it follows from Equations (11.1) and (11.15) that

$$\frac{\partial\psi}{\partial t} = -i\,\omega\,\psi_0\,e^{-i(\omega t-kx)} = -i\,\frac{E}{\hbar}\,\psi, \tag{11.19}$$

which can be rearranged to give

$$E\psi = i\hbar\,\frac{\partial\psi}{\partial t}. \tag{11.20}$$

Likewise, from Equations (11.3) and (11.15),

$$\frac{\partial^2 \psi}{\partial x^2} = -k^2 \psi_0 \, e^{-i(kx-\omega t)} = -\frac{p^2}{\hbar^2} \psi, \tag{11.21}$$

which can be rearranged to give

$$K \psi = \frac{p^2}{2m} \psi = -\frac{\hbar^2}{2m} \frac{\partial^2 \psi}{\partial x^2}. \tag{11.22}$$

Thus, combining Equations (11.18), (11.20), and (11.22), we obtain

$$i\hbar \frac{\partial \psi}{\partial t} = -\frac{\hbar^2}{2m} \frac{\partial^2 \psi}{\partial x^2} + U(x)\psi. \tag{11.23}$$

This equation, which is known as *Schrödinger's equation*—because it was first formulated by Erwin Schrödinder in 1926—is the fundamental equation of wave mechanics (Dirac 1982).

For a massive particle moving in free space (i.e., $U = 0$), the complex wavefunction (11.15) is a solution of Schrödinger's equation, (11.23), provided

$$\omega = \frac{\hbar}{2m} k^2. \tag{11.24}$$

The previous expression can be thought of as the dispersion relation (see Section 4.2) for matter waves in free space. The associated phase velocity (see Section 6.3) is

$$v_p = \frac{\omega}{k} = \frac{\hbar k}{2m} = \frac{p}{2m}, \tag{11.25}$$

where use has been made of Equation (11.3). However, this phase velocity is only half the classical velocity, $v = p/m$, of a massive (non-relativistic) particle.

11.6 PROBABILITY INTERPRETATION OF WAVEFUNCTION

After many false starts, physicists in the early 20th century eventually came to the conclusion that the only physical interpretation of a particle wavefunction that is consistent with experimental observations is probabilistic in nature (Dirac 1982). To be more exact, if $\psi(x, t)$ is the complex wavefunction of a given particle, moving in one dimension along the x-axis, then the probability of finding the particle between x and $x + dx$ at time t is

$$P(x, t) = |\psi(x, t)|^2 \, dx. \tag{11.26}$$

A probability is a real number lying in the range 0 to 1. An event that has a probability 0 is impossible. On the other hand, an event that has a probability 1 is certain to occur. An event that has a probability 1/2 (say) is such that in a very large number of identical trials the event occurs in half of the trials. We can interpret

$$P(t) = \int_{-\infty}^{\infty} |\psi(x, t)|^2 \, dx \tag{11.27}$$

as the probability of the particle being found anywhere between $x = -\infty$ and $x = +\infty$ at time t. This follows, via induction, from the fundamental result in probability theory that the probability of the occurrence of one or other of two mutually exclusive events (such as the particle being found in two non-overlapping regions) is the sum (or integral) of the probabilities of the individual events (Reif 2008). (For example, the probability of throwing a 1 on a six-sided die is 1/6. Likewise, the probability of throwing a 2 is $1/6$. Hence, the probability of throwing a 1 or a 2 is $1/6 + 1/6 = 1/3$.) Assuming that the particle exists, it is certain that it will be found somewhere between $x = -\infty$ and $x = +\infty$ at time t. Because a certain event has probability 1, our probability interpretation of the wavefunction is only tenable provided

$$\int_{-\infty}^{\infty} |\psi(x,t)|^2 \, dx = 1 \qquad (11.28)$$

at all times. A wavefunction that satisfies the previous condition—which is known as the *normalization condition*—is said to be properly normalized.

Suppose that we have a wavefunction, $\psi(x,t)$, which is such that it satisfies the normalization condition (11.28) at time $t = 0$. Furthermore, let the wavefunction evolve in time according to Schrödinger's equation, (11.23). Our probability interpretation of the wavefunction only makes sense if the normalization condition remains satisfied at all subsequent times. This follows because if the particle is certain to be found somewhere on the x-axis (which is the interpretation put on the normalization condition) at time $t = 0$ then it is equally certain to be found somewhere on the x-axis at a later time (because we are not considering any physical process by which particles can be created or destroyed). Thus, it is necessary for us to demonstrate that Schrödinger's equation preserves the normalization of the wavefunction.

Taking Schrödinger's equation, and multiplying it by ψ^* (the complex conjugate of the wavefunction), we obtain

$$i\hbar \frac{\partial \psi}{\partial t} \psi^* = -\frac{\hbar^2}{2m} \frac{\partial^2 \psi}{\partial x^2} \psi^* + U(x) |\psi|^2. \qquad (11.29)$$

The complex conjugate of the previous expression yields

$$-i\hbar \frac{\partial \psi^*}{\partial t} \psi = -\frac{\hbar^2}{2m} \frac{\partial^2 \psi^*}{\partial x^2} \psi + U(x) |\psi|^2. \qquad (11.30)$$

Here, use has been made of the readily demonstrated results $(\psi^*)^* = \psi$ and $i^* = -i$, as well as the fact that $U(x)$ is real. Taking the difference between the previous two expressions, we obtain

$$i\hbar \left(\frac{\partial \psi}{\partial t} \psi^* + \frac{\partial \psi^*}{\partial t} \psi \right) = -\frac{\hbar^2}{2m} \left(\frac{\partial^2 \psi}{\partial x^2} \psi^* - \frac{\partial^2 \psi^*}{\partial x^2} \psi \right), \qquad (11.31)$$

which can be written

$$i\hbar \frac{\partial |\psi|^2}{\partial t} = -\frac{\hbar^2}{2m} \frac{\partial}{\partial x} \left(\frac{\partial \psi}{\partial x} \psi^* - \frac{\partial \psi^*}{\partial x} \psi \right). \qquad (11.32)$$

Integrating in x, we get

$$i\hbar \frac{d}{dt} \int_{-\infty}^{\infty} |\psi|^2 \, dx = -\frac{\hbar^2}{2m} \left[\frac{\partial \psi}{\partial x} \psi^* - \frac{\partial \psi^*}{\partial x} \psi \right]_{-\infty}^{\infty}. \qquad (11.33)$$

Finally, assuming that the wavefunction is localized in space; that is,

$$|\psi(x,t)| \to 0 \quad \text{as} \quad |x| \to \infty, \qquad (11.34)$$

we obtain

$$\frac{d}{dt} \int_{-\infty}^{\infty} |\psi|^2 \, dx = 0. \qquad (11.35)$$

It follows, from the preceding analysis, that if a localized wavefunction is properly normalized at $t = 0$ (i.e., if $\int_{-\infty}^{\infty} |\psi(x,0)|^2 \, dx = 1$) then it will remain properly normalized as it evolves in time according to Schrödinger's equation. Incidentally, a wavefunction that is not localized cannot be properly normalized, because its normalization integral $\int_{-\infty}^{\infty} |\psi|^2 \, dx$ is necessarily infinite. For such a wavefunction, $|\psi(x,t)|^2 \, dx$ gives the relative, rather than the absolute, probability of finding the particle between x and $x + dx$ at time t. In other words, [cf., Equation (11.26)]

$$P(x,t) \propto |\psi(x,t)|^2 \, dx. \tag{11.36}$$

11.7 WAVE PACKETS

As we have seen, the wavefunction of a massive particle of momentum p and energy E, moving in free space along the x-axis, can be written

$$\psi(x,t) = \bar{\psi} \, e^{-i(\omega t - kx)}, \tag{11.37}$$

where $k = p/\hbar$, $\omega = E/\hbar$, and $\bar{\psi}$ is a complex constant. Here, ω and k are linked via the matter-wave dispersion relation (11.24). Expression (11.37) represents a plane wave that propagates in the x-direction with the phase velocity $v_p = \omega/k$. However, it follows from Equation (11.25) that this phase velocity is only half of the classical velocity of a massive particle.

According to the discussion in the previous section, the most reasonable physical interpretation of the wavefunction is that $|\psi(x,t)|^2 \, dx$ is proportional to (assuming that the wavefunction is not properly normalized) the probability of finding the particle between x and $x + dx$ at time t. However, the modulus squared of the wavefunction (11.37) is $|\bar{\psi}|^2$, which is a constant that depends on neither x nor t. In other words, the previous wavefunction represents a particle that is equally likely to be found anywhere on the x-axis at all times. Hence, the fact that this wavefunction propagates at a phase velocity that does not correspond to the classical particle velocity has no observable consequences.

How can we write the wavefunction of a particle that is localized in x? In other words, a particle that is more likely to be found at some positions on the x-axis than at others. It turns out that we can achieve this goal by forming a linear combination of plane waves of different wavenumbers; that is,

$$\psi(x,t) = \int_{-\infty}^{\infty} \bar{\psi}(k) \, e^{-i(\omega t - kx)} \, dk. \tag{11.38}$$

Here, $\bar{\psi}(k)$ represents the complex amplitude of plane waves of wavenumber k within this combination. In writing the previous expression, we are relying on the assumption that matter waves are superposable. In other words, it is possible to add two valid wave solutions to form a third valid wave solution. The ultimate justification for this assumption is that matter waves satisfy the linear wave equation (11.23).

There is a fundamental mathematical theorem, known as Fourier's theorem (see Section 8.2 and Exercise 11.11), which states that if

$$f(x) = \int_{-\infty}^{\infty} \bar{f}(k) \, e^{ikx} \, dk, \tag{11.39}$$

then

$$\bar{f}(k) = \frac{1}{2\pi} \int_{-\infty}^{\infty} f(x) \, e^{-ikx} \, dx. \tag{11.40}$$

Here, $\bar{f}(k)$ is known as the Fourier transform of the function $f(x)$. We can use Fourier's theorem to find the k-space function $\bar{\psi}(k)$ that generates any given x-space wavefunction $\psi(x)$ at a given time.

For instance, suppose that at $t = 0$ the wavefunction of our particle takes the form

$$\psi(x,0) \propto \exp\left[i\,k_0\,x - \frac{(x - x_0)^2}{4\,(\Delta x)^2} \right]. \tag{11.41}$$

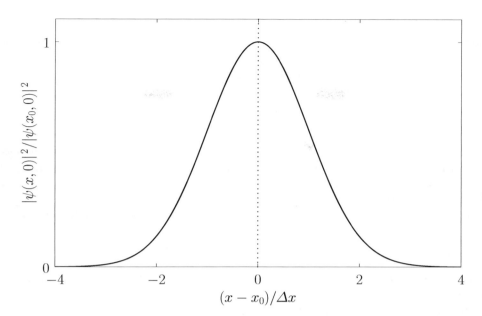

FIGURE 11.3   A one-dimensional Gaussian probability distribution.

Thus, the initial probability distribution for the particle's x-coordinate is

$$|\psi(x, 0)|^2 \propto \exp\left[-\frac{(x - x_0)^2}{2\,(\Delta x)^2}\right].\tag{11.42}$$

This particular distribution is called a Gaussian distribution (see Section 8.2), and is plotted in Figure 11.3. It can be seen that a measurement of the particle's position is most likely to yield the value x_0, and very unlikely to yield a value which differs from x_0 by more than $3\,\Delta x$. Thus, Equation (11.41) is the wavefunction of a particle that is initially localized in some region of x-space, centered on $x = x_0$, whose width is of order Δx. This type of wavefunction is known as a *wave packet*. However, a wave packet is just another name for a wave pulse. (See Chapter 8.)

According to Equation (11.38),

$$\psi(x, 0) = \int_{-\infty}^{\infty} \bar\psi(k)\,e^{ikx}\,dk.\tag{11.43}$$

Hence, we can employ Fourier's theorem to invert this expression to give

$$\bar\psi(k) \propto \int_{-\infty}^{\infty} \psi(x, 0)\,e^{-ikx}\,dx.\tag{11.44}$$

Making use of Equation (11.41), we obtain

$$\bar\psi(k) \propto e^{-i(k-k_0)\,x_0} \int_{-\infty}^{\infty} \exp\left[-i\,(k - k_0)\,(x - x_0) - \frac{(x - x_0)^2}{4\,(\Delta x)^2}\right] dx.\tag{11.45}$$

Changing the variable of integration to $y = (x - x_0)/(2\,\Delta x)$, the previous expression reduces to

$$\bar\psi(k) \propto e^{-i\,k\,x_0 - \beta^2/4} \int_{-\infty}^{\infty} e^{-(y-y_0)^2}\,dy,\tag{11.46}$$

where $\beta = 2\,(k - k_0)\,\Delta x$ and $y_0 = -\mathrm{i}\,\beta/2$. The integral in the previous equation is now just a number, as can easily be seen by making the second change of variable $z = y - y_0$. Hence, we deduce that

$$\bar{\psi}(k) \propto \exp\left[-\mathrm{i}\,k\,x_0 - \frac{(k - k_0)^2}{4\,(\Delta k)^2}\right],\qquad(11.47)$$

where

$$\Delta k = \frac{1}{2\,\Delta x}.\qquad(11.48)$$

If $|\psi(x, 0)|^2\,dx$ is proportional to the probability of a measurement of the particle's position yielding a value in the range x to $x + dx$ at time $t = 0$ then it stands to reason that $|\bar{\psi}(k)|^2\,dk$ is proportional to the probability of a measurement of the particle's wavenumber yielding a value in the range k to $k + dk$. (Recall that $p = \hbar k$, so a measurement of the particle's wavenumber, k, is equivalent to a measurement of the particle's momentum, p.) According to Equation (11.47),

$$|\bar{\psi}(k)|^2 \propto \exp\left[-\frac{(k - k_0)^2}{2\,(\Delta k)^2}\right].\qquad(11.49)$$

This probability distribution is a Gaussian in k-space. [See Equation (11.42) and Figure 11.3.] Hence, a measurement of k is most likely to yield the value k_0, and very unlikely to yield a value that differs from k_0 by more than $3\,\Delta k$.

We have just seen that a wave packet with a Gaussian probability distribution of characteristic width Δx in x-space [see Equation (11.42)] is equivalent to a wave packet with a Gaussian probability distribution of characteristic width Δk in k-space [see Equation (11.49)], where

$$\Delta x\,\Delta k = \frac{1}{2}.\qquad(11.50)$$

This illustrates an important property of wave packets. Namely, in order to construct a packet that is highly localized in x-space (i.e., with small Δx) we need to combine plane waves with a very wide range of different k-values (i.e., with large Δk). Conversely, if we only combine plane waves whose wavenumbers differ by a small amount (i.e., if Δk is small) then the resulting wave packet is highly extended in x-space (i.e., Δx is large).

According to Section 9.2, a wave packet made up of a superposition of plane waves that is strongly peaked around some central wavenumber k_0 propagates at the group velocity,

$$v_g = \frac{d\omega(k_0)}{dk},\qquad(11.51)$$

rather than the phase velocity, $v_p = (\omega/k)_{k_0}$, assuming that all of the constituent plane waves satisfy a dispersion relation of the form $\omega = \omega(k)$. For the case of matter waves, the dispersion relation is (11.24). Thus, the associated group velocity is

$$v_g = \frac{\hbar\,k_0}{m} = \frac{p}{m},\qquad(11.52)$$

where $p = \hbar\,k_0$. This velocity is identical to the classical velocity of a (non-relativistic) massive particle. We conclude that the matter-wave dispersion relation (11.24) is perfectly consistent with classical physics, as long as we recognize that particles must be identified with wave packets (which propagate at the group velocity) rather than plane waves (which propagate at the phase velocity).

In Section 9.2, it was demonstrated that the spatial extent of a wave packet of initial extent $(\Delta x)_0$ grows, as the packet evolves in time, like

$$\Delta x \simeq (\Delta x)_0 + \frac{d^2\omega(k_0)}{dk^2}\,\frac{t}{(\Delta x)_0},\qquad(11.53)$$

where k_0 is the packet's central wavenumber. Thus, it follows from the matter-wave dispersion relation, (11.24), that the width of a particle wave packet grows in time as

$$\Delta x \simeq (\Delta x)_0 + \frac{\hbar}{m} \frac{t}{(\Delta x)_0}. \tag{11.54}$$

For example, if an electron wave packet is initially localized in a region of atomic dimensions (i.e., $\Delta x \sim 10^{-10}$ m) then the width of the packet doubles in about 10^{-16} s.

11.8 HEISENBERG'S UNCERTAINTY PRINCIPLE

According to the analysis contained in the previous section, a particle wave packet that is initially localized in x-space, with characteristic width Δx, is also localized in k-space, with characteristic width $\Delta k = 1/(2\Delta x)$. However, as time progresses, the width of the wave packet in x-space increases [see Equation (11.54)], while that of the packet in k-space stays the same [because $\bar{\psi}(k)$ is given by Equation (11.44) at all times]. Hence, in general, we can say that

$$\Delta x \, \Delta k \gtrsim \frac{1}{2}. \tag{11.55}$$

Furthermore, we can interpret Δx and Δk as characterizing our uncertainty regarding the values of the particle's position and wavenumber, respectively.

A measurement of a particle's wavenumber, k, is equivalent to a measurement of its momentum, p, because $p = \hbar k$. Hence, an uncertainty in k of order Δk translates to an uncertainty in p of order $\Delta p = \hbar \Delta k$. It follows, from the previous inequality, that

$$\Delta x \, \Delta p \gtrsim \frac{\hbar}{2}. \tag{11.56}$$

This is the famous *Heisenberg uncertainty principle*, first proposed by Werner Heisenberg in 1927 (Dirac 1982). According to this principle, it is impossible to simultaneously measure the position and momentum of a particle (exactly). Indeed, a good knowledge of the particle's position implies a poor knowledge of its momentum, and vice versa. The uncertainty principle is a direct consequence of representing particles as waves.

It is apparent, from Equation (11.54), that a particle wave packet of initial spatial extent $(\Delta x)_0$ spreads out in such a manner that its spatial extent becomes

$$\Delta x \sim \frac{\hbar t}{m (\Delta x)_0} \tag{11.57}$$

at large t. It is readily demonstrated that this spreading of the wave packet is a consequence of the uncertainty principle. Indeed, because the initial uncertainty in the particle's position is $(\Delta x)_0$, it follows that the uncertainty in its momentum is of order $\hbar/(\Delta x)_0$. This translates to an uncertainty in velocity of $\Delta v = \hbar/[m(\Delta x)_0]$. Thus, if we imagine that part of the wave packet propagates at $v_0 + \Delta v/2$, and another part at $v_0 - \Delta v/2$, where v_0 is the mean propagation velocity, then it follows that the wave packet will spread out as time progresses. Indeed, at large t, we expect the width of the wave packet to be

$$\Delta x \sim \Delta v \, t \sim \frac{\hbar t}{m (\Delta x)_0}, \tag{11.58}$$

which is identical to Equation (11.57). Evidently, the spreading of a particle wave packet, as time progresses, should be interpreted as representing an increase in our uncertainty regarding the particle's position, rather than an increase in the spatial extent of the particle itself.

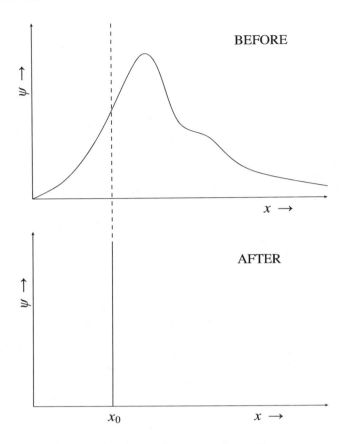

FIGURE 11.4   Collapse of the wavefunction upon measurement of x.

11.9   WAVEFUNCTION COLLAPSE

Consider a spatially extended wavefunction, $\psi(x,t)$. According to our usual interpretati
$|\psi(x,t)|^2\,dx$ is proportional to the probability of a measurement of the particle's position yie
ing a value in the range x to $x + dx$ at time t. Thus, if the wavefunction is extended then there
wide range of likely values that such a measurement could give. Suppose, however, that we m
a measurement of the particle's position, and obtain the value x_0. We now know that the part
is located at $x = x_0$. If we make another measurement, immediately after the first one, then w
value would we expect to obtain? Common sense tells us that we should obtain the same val
x_0, because the particle cannot have shifted position appreciably in an infinitesimal time inter
Thus, immediately after the first measurement, a measurement of the particle's position is certai
give the value x_0, and has no chance of giving any other value. This implies that the wavefunct
must have collapsed to some sort of "spike" function, centered on $x = x_0$. This idea is illustra
in Figure 11.4. As soon as the wavefunction collapses, it starts to expand again, as described in
previous section. Thus, the second measurement must be made reasonably quickly after the f
one, otherwise the same result will not necessarily be obtained.

The preceding discussion illustrates an important point in wave mechanics. That is, the wa
function of a massive particle changes discontinuously (in time) whenever a measurement of
particle's position is made. We conclude that there are two types of time evolution of the wa
function in wave mechanics. First, there is a smooth evolution that is governed by Schröding

equation. This evolution takes place between measurements. Second, there is a discontinuous evolution that takes place each time a measurement is made.

11.10   STATIONARY STATES

Consider separable solutions to Schrödinger's equation of the form

$$\psi(x, t) = \psi(x)\, e^{-i\omega t}. \tag{11.59}$$

According to Equation (11.20), such solutions have definite energies $E = \hbar\omega$. For this reason, they are usually written

$$\psi(x, t) = \psi(x)\, e^{-i(E/\hbar)\, t}. \tag{11.60}$$

The probability of finding the particle between x and $x + dx$ at time t is

$$P(x, t) = |\psi(x, t)|^2\, dx = |\psi(x)|^2\, dx. \tag{11.61}$$

This probability is time independent. For this reason, states whose wavefunctions are of the form (11.60) are known as *stationary states*. Moreover, $\psi(x)$ is called a stationary wavefunction. Substituting (11.60) into Schrödinger's equation, (11.23), we obtain the following differential equation for $\psi(x)$;

$$-\frac{\hbar^2}{2m}\frac{d^2\psi}{dx^2} + U(x)\,\psi = E\,\psi. \tag{11.62}$$

This equation is called the *time-independent Schrödinger equation*.

Consider a particle trapped in a one-dimensional square potential well, of infinite depth, which is such that

$$U(x) = \begin{cases} 0 & 0 \le x \le a \\ \infty & \text{otherwise} \end{cases}. \tag{11.63}$$

The particle is excluded from the region $x < 0$ or $x > a$, so $\psi = 0$ in this region (i.e., there is zero probability of finding the particle outside the well). Within the well, a particle of definite energy E has a stationary wavefunction, $\psi(x)$, that satisfies

$$-\frac{\hbar^2}{2m}\frac{d^2\psi}{dx^2} = E\,\psi. \tag{11.64}$$

The boundary conditions are

$$\psi(0) = \psi(a) = 0. \tag{11.65}$$

This follows because $\psi = 0$ in the region $x < 0$ or $x > a$, and $\psi(x)$ must be continuous [because a discontinuous wavefunction would generate a singular term (i.e., the term involving $d^2\psi/dx^2$) in the time-independent Schrödinger equation, (11.62), that could not be balanced, even by an infinite potential].

Let us search for solutions to Equation (11.64) of the form

$$\psi(x) = \psi_0\, \sin(k\, x), \tag{11.66}$$

where ψ_0 is a constant. It follows that

$$\frac{\hbar^2 k^2}{2m} = E. \tag{11.67}$$

The solution (11.66) automatically satisfies the boundary condition $\psi(0) = 0$. The second boundary condition, $\psi(a) = 0$, leads to a quantization of the wavenumber; that is,

$$k = n\,\frac{\pi}{a}, \tag{11.68}$$

where $n = 1, 2, 3$, et cetera. (A "quantized" quantity is one that can only take certain discrete values.) According to Equation (11.67), the energy is also quantized. In fact, $E = E_n$, where

$$E_n = n^2 \frac{\hbar^2 \pi^2}{2 m a^2}. \tag{11.69}$$

Thus, the allowed wavefunctions for a particle trapped in a one-dimensional square potential well of infinite depth are

$$\psi_n(x, t) = A_n \sin\left(n \pi \frac{x}{a}\right) \exp\left(-i n^2 \frac{E_1}{\hbar} t\right), \tag{11.70}$$

where n is a positive integer, and A_n a constant. We cannot have $n = 0$, because, in this case, we obtain a null wavefunction; that is, $\psi = 0$, everywhere. Furthermore, if n takes a negative integer value then it generates exactly the same wavefunction as the corresponding positive integer value (assuming $A_{-n} = -A_n$).

The constant A_n, appearing in the previous wavefunction, can be determined from the constraint that the wavefunction be properly normalized. For the case under consideration, the normalization condition (11.28) reduces to

$$\int_0^a |\psi(x)|^2 \, dx = 1. \tag{11.71}$$

It follows from Equation (11.70) that $|A_n|^2 = 2/a$. Hence, the properly normalized version of the wavefunction (11.70) is

$$\psi_n(x, t) = \left(\frac{2}{a}\right)^{1/2} \sin\left(n \pi \frac{x}{a}\right) \exp\left(-i n^2 \frac{E_1}{\hbar} t\right). \tag{11.72}$$

Figure 11.5 shows the first four properly normalized stationary wavefunctions for a particle trapped in a one-dimensional square potential well of infinite depth; that is, $\psi_n(x) = (2/a)^{1/2} \sin(n \pi x/a)$, for $n = 1$ to 4.

The stationary wavefunctions that we have just found are, in essence, standing wave solutions to Schrödinger's equation. Indeed, the wavefunctions are very similar in form to the classical standing wave solutions discussed in Chapters 4 and 5.

At first sight, it seems rather strange that the lowest possible energy for a particle trapped in a one-dimensional potential well is not zero, as would be the case in classical mechanics, but rather $E_1 = \hbar^2 \pi^2/(2 m a^2)$. In fact, as explained in the following, this residual energy is a direct consequence of Heisenberg's uncertainty principle. A particle trapped in a one-dimensional well of width a is likely to be found anywhere inside the well. Thus, the uncertainty in the particle's position is $\Delta x \sim a$. It follows from the uncertainty principle, (11.56), that

$$\Delta p \gtrsim \frac{\hbar}{2 \Delta x} \sim \frac{\hbar}{a}. \tag{11.73}$$

In other words, the particle cannot have zero momentum. In fact, the particle's momentum must be at least $p \sim \hbar/a$. However, for a free particle, $E = p^2/2 m$. Hence, the residual energy associated with the particle's residual momentum is

$$E \sim \frac{p^2}{m} \sim \frac{\hbar^2}{m a^2} \sim E_1. \tag{11.74}$$

This type of residual energy, which often occurs in quantum mechanical systems, and has no equivalent in classical mechanics, is called *zero point energy*.

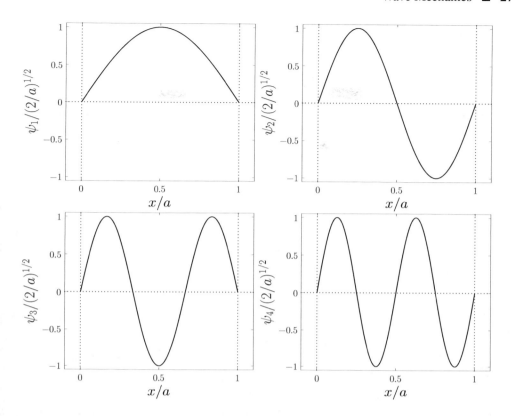

FIGURE 11.5 First four stationary wavefunctions for a particle trapped in a one-dimensional square potential well of infinite depth.

11.11 PARTICLE IN FINITE SQUARE POTENTIAL WELL

Consider a particle of mass m trapped in a one-dimensional, square, potential well of width a and finite depth $V > 0$. Suppose that the potential takes the form

$$U(x) = \begin{cases} -V & |x| \leq a/2 \\ 0 & \text{otherwise} \end{cases}. \qquad (11.75)$$

Here, we have adopted the standard convention that $U(x) \to 0$ as $|x| \to \infty$. This convention is useful because, just as in classical mechanics, a particle whose overall energy, E, is negative is bound in the well (i.e., it cannot escape to infinity), whereas a particle whose overall energy is positive is unbound (Fitzpatrick 2012). Because we are interested in bound particles, we shall assume that $E < 0$. We shall also assume that $E + V > 0$, in order to allow the particle to have a positive kinetic energy inside the well.

Let us search for a stationary state

$$\psi(x, t) = \psi(x) \, e^{-i (E/\hbar) t}, \qquad (11.76)$$

whose stationary wavefunction, $\psi(x)$, satisfies the time-independent Schrödinger equation, (11.62). Solutions to Equation (11.62) in the symmetric [i.e., $U(-x) = U(x)$] potential (11.75) are either totally symmetric [i.e., $\psi(-x) = \psi(x)$] or totally antisymmetric [i.e., $\psi(-x) = -\psi(x)$]. Moreover, the solutions must satisfy the boundary condition

$$\psi \to 0 \quad \text{as} \quad |x| \to \infty, \tag{11.77}$$

otherwise they would not correspond to bound states.

Let us, first of all, search for a totally-symmetric solution. In the region to the left of the well (i.e., $x < -a/2$), the solution of the time-independent Schrödinger equation that satisfies the boundary condition $\psi \to 0$ as $x \to -\infty$ is

$$\psi(x) = A\,e^{qx}, \tag{11.78}$$

where

$$q = \sqrt{\frac{2\,m\,(-E)}{\hbar^2}}, \tag{11.79}$$

and A is a constant. By symmetry, the solution in the region to the right of the well (i.e., $x > a/2$) is

$$\psi(x) = A\,e^{-qx}. \tag{11.80}$$

The solution inside the well (i.e., $|x| \le a/2$) that satisfies the symmetry constraint $\psi(-x) = \psi(x)$ is

$$\psi(x) = B\,\cos(k\,x), \tag{11.81}$$

where

$$k = \sqrt{\frac{2\,m\,(V+E)}{\hbar^2}}, \tag{11.82}$$

and B is a constant. The appropriate matching conditions at the edges of the well (i.e., $x = \pm a/2$) are that $\psi(x)$ and $d\psi(x)/dx$ both be continuous [because a discontinuity in the wavefunction, or its first derivative, would generate a singular term in the time-independent Schrödinger equation (i.e., the term involving $d^2\psi/dx^2$) that could not be balanced]. The matching conditions yield

$$q = k\,\tan(k\,a/2). \tag{11.83}$$

Let $y = k\,a/2$. It follows that

$$E = E_0\,y^2 - V, \tag{11.84}$$

where

$$E_0 = \frac{2\,\hbar^2}{m\,a^2}. \tag{11.85}$$

Moreover, Equation (11.83) becomes

$$\frac{\sqrt{\lambda - y^2}}{y} = \tan y, \tag{11.86}$$

with

$$\lambda = \frac{V}{E_0}. \tag{11.87}$$

Here, y must lie in the range $0 < y < \lambda^{1/2}$, in order to ensure that E lies in the range $-V < E < 0$.

The solutions of Equation (11.86) correspond to the intersection of the curve $(\lambda - y^2)^{1/2}/y$ with the curve $\tan y$. Figure 11.6 shows these two curves plotted for a particular value of λ. In this case, the curves intersect twice, indicating the existence of two totally-symmetric bound states in the well. It is apparent, from the figure, that as λ increases (i.e., as the well becomes deeper) there are more and more bound states. However, it is also apparent that there is always at least one totally-symmetric bound state, no matter how small λ becomes (i.e., no matter how shallow the well becomes). In

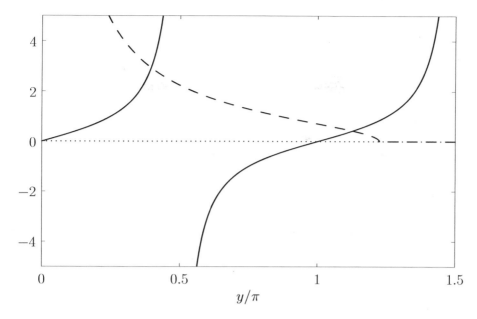

FIGURE 11.6   The curves $\tan y$ (solid) and $(\lambda - y^2)^{1/2}/y$ (dashed), calculated for $\lambda = 1.5\,\pi^2$. The latter curve takes the value 0 when $y > \lambda^{1/2}$.

the limit $\lambda \gg 1$ (i.e., the limit in which the well is very deep), the solutions to Equation (11.86) asymptote to the roots of $\tan y = \infty$. This gives $y = (2\,n - 1)\,\pi/2$, where n is a positive integer, or

$$k = (2\,n - 1)\,\frac{\pi}{a}. \qquad (11.88)$$

These solutions are equivalent to the odd-n infinite-depth potential well solutions specified by Equation (11.68).

For the case of a totally-antisymmetric bound state, similar analysis to the preceding yields (see Exercise 11.12)

$$-\frac{y}{\sqrt{\lambda - y^2}} = \tan y. \qquad (11.89)$$

The solutions of this equation correspond to the intersection of the curve $\tan y$ with the curve $-y/(\lambda - y^2)^{1/2}$. Figure 11.7 shows these two curves plotted for the same value of λ as that used in Figure 11.6. In this case, the curves intersect once, indicating the existence of a single totally-antisymmetric bound state in the well. It is, again, apparent, from the figure, that as λ increases (i.e., as the well becomes deeper) there are more and more bound states. However, it is also apparent that when λ becomes sufficiently small [i.e., $\lambda < (\pi/2)^2$] then there is no totally antisymmetric bound state. In other words, a very shallow potential well always possesses a totally-symmetric bound state, but does not generally possess a totally-antisymmetric bound state. In the limit $\lambda \gg 1$ (i.e., the limit in which the well becomes very deep), the solutions to Equation (11.89) asymptote to the roots of $\tan y = 0$. This gives $y = n\,\pi$, where n is a positive integer, or

$$k = 2\,n\,\frac{\pi}{a}. \qquad (11.90)$$

These solutions are equivalent to the even-n infinite-depth potential well solutions specified by Equation (11.68).

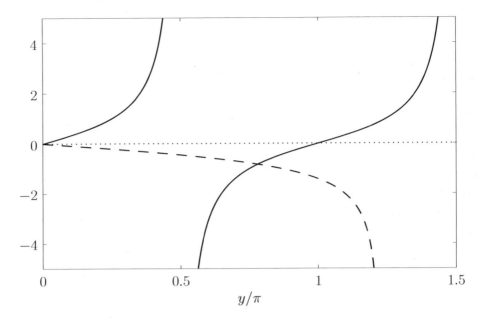

FIGURE 11.7 The curves $\tan y$ (solid) and $-y/(\lambda - y^2)^{1/2}$ (dashed), calculated for $\lambda = 1.5\,\pi^2$.

Probably the most surprising aspect of the bound states that we have just described is the p
sibility of finding the particle outside the well; that is, in the region $|x| > a/2$ where $U(x) >$
This follows from Equation (11.80) and (11.81) because the ratio $A/B = \exp(q\,a/2)\cos(k\,a/2$
not necessarily zero. Such behavior is strictly forbidden in classical mechanics, according to wh
a particle of energy E is restricted to regions of space where $E > U(x)$ (Fitzpatrick 2012). In fa
in the case of the ground state (i.e., the lowest energy symmetric state) it is possible to demonstr
that the probability of a measurement finding the particle outside the well is (see Exercise 11.13

$$P_{\text{out}} \simeq 1 - 2\lambda \qquad (11.$$

for a shallow well (i.e., $\lambda \ll 1$), and

$$P_{\text{out}} \simeq \frac{\pi^2}{4}\frac{1}{\lambda^{3/2}} \qquad (11.$$

for a deep well (i.e., $\lambda \gg 1$). It follows that the particle is very likely to be found outside a shall
well, and there is a small, but finite, probability of it being found outside a deep well. In fact,
probability of finding the particle outside the well only goes to zero in the case of an infinitely d
well (i.e., $\lambda \to \infty$).

11.12   SQUARE POTENTIAL BARRIER

Consider a particle of mass m and energy $E > 0$ interacting with the simple, one-dimension
potential barrier

$$U(x) = \begin{cases} V & \text{for } 0 \le x \le a \\ 0 & \text{otherwise} \end{cases} \qquad (11.$$

where $V > 0$. In the regions to the left and to the right of the barrier, the stationary wavefuncti
$\psi(x)$, satisfies

$$\frac{d^2\psi}{dx^2} = -k^2\psi, \qquad (11.$$

where

$$k = \sqrt{\frac{2\,m\,E}{\hbar^2}}.$$
(11.95)

Let us adopt the following solution of the previous equation to the left of the barrier (i.e., $x < 0$):

$$\psi(x) = e^{ikx} + R\,e^{-ikx}.$$
(11.96)

This solution consists of a plane wave of unit amplitude traveling to the right [because the full wave-function is multiplied by a factor $\exp(-i\,E\,t/\hbar)$], and a plane wave of complex amplitude R traveling to the left. We interpret the first plane wave as an incident particle, and the second as a particle reflected by the potential barrier. Hence, $|R|^2$ is the probability of reflection. (See Sections 6.7 and 11.6.)

Let us adopt the following solution to Equation (11.94) to the right of the barrier (i.e. $x > a$):

$$\psi(x) = T\,e^{ikx}.$$
(11.97)

This solution consists of a plane wave of complex amplitude T traveling to the right. We interpret the plane wave as a particle transmitted through the barrier. Hence, $|T|^2$ is the probability of transmission.

Let us, first of all, consider the situation in which $E > V$. In this case, according to classical mechanics, the particle slows down as it passes through the barrier, but is otherwise unaffected. In other words, the classical probability of reflection is zero, and the classical probability of transmission is unity. However, this is not necessarily the case in wave mechanics. In fact, inside the barrier (i.e., $0 \le x \le a$), $\psi(x)$ satisfies

$$\frac{d^2\psi}{dx^2} = -q^2\psi,$$
(11.98)

where

$$q = \sqrt{\frac{2\,m\,(E-V)}{\hbar^2}}.$$
(11.99)

The general solution to Equation (11.98) takes the form

$$\psi(x) = A\,e^{iqx} + B\,e^{-iqx}.$$
(11.100)

Continuity of ψ and $d\psi/dx$ at the left edge of the barrier (i.e., $x = 0$) yields

$$1 + R = A + B,$$
(11.101)

$$k\,(1-R) = q\,(A-B).$$
(11.102)

Likewise, continuity of ψ and $d\psi/dx$ at the right edge of the barrier (i.e., $x = a$) gives

$$A\,e^{iqa} + B\,e^{-iqa} = T\,e^{ika},$$
(11.103)

$$q\left(A\,e^{iqa} - B\,e^{-iqa}\right) = kT\,e^{ika}.$$
(11.104)

After considerable algebra, the previous four equations yield

$$|T|^2 = 1 - |R|^2 = \frac{4\,k^2\,q^2}{4\,k^2\,q^2 + (k^2 - q^2)^2\,\sin^2(q\,a)}.$$
(11.105)

The fact that $|R|^2 + |T|^2 = 1$ ensures that the probabilities of reflection and transmission sum to unity, as must be the case, because reflection and transmission are the only possible outcomes for a particle incident on the barrier.

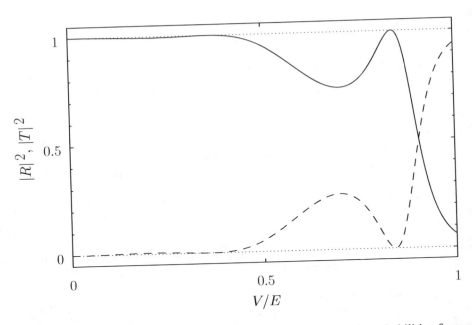

FIGURE 11.8 Transmission (solid curve) and reflection (dashed curve) probabilities for a square potential barrier of width $a = 1.25\,\lambda$, where λ is the free-space de Broglie wavelength, as a function of the ratio of the height of the barrier, V, to the energy, E, of the incident particle.

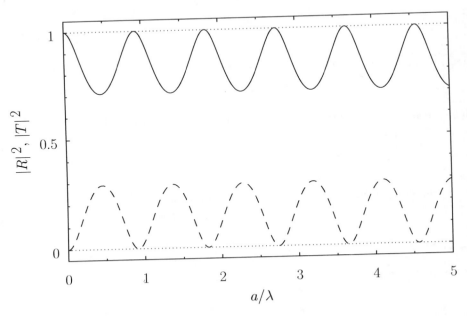

FIGURE 11.9 Transmission (solid curve) and reflection (dashed curve) probabilities for a particle of energy E incident on a square potential barrier of height $V = 0.75\,E$ as a function of the ratio of the width of the barrier, a, to the free-space de Broglie wavelength, λ.

The reflection and transmission probabilities obtained from Equation (11.105) are plotted in Figures 11.8 and 11.9. It can be seen, from Figure 11.8, that the classical result, $|R|^2 = 0$ and $|T|^2 = 1$, is obtained in the limit where the height of the barrier is relatively small (i.e., $V \ll E$). However, if V is of order E then there is a substantial probability that the incident particle will be reflected by the barrier. According to classical physics, reflection is impossible when $V < E$.

It can also be seen, from Figure 11.9, that at certain barrier widths the probability of reflection goes to zero. It turns out that this is true irrespective of the energy of the incident particle. It is evident, from Equation (11.105), that these special barrier widths correspond to

$$q\,a = n\,\pi, \tag{11.106}$$

where $n = 1, 2, 3, \cdots$. In other words, the special barrier widths are integer multiples of half the de Broglie wavelength of the particle inside the barrier. There is no reflection at the special barrier widths because, at these widths, the backward traveling wave reflected from the left edge of the barrier interferes destructively with the similar wave reflected from the right edge of the barrier to give zero net reflected wave. (See Section 6.7.)

Let us now consider the situation in which $E < V$. In this case, according to classical mechanics, the particle is unable to penetrate the barrier, so the coefficient of reflection is unity, and the coefficient of transmission zero. However, this is not necessarily the case in wave mechanics. In fact, inside the barrier (i.e., $0 \le x \le a$), $\psi(x)$ satisfies

$$\frac{d^2\psi}{dx^2} = q^2\,\psi, \tag{11.107}$$

where

$$q = \sqrt{\frac{2\,m\,(V - E)}{\hbar^2}}. \tag{11.108}$$

The general solution to Equation (11.107) takes the form

$$\psi(x) = A\,e^{q\,x} + B\,e^{-q\,x}. \tag{11.109}$$

Continuity of ψ and $d\psi/dx$ at the left edge of the barrier (i.e., $x = 0$) yields

$$1 + R = A + B, \tag{11.110}$$

$$i\,k\,(1 - R) = q\,(A - B). \tag{11.111}$$

Likewise, continuity of ψ and $d\psi/dx$ at the right edge of the barrier (i.e., $x = a$) gives

$$A\,e^{q\,a} + B\,e^{-q\,a} = T\,e^{i\,k\,a}, \tag{11.112}$$

$$q\,(A\,e^{q\,a} - B\,e^{-q\,a}) = i\,k\,T\,e^{i\,k\,a}. \tag{11.113}$$

After considerable algebra, the preceding four equations yield

$$|T|^2 = 1 - |R|^2 = \frac{4\,k^2\,q^2}{4\,k^2\,q^2 + (k^2 + q^2)^2\,\sinh^2(q\,a)}. \tag{11.114}$$

The fact that $|R|^2 + |T|^2 = 1$ again ensures that the probabilities of reflection and transmission sum to unity, as must be the case, because reflection and transmission are the only possible outcomes for a particle incident on the barrier.

The reflection and transmission probabilities obtained from Equation (11.114) are plotted in Figures 11.10 and 11.11. It can be seen, from these two figures, that the classical result, $|R|^2 = 1$

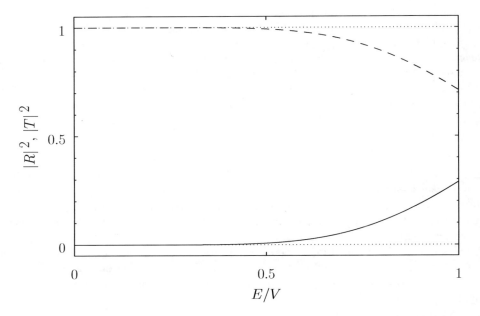

FIGURE 11.10 Transmission (solid curve) and reflection (dashed curve) probabilities for a square potential barrier of width $a = 0.5\,\lambda$, where λ is the free-space de Broglie wavelength, as a function of the ratio of the energy, E, of the incoming particle to the height, V, of the barrier.

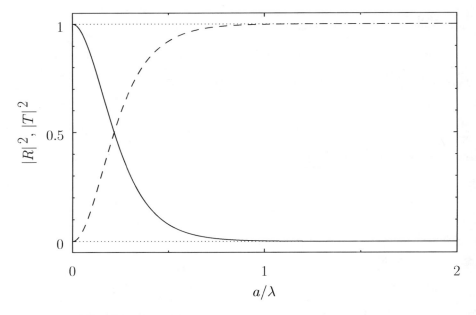

FIGURE 11.11 Transmission (solid curve) and reflection (dashed curve) probabilities for a particle of energy E incident on a square potential barrier of height $V = (4/3)\,E$ as a function of the ratio of the width of the barrier, a, to the free-space de Broglie wavelength, λ.

and $|T|^2 = 0$, is obtained for relatively thin barriers (i.e., $q a \sim 1$) in the limit where the height of the barrier is relatively large (i.e., $V \gg E$). However, if V is of order E then there is a substantial probability that the incident particle will be transmitted by the barrier. According to classical physics, transmission is impossible when $V > E$.

It can also be seen, from Figure 11.11, that the transmission probability decays exponentially as the width of the barrier increases. Nevertheless, even for very wide barriers (i.e., $q a \gg 1$), there is a small but finite probability that a particle incident on the barrier will be transmitted. This phenomenon, which is inexplicable within the context of classical physics, is called *tunneling*. For the case of a very high barrier, such that $V \gg E$, the tunneling probability reduces to

$$|T|^2 \simeq \frac{4 E}{V} e^{-2a/\lambda}, \tag{11.115}$$

where $\lambda = (\hbar^2/2 m V)^{1/2}$ is the de Broglie wavelength inside the barrier. Here, it is assumed that $a \gg \lambda$. Thus, even in the limit that the barrier is very high, there is an exponentially small, but nevertheless non-zero, tunneling probability. Quantum mechanical tunneling plays an important role in the physics of electron field emission and α-decay (Park 1974). (See Sections 11.14 and 11.15.)

11.13 WKB APPROXIMATION

Consider a particle of mass m and energy $E > 0$ moving through some slowly-varying potential, $U(x)$. The particle's wavefunction satisfies

$$\frac{d^2\psi(x)}{dx^2} = -k^2(x)\,\psi(x), \tag{11.116}$$

where

$$k^2(x) = \frac{2 m [E - U(x)]}{\hbar^2}. \tag{11.117}$$

Let us try a solution to Equation (11.116) of the form

$$\psi(x) = \psi_0 \exp\left(\int_0^x i k(x')\,dx'\right), \tag{11.118}$$

where ψ_0 is a complex constant. Note that this solution represents a particle propagating in the positive x-direction [because the full wavefunction is multiplied by $\exp(-i\omega t)$, where $\omega = E/\hbar > 0$] with the continuously-varying wavenumber $k(x)$. It follows that

$$\frac{d\psi(x)}{dx} = i k(x)\,\psi(x), \tag{11.119}$$

and

$$\frac{d^2\psi(x)}{dx^2} = i k'(x)\,\psi(x) - k^2(x)\,\psi(x), \tag{11.120}$$

where $k' \equiv dk/dx$. A comparison of Equations (11.116) and (11.120) reveals that Equation (11.118) represents an approximate solution to Equation (11.116) provided that the first term on the right-hand side of Equation (11.120) is negligible compared to the second. This yields the validity criterion $|k'| \ll k^2$, or

$$\frac{k}{|k'|} \gg k^{-1}. \tag{11.121}$$

In other words, the variation lengthscale of $k(x)$, which is approximately the same as the variation lengthscale of $U(x)$, must be much greater than the particle's de Broglie wavelength (which is of order k^{-1}). Let us suppose that this is the case. Incidentally, the approximation involved in dropping the first term on the right-hand side of Equation (11.120) is generally known as the *WKB approximation*, after G. Wentzel, H.A. Kramers, and L. Brillouin. (See Section 6.10.) Similarly, Equation (11.118) is termed a WKB solution.

According to the WKB solution, (11.118), the probability density remains constant; that is,

$$|\psi(x)|^2 = |\psi_0|^2; \tag{11.122}$$

as long as the particle moves through a region in which $E > U(x)$, and $k(x)$ is consequently real (i.e., an allowed region according to classical physics). Suppose, however, that the particle encounters a potential barrier (i.e., a region from which the particle is excluded according to classical physics). By definition, $E < U(x)$ inside such a barrier, and $k(x)$ is consequently imaginary. Let the barrier extend from $x = x_1$ to x_2, where $0 < x_1 < x_2$. The WKB solution inside the barrier is written

$$\psi(x) = \psi_1 \exp\left(-\int_{x_1}^{x} |k(x')| \, dx'\right), \tag{11.123}$$

where

$$\psi_1 = \psi_0 \exp\left(\int_{0}^{x_1} i\,k(x')\,dx'\right). \tag{11.124}$$

Here, we have neglected the unphysical exponentially-growing solution.

According to the WKB solution, (11.123), the probability density decays exponentially inside the barrier; that is,

$$|\psi(x)|^2 = |\psi_1|^2 \exp\left(-2\int_{x_1}^{x} |k(x')| \, dx'\right), \tag{11.125}$$

where $|\psi_1|^2$ is the probability density at the left-hand side of the barrier (i.e., $x = x_1$). It follows that the probability density at the right-hand side of the barrier (i.e., $x = x_2$) is

$$|\psi_2|^2 = |\psi_1|^2 \exp\left(-2\int_{x_1}^{x_2} |k(x')| \, dx'\right). \tag{11.126}$$

Note that $|\psi_2|^2 < |\psi_1|^2$. Of course, in the region to the right of the barrier (i.e., $x > x_2$), the probability density takes the constant value $|\psi_2|^2$.

We can interpret the ratio of the probability densities to the right and to the left of the potential barrier as the probability, $|T|^2$, that a particle incident from the left will tunnel through the barrier and emerge on the other side; that is,

$$|T|^2 = \frac{|\psi_2|^2}{|\psi_1|^2} = \exp\left(-2\int_{x_1}^{x_2} |k(x')| \, dx'\right). \tag{11.127}$$

It is easily demonstrated that the probability of a particle incident from the right tunneling through the barrier is the same.

Note that the criterion (11.121) for the validity of the WKB approximation implies that the previous transmission probability is very small. Hence, the WKB approximation only applies to situations in which there is very little chance of a particle tunneling through the potential barrier in question. Unfortunately, the validity criterion (11.121) breaks down completely at the edges of the barrier (i.e., at $x = x_1$ and x_2), because $k(x) = 0$ at these points. However, it can be demonstrated that the contribution of those regions, around $x = x_1$ and x_2, in which the WKB approximation breaks down to the integral in Equation (11.127) is fairly negligible (Schiff 1955). Hence, the previous expression for the tunneling probability is a reasonable approximation provided that the incident particle's de Broglie wavelength is much smaller than the spatial extent of the potential barrier.

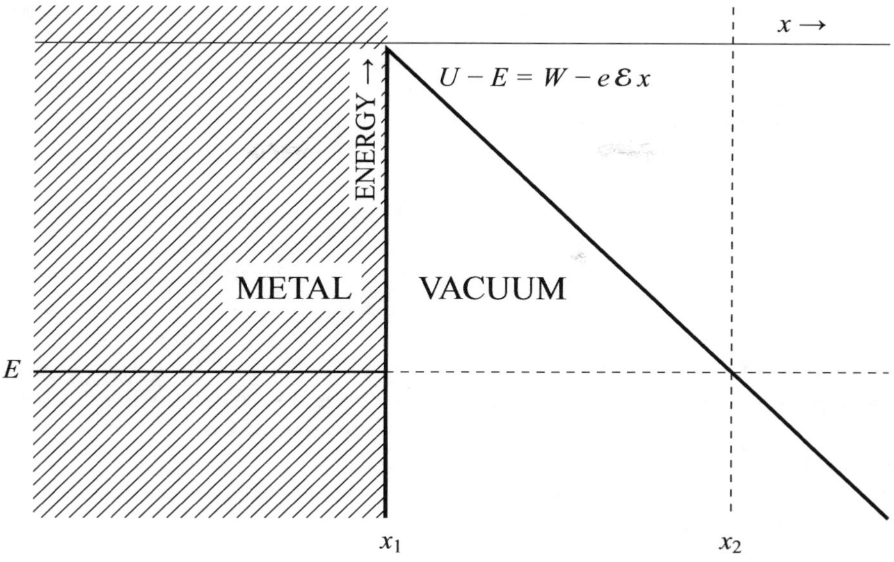

FIGURE 11.12   The potential barrier for an electron in a metal surface subject to an external electric field.

11.14   COLD EMISSION

Suppose that an unheated metal surface is subject to a large uniform external electric field, of strength $\mathcal{E}$, which is directed such that it accelerates electrons away from the surface. We have already seen (in Section 11.2) that electrons just below the surface of a metal can be regarded as being in a potential well of depth W, where W is called the workfunction of the surface. Adopting a simple one-dimensional treatment of the problem, let the metal lie at $x < 0$, and the surface at $x = 0$. Now, the applied electric field is shielded from the interior of the metal (Fitzpatrick 2008). Hence, the energy, E, say, of an electron just below the surface is unaffected by the field. In the absence of the electric field, the potential barrier just above the surface is simply $U(x) - E = W$. The electric field modifies this to $U(x) - E = W - e\,\mathcal{E}\,x$, where e is the magnitude of the electron charge. The potential barrier is sketched in Figure 11.12.

It can be seen, from Figure 11.12, that an electron just below the surface of the metal is confined by a triangular potential barrier that extends from $x = x_1$ to x_2, where $x_1 = 0$ and $x_2 = W/(e\,\mathcal{E})$. Making use of the WKB approximation (see Section 11.13), the probability of such an electron tunneling through the barrier, and consequently being emitted from the surface, is

$$|T|^2 = \exp\left(-\frac{2\sqrt{2\,m_e}}{\hbar} \int_{x_1}^{x_2} \sqrt{U(x) - E}\,dx\right), \tag{11.128}$$

or

$$|T|^2 = \exp\left(-\frac{2\sqrt{2\,m_e}}{\hbar} \int_{0}^{W/e\mathcal{E}} \sqrt{W - e\,\mathcal{E}\,x}\,dx\right), \tag{11.129}$$

where m_e is the electron mass. This reduces to

$$|T|^2 = \exp\left(-2\sqrt{2}\,\frac{m_e^{1/2}\,W^{3/2}}{\hbar\,e\,\mathcal{E}} \int_{0}^{1} \sqrt{1 - y}\,dy\right), \tag{11.130}$$

or

$$|T|^2 = \exp\left(-\frac{4\sqrt{2}}{3}\frac{m_e^{1/2}\,W^{3/2}}{\hbar\,e\,\mathcal{E}}\right). \tag{11.131}$$

The previous result is known as the *Fowler–Nordheim formula*. Note that the probability of emission increases exponentially as the electric field-strength above the surface of the metal increases.

The cold emission of electrons from a metal surface is the basis of an important device known as a *scanning tunneling microscope*, or an STM. An STM consists of a very sharp conducting probe that is scanned over the surface of a metal (or any other solid conducting medium). A large voltage difference is applied between the probe and the surface. Now, the surface electric field-strength immediately below the probe tip is proportional to the applied potential difference, and inversely proportional to the spacing between the tip and the surface. Electrons tunneling between the surface and the probe tip give rise to a weak electric current. The magnitude of this current is proportional to the tunneling probability, (11.131). It follows that the current is an extremely sensitive function of the surface electric field-strength, and, hence, of the spacing between the tip and the surface (assuming that the potential difference is held constant). An STM can, thus, be used to construct a very accurate contour map of the surface under investigation. In fact, STMs are capable of achieving sufficient resolution to image individual atoms.

11.15   ALPHA DECAY

Many types of heavy atomic nuclei spontaneously decay to produce daughter nuclei via the emission of α-particles (i.e., helium nuclei) of some characteristic energy. This process is known as α-decay. Let us investigate the α-decay of a particular type of atomic nucleus of radius R, charge-number Z, and mass-number A. Such a nucleus thus decays to produce a daughter nucleus of charge-number $Z_1 = Z - 2$ and mass-number $A_1 = A - 4$, and an α-particle of charge-number $Z_2 = 2$ and mass-number $A_2 = 4$. Let the characteristic energy of the α-particle be E. Incidentally, nuclear radii are found to satisfy the empirical formula

$$R = 1.5 \times 10^{-15}\,A^{1/3}\,\text{m} = 2.0 \times 10^{-15}\,Z_1^{1/3}\,\text{m} \tag{11.132}$$

for $Z \gg 1$ (Park 1974).

In 1928, George Gamov proposed a very successful theory of α-decay, according to which the α-particle moves freely inside the nucleus, and is emitted after tunneling through the potential barrier between itself and the daughter nucleus. In other words, the α-particle, whose energy is E, is trapped in a potential well of radius R by the potential barrier

$$U(r) = \frac{Z_1\,Z_2\,e^2}{4\pi\,\epsilon_0\,r} \tag{11.133}$$

for $r > R$. Here, e is the magnitude of the electron charge.

Making use of the WKB approximation (and neglecting the fact that r is a radial, rather than a Cartesian, coordinate), the probability of the α-particle tunneling through the barrier is

$$|T|^2 = \exp\left(-\frac{2\sqrt{2m}}{\hbar}\int_{r_1}^{r_2}\sqrt{U(r)-E}\,dr\right), \tag{11.134}$$

where $r_1 = R$ and $r_2 = Z_1\,Z_2\,e^2/(4\pi\,\epsilon_0\,E)$. Here, $m = 4\,m_p$ is the α-particle mass, and m_p is the proton mass. The previous expression reduces to

$$|T|^2 = \exp\left[-2\sqrt{2}\beta\int_1^{E_c/E}\left(\frac{1}{y}-\frac{E}{E_c}\right)^{1/2}dy\right], \tag{11.135}$$

where

$$\beta = \left(\frac{Z_1 Z_2 e^2 m R}{4\pi \epsilon_0 \hbar^2}\right)^{1/2} = 0.74 Z_1^{2/3} \tag{11.136}$$

is a dimensionless constant, and

$$E_c = \frac{Z_1 Z_2 e^2}{4\pi \epsilon_0 R} = 1.44 Z_1^{2/3} \text{ MeV} \tag{11.137}$$

is the characteristic energy the α-particle would need in order to escape from the nucleus without tunneling. Of course, $E \ll E_c$. It is easily demonstrated that

$$\int_1^{1/\epsilon} \left(\frac{1}{y} - \epsilon\right)^{1/2} dy \simeq \frac{\pi}{2\sqrt{\epsilon}} - 2 \tag{11.138}$$

when $\epsilon \ll 1$. Hence.

$$|T|^2 \simeq \exp\left[-2\sqrt{2}\beta\left(\frac{\pi}{2}\sqrt{\frac{E_c}{E}} - 2\right)\right]. \tag{11.139}$$

Now, the α-particle moves inside the nucleus at the characteristic velocity $v = \sqrt{2 E/m}$. It follows that the particle bounces backward and forward within the nucleus at the frequency $\nu \simeq v/R$, giving

$$\nu \simeq 2 \times 10^{28} \text{ yr}^{-1} \tag{11.140}$$

for a 1 MeV α-particle trapped inside a typical heavy nucleus of radius 10^{-14} m. Thus, the α-particle effectively attempts to tunnel through the potential barrier ν times a second. If each of these attempts has a probability $|T|^2$ of succeeding then the probability of decay per unit time is $\nu |T|^2$. Hence, if there are $N(t) \gg 1$ intact nuclei at time t then there are only $N + dN$ at time $t + dt$, where

$$dN = -N \nu |T|^2 dt. \tag{11.141}$$

This expression can be integrated to give

$$N(t) = N(0) \exp\left(-\nu |T|^2 t\right). \tag{11.142}$$

The *half-life*, τ, is defined as the time which must elapse in order for half of the nuclei originally present to decay. It follows from the previous formula that

$$\tau = \frac{\ln 2}{\nu |T|^2}. \tag{11.143}$$

Note that the half-life is independent of $N(0)$.

Finally, making use of the previous results, we obtain

$$\log_{10}[\tau(\text{yr})] = -C_1 - C_2 Z_1^{2/3} + C_3 \frac{Z_1}{\sqrt{E(\text{MeV})}}, \tag{11.144}$$

where

$$C_1 = 28.5, \tag{11.145}$$
$$C_2 = 1.83, \tag{11.146}$$
$$C_3 = 1.73. \tag{11.147}$$

The half-life, τ, the daughter charge-number, $Z_1 = Z-2$, and the α-particle energy, E, for atomic

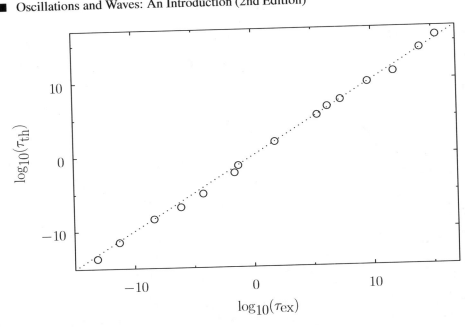

FIGURE 11.13 The experimentally determined half-life, τ_{ex}, of various atomic nuclei that decay via α-emission versus the best-fit theoretical half-life $\log_{10}(\tau_{th}) = -28.9 - 1.60\,Z_1^{2/3} + 1.61\,Z_1/\sqrt{E}$. Both half-lives are measured in years. Here, $Z_1 = Z - 2$, where Z is the charge-number of the nucleus and E the characteristic energy of the emitted α-particle in MeV. In order of increasing half-life, the points correspond to the following nuclei: Rn 215, Po 214, Po 216, Po 197, Fm 250, Ac 225, U 230, U 232, U 234, Gd 150, U 236, U 238, Pt 190, Gd 152, Nd 144. (Data obtained from International Atomic Energy Agency, Nuclear Data Center.)

nuclei that undergo α-decay are indeed found to satisfy a relationship of the form (11.144). The best fit to the data (see Figure 11.13) is obtained using

$$C_1 = 28.9, \tag{11.148}$$

$$C_2 = 1.60, \tag{11.149}$$

$$C_3 = 1.61. \tag{11.150}$$

It can be seen that these values are remarkably similar to those calculated previously.

11.16   THREE-DIMENSIONAL WAVE MECHANICS

Up to now, we have only discussed wave mechanics for a particle moving in one dimension. However, the generalization to a particle moving in three dimensions is fairly straightforward. A massive particle moving in three dimensions has a complex wavefunction of the form [cf., Equation (11.15

$$\psi(x, y, z, t) = \psi_0\,e^{-i\,(\omega t - \mathbf{k} \cdot \mathbf{r})}, \tag{11.15}$$

where ψ_0 is a complex constant, and $\mathbf{r} = (x, y, z)$. Here, the wavevector, $\mathbf{k}$, and the angular frequency, ω, are related to the particle momentum, $\mathbf{p}$, and energy, E, according to [cf., Equation (11.1

$$\mathbf{p} = \hbar\,\mathbf{k}, \tag{11.15}$$

and [cf., Equation (11.1)]

$$E = \hbar \omega, \tag{11.153}$$

respectively. Generalizing the analysis of Section 11.5, the three-dimensional version of Schrödinger's equation is [cf., Equation (11.23)]

$$i\hbar \frac{\partial \psi}{\partial t} = -\frac{\hbar^2}{2m} \nabla^2 \psi + U(\mathbf{r})\psi, \tag{11.154}$$

where the differential operator

$$\nabla^2 \equiv \frac{\partial^2}{\partial x^2} + \frac{\partial^2}{\partial y^2} + \frac{\partial^2}{\partial z^2} \tag{11.155}$$

is known as the *Laplacian*. The interpretation of a three-dimensional wavefunction is that the probability of simultaneously finding the particle between x and $x+dx$, between y and $y+dy$, and between z and $z + dz$, at time t is [cf., Equation (11.26)]

$$P(x, y, z, t) = |\psi(x, y, z, t)|^2 \, dx \, dy \, dz. \tag{11.156}$$

Moreover, the normalization condition for the wavefunction becomes [cf., Equation (11.28)]

$$\int_{-\infty}^{\infty} \int_{-\infty}^{\infty} \int_{-\infty}^{\infty} |\psi(x, y, z, t)|^2 \, dx \, dy \, dz = 1. \tag{11.157}$$

It can be demonstrated that Schrödinger's equation, (11.154), preserves the normalization condition, (11.157), of a localized wavefunction (Gasiorowicz 1996). Heisenberg's uncertainty principle generalizes to [cf., Equation (11.56)]

$$\Delta x \, \Delta p_x \gtrsim \frac{\hbar}{2}, \tag{11.158}$$

$$\Delta y \, \Delta p_y \gtrsim \frac{\hbar}{2}, \tag{11.159}$$

$$\Delta z \, \Delta p_z \gtrsim \frac{\hbar}{2}. \tag{11.160}$$

Finally, a stationary state of energy E is written [cf., Equation (11.60)]

$$\psi(x, y, z, t) = \psi(x, y, z) \, e^{-i(E/\hbar)t}, \tag{11.161}$$

where the stationary wavefunction, $\psi(x, y, z)$, satisfies [cf., Equation (11.62)]

$$-\frac{\hbar^2}{2m} \nabla^2 \psi + U(\mathbf{r})\psi = E\psi. \tag{11.162}$$

11.17 PARTICLE IN BOX

As an example of a three-dimensional problem in wave mechanics, consider a particle trapped in a square potential well of infinite depth, such that

$$U(x, y, z) = \begin{cases} 0 & 0 \le x \le a, \, 0 \le y \le a, \, 0 \le z \le a \\ \infty & \text{otherwise} \end{cases}. \tag{11.163}$$

Within the well, the stationary wavefunction, $\psi(x, y, z)$, satisfies

$$-\frac{\hbar^2}{2m} \nabla^2 \psi = E\psi, \tag{11.164}$$

subject to the boundary conditions

$$\psi(0, y, z) = \psi(x, 0, z) = \psi(x, y, 0) = 0, \tag{11.165}$$

and

$$\psi(a, y, z) = \psi(x, a, z) = \psi(x, y, a) = 0, \tag{11.166}$$

because $\psi = 0$ outside the well. Let us try a separable wavefunction of the form

$$\psi(x, y, z) = \psi_0 \sin(k_x x) \sin(k_y y) \sin(k_z z). \tag{11.167}$$

This expression automatically satisfies the boundary conditions (11.165). The remaining boundary conditions, (11.166), are satisfied provided

$$k_x = l_x \frac{\pi}{a}, \tag{11.168}$$

$$k_y = l_y \frac{\pi}{a}, \tag{11.169}$$

$$k_z = l_z \frac{\pi}{a}, \tag{11.170}$$

where l_x, l_y, and l_z are (independent) positive integers. Substitution of the wavefunction (11.167) into Equation (11.164) yields

$$E = \frac{\hbar^2}{2m} (k_x^2 + k_y^2 + k_z^2). \tag{11.171}$$

Thus, it follows from Equations (11.168)–(11.170) that the particle energy is quantized, and that the allowed *energy levels* are

$$E_{l_x, l_y, l_z} = \frac{\hbar^2}{2ma^2} (l_x^2 + l_y^2 + l_z^2). \tag{11.172}$$

The properly normalized [see Equation (11.157)] stationary wavefunctions corresponding to these energy levels are

$$\psi_{l_x, l_y, l_z}(x, y, z) = \left(\frac{2}{a}\right)^{3/2} \sin\left(l_x \pi \frac{x}{a}\right) \sin\left(l_y \pi \frac{y}{a}\right) \sin\left(l_z \pi \frac{z}{a}\right). \tag{11.173}$$

As is the case for a particle trapped in a one-dimensional potential well, the lowest energy level for a particle trapped in a three-dimensional well is not zero, but rather

$$E_{1,1,1} = 3 E_1. \tag{11.174}$$

Here,

$$E_1 = \frac{\hbar^2}{2ma^2}. \tag{11.175}$$

is the *ground state* (i.e., the lowest energy state) energy in the one-dimensional case. It follows from Equation (11.172) that distinct permutations of l_x, l_y, and l_z that do not alter the value of $l_x^2 + l_y^2 + l_z^2$ also do not alter the energy. In other words, in three dimensions, it is possible for distinct wavefunctions to be associated with the same energy level. In this situation, the energy level is said to be *degenerate*. The ground-state energy level, $3 E_1$, is non-degenerate, because the only combination of (l_x, l_y, l_z) that gives this energy is $(1, 1, 1)$. However, the next highest energy level, $6 E_1$, is degenerate, because it is obtained when (l_x, l_y, l_y) take the values $(2, 1, 1)$, or $(1, 2, 1)$, or $(1, 1, 2)$. In fact, a non-degenerate energy level corresponds to a case where the three quantum numbers (i.e., l_x, l_y, and l_z) all have the same value, whereas a threefold degenerate energy level corresponds to a case where only two of the quantum numbers have the same value, and, finally, a sixfold degenerate energy level corresponds to a case where the quantum numbers are all different.

11.18   DEGENERATE ELECTRON GAS

Consider N electrons trapped in a cubic box of dimension a. Let us treat the electrons as essentially non-interacting particles. The total energy of a system consisting of many non-interacting particles is simply the sum of the single-particle energies of the individual particles. Furthermore, because the electrons are indistinguishable fermions, they are subject to the so-called *Pauli exclusion principle* (Park 1974). The exclusion principle states that no two electrons in our system can occupy the same single-particle energy level. Now, from Section 11.17, the single-particle energy levels for a particle in a box are characterized by the three quantum numbers, l_x, l_y, and l_z. Thus, we conclude that no two electrons in our system can have the same set of values of l_x, l_y, and l_z. It turns out that this is not quite true, because electrons possess an intrinsic angular momentum called *spin* (Park 1974). The spin states of an electron are governed by an additional quantum number that can take one of two different values. Hence, when spin is taken into account, we conclude that a maximum of two electrons (with different spin quantum numbers) can occupy a single-particle energy level corresponding to a particular set of values of l_x, l_y, and l_z. It is clear, from Equation (11.172), that the associated particle energy is proportional to $l^2 = l_x^2 + l_y^2 + l_z^2$.

Suppose that our electrons are cold; that is, they have comparatively little thermal energy. In this case, we would expect them to fill the lowest single-particle energy levels available to them. We can imagine the single-particle energy levels as existing in a sort of three-dimensional quantum number space whose Cartesian coordinates are l_x, l_y, and l_z. Thus, the energy levels are uniformly distributed in this space on a cubic lattice. Moreover, the distance between nearest-neighbor energy levels is unity. This implies that the number of energy levels per unit volume is also unity. Finally, the energy of a given energy level is proportional to its distance, $l^2 = l_x^2 + l_y^2 + l_z^2$, from the origin.

Because we expect cold electrons to occupy the lowest energy levels available to them, but only two electrons can occupy a given energy level, it follows that if the number of electrons, N, is very large then the filled energy levels will be approximately distributed in a sphere centered on the origin of quantum number space. The number of energy levels contained in a sphere of radius l is approximately equal to the volume of the sphere, because the number of energy levels per unit volume is unity. It turns out that this is not quite correct, because we have forgotten that the quantum numbers l_x, l_y, and l_z can only take positive values. Hence, the filled energy levels actually only occupy one octant of a sphere. The radius, l_F, of the octant of filled energy levels in quantum number space can be calculated by equating the number of energy levels it contains to the number of electrons, N. Thus, we can write

$$N = 2 \times \frac{1}{8} \times \frac{4\pi}{3} l_F^3. \tag{11.176}$$

Here, the factor 2 is to take into account the two spin states of an electron, and the factor $1/8$ is to take account of the fact that l_x, l_y, and l_z can only take positive values. Thus,

$$l_F = \left(\frac{3N}{\pi}\right)^{1/3}. \tag{11.177}$$

According to Equation (11.172), the energy of the most energetic electrons—which is known as the *Fermi energy*—is given by

$$E_F = \frac{l_F^2 \pi^2 \hbar^2}{2 m_e a^2} = \frac{\pi^2 \hbar^2}{2 m_e a^2} \left(\frac{3N}{\pi}\right)^{2/3}, \tag{11.178}$$

where m_e is the electron mass. This expression can also be written as

$$E_F = \frac{\pi^2 \hbar^2}{2 m_e} \left(\frac{3n}{\pi}\right)^{2/3}, \tag{11.179}$$

where $n = N/a^3$ is the number of electrons per unit volume (in real space). Note that the Fermi energy only depends on the number density of the confined electrons.

The mean energy of the electrons is given by

$$\overline{E} = E_F \int_0^{l_F} l^2 \, 4\pi \, l^2 \, dl \bigg| \frac{4}{3} \pi \, l_F^5 = \frac{3}{5} E_F, \tag{11.180}$$

because $E \propto l^2$, and the energy levels are uniformly distributed in quantum-number space within an octant of radius l_F. According to classical physics, the mean thermal energy of the electrons is $(3/2) k_B T$, where T is the electron temperature, and k_B the Boltzmann constant (Reif 2008). Thus, if $k_B T \ll E_F$ then our original assumption that the electrons are cold is valid. Note that, in this case, the electron energy is much larger than that predicted by classical physics—electrons in this state are termed *degenerate*. On the other hand, if $k_B T \gg E_F$ then the electrons are hot, and are essentially governed by classical physics—electrons in this state are termed *non-degenerate*.

The total energy of a degenerate electron gas is

$$E_{\text{total}} = N \overline{E} = \frac{3}{5} N E_F. \tag{11.181}$$

Hence, the gas pressure takes the form (Reif 2008)

$$P = -\frac{\partial E_{\text{total}}}{\partial V} = \frac{2}{5} n E_F, \tag{11.182}$$

because $E_F \propto a^{-2} = V^{-2/3}$. [See Equation (11.178).] Now, the pressure predicted by classical physics is $P = n k_B T$ (Reif 2008). Thus, a degenerate electron gas has a much higher pressure than that which would be predicted by classical physics. This is an entirely quantum mechanical effect, and is due to the fact that identical fermions cannot get significantly closer together than a de Broglie wavelength without violating the Pauli exclusion principle. Note that, according to Equation (11.179), the mean spacing between degenerate electrons is

$$d \sim n^{-1/3} \sim \frac{h}{\sqrt{m_e E}} \sim \frac{h}{p} \sim \lambda, \tag{11.183}$$

where λ is the de Broglie wavelength. Thus, an electron gas is non-degenerate when the mean spacing between the electrons is much greater than the de Broglie wavelength, and becomes degenerate as the mean spacing approaches the de Broglie wavelength.

In turns out that the conduction (i.e., free) electrons inside metals are highly degenerate (because the number of electrons per unit volume is very large, and $E_F \propto n^{2/3}$). Indeed, most metals are hard to compress as a direct consequence of the high degeneracy pressure of their conduction electrons. To be more exact, resistance to compression is usually measured in terms of a quantity known as the *bulk modulus*, which is defined

$$B = -V \frac{\partial P}{\partial V} \tag{11.184}$$

Now, for a fixed number of electrons, $P \propto V^{-5/3}$ [see Equations (11.178) and (11.182)]. Hence,

$$B = \frac{5}{3} P = \frac{\pi^3 \hbar^2}{9 m_e} \left(\frac{3 n}{\pi} \right)^{5/3}. \tag{11.185}$$

For example, the number density of free electrons in magnesium is $n \sim 8.6 \times 10^{28} \, \text{m}^{-3}$. This leads to the following estimate for the bulk modulus; $B \sim 6.4 \times 10^{10} \, \text{N m}^{-2}$. The actual bulk modulus is $B = 4.5 \times 10^{10} \, \text{N m}^{-2}$.

11.19 WHITE-DWARF STAR

A main-sequence hydrogen-burning star, such as the Sun, is maintained in equilibrium via the balance of the gravitational attraction tending to make it collapse, and the thermal pressure tending to make it expand. Of course, the thermal energy of the star is generated by nuclear reactions occurring deep inside its core. Eventually, however, the star will run out of burnable fuel, and, therefore, start to collapse, as it radiates away its remaining thermal energy. What is the ultimate fate of such a star?

A burnt-out star is basically a gas of electrons and ions. As the star collapses, its density increases, and so the mean separation between its constituent particles decreases. Eventually, the mean separation becomes of order of the de Broglie wavelength of the electrons, and the electron gas becomes degenerate. Note that the de Broglie wavelength of the ions is much smaller than that of the electrons (because the ions are much more massive), so the ion gas remains non-degenerate. Now, even at zero temperature, a degenerate electron gas exerts a substantial pressure, because the Pauli exclusion principle prevents the mean electron separation from becoming significantly smaller than the typical de Broglie wavelength. (See Section 11.18.) Thus, it is possible for a burnt-out star to maintain itself against complete collapse under gravity via the degeneracy pressure of its constituent electrons. Such stars are termed *white-dwarfs*. Let us investigate the physics of white-dwarfs in more detail.

The total energy of a white-dwarf star can be written

$$\mathcal{E} = K + U,$$
(11.186)

where K is the kinetic energy of the degenerate electrons (the kinetic energy of the ions is negligible), and U is the gravitational potential energy. Let us assume, for the sake of simplicity, that the density of the star is uniform. In this case, the gravitational potential energy takes the form (Fitzpatrick 2012)

$$U = -\frac{3}{5}\frac{G M^2}{R},$$
(11.187)

where G is the gravitational constant, M is the stellar mass, and R is the stellar radius.

From the previous section, the kinetic energy of a degenerate electron gas is simply

$$K = N\overline{E} = \frac{3}{5}N E_F = \frac{3}{5}N\frac{\pi^2\hbar^2}{2 m_e}\left(\frac{3 N}{\pi V}\right)^{2/3},$$
(11.188)

where N is the number of electrons, V the volume of the star, and m_e the electron mass.

The interior of a white-dwarf star is composed of atoms like C^{12} and O^{16} which contain equal numbers of protons, neutrons, and electrons. Thus,

$$M = 2 N m_p,$$
(11.189)

where m_p is the proton mass.

Equations (11.186)–(11.189) can be combined to give

$$\mathcal{E} = \frac{A}{R^2} - \frac{B}{R},$$
(11.190)

where

$$A = \frac{3}{20}\left(\frac{9\pi}{8}\right)^{2/3}\frac{\hbar^2}{m_e}\left(\frac{M}{m_p}\right)^{5/3},$$
(11.191)

$$B = \frac{3}{5}G M^2.$$
(11.192)

The equilibrium radius of the star, R_*, is that which minimizes the total energy $\mathcal{E}$. In fact, it is easily demonstrated that

$$R_* = \frac{2A}{B},$$ (11.193)

which yields

$$R_* = \frac{(9\pi)^{2/3}}{8} \frac{\hbar^2}{G m_e m_p^{5/3} M^{1/3}}.$$ (11.194)

The previous formula can also be written

$$\frac{R_*}{R_\odot} = 0.010 \left(\frac{M_\odot}{M}\right)^{1/3},$$ (11.195)

where $R_\odot = 7 \times 10^5$ km is the solar radius, and $M_\odot = 2 \times 10^{30}$ kg the solar mass. It follows that the radius of a typical solar-mass white-dwarf is about 7000 km; that is, about the same as the radius of the Earth. The first white-dwarf to be discovered (in 1862) was the companion of Sirius. Nowadays, thousands of white-dwarfs have been observed, all with properties similar to those described previously.

Note from Equations (11.188), (11.189), and (11.195) that $\bar{E} \propto M^{4/3}$. In other words, the mean energy of the electrons inside a white-dwarf increases as the stellar mass increases. Hence, for a sufficiently massive white-dwarf, the electrons can become relativistic. It turns out that the degeneracy pressure for relativistic electrons only scales as R^{-1}, rather that R^{-2} (Park 1974), and, thus is unable to balance the gravitational pressure [which also scales as R^{-1}; see Equation (11.190)]. It follows that electron degeneracy pressure is only able to halt the collapse of a burnt-out star provided that the stellar mass does not exceed some critical value, known as the *Chandrasekhar limit*, which turns out to be about 1.4 times the mass of the Sun (Park 1974). Stars whose mass exceed the Chandrasekhar limit inevitably collapse to produce extremely compact objects, such as neutron stars (which are held up by the degeneracy pressure of their constituent neutrons), or black holes.

EXERCISES

11.1 Use the standard power law expansions,

$$e^x = 1 + x + \frac{x^2}{2!} + \frac{x^3}{3!} + \frac{x^4}{4!} + \frac{x^5}{5!} + \frac{x^6}{6!} + \frac{x^7}{7!} + \cdots,$$

$$\sin x = x - \frac{x^3}{3!} + \frac{x^5}{5!} - \frac{x^7}{7!} + \cdots,$$

$$\cos x = 1 - \frac{x^2}{2!} + \frac{x^4}{4!} - \frac{x^6}{6!} + \cdots,$$

which are valid for complex x, to prove Euler's theorem,

$$e^{i\theta} = \cos\theta + i\,\sin\theta,$$

where θ is real.

11.2 Equations (8.27) and (8.28) can be combined with Euler's theorem to give

$$\delta(k) = \frac{1}{2\pi} \int_{-\infty}^{\infty} e^{ikx}\,dx,$$

where $\delta(k)$ is a Dirac delta function. Use this result to prove Fourier's theorem; that is, if

$$f(x) = \int_{-\infty}^{\infty} \bar{f}(k) e^{ikx} dk,$$

then

$$\bar{f}(k) = \frac{1}{2\pi} \int_{-\infty}^{\infty} f(x) e^{-ikx} dx.$$

11.3 A He–Ne laser emits radiation of wavelength $\lambda = 633$ nm. How many photons are emitted per second by a laser with a power of 1 mW? What force does such a laser exert on a body that completely absorbs its radiation?

11.4 The ionization energy of a hydrogen atom in its ground state is $E_{ion} = 13.6$ eV. Calculate the frequency (in hertz), wavelength, and wavenumber of the electromagnetic radiation that will just ionize the atom.

11.5 The maximum energy of photoelectrons from aluminum is 2.3 eV for radiation of wavelength 200 nm, and 0.90 eV for radiation of wavelength 258 nm. Use this data to calculate Planck's constant (divided by 2π) and the work function of aluminum. [Adapted from Gasiorowicz 1996.]

11.6 Show that the de Broglie wavelength of an electron accelerated across a potential difference V is given by

$$\lambda = 1.23 \times 10^{-9} V^{-1/2} \text{ m},$$

where V is measured in volts. [From Pain 1999.]

11.7 If the atoms in a regular crystal are separated by 3×10^{-10} m demonstrate that an accelerating voltage of about 2 kV would be required to produce a useful electron diffraction pattern from the crystal. [Modified from Pain 1999.]

11.8 A particle of mass m has a wavefunction

$$\psi(x, t) = A \exp\left[-a\left(m x^2/\hbar + i t\right)\right],$$

where A and a are positive real constants. For what potential $U(x)$ does $\psi(x, t)$ satisfy Schrödinger's equation?

11.9 Show that the wavefunction of a particle of mass m trapped in a one-dimensional square potential well of width a, and infinite depth, returns to its original form after a quantum revival time; $T = 4 m a^2/\pi \hbar$.

11.10 Show that the normalization constant for the stationary wavefunction

$$\psi(x, y, z) = A \sin\left(l_x \pi \frac{x}{a}\right) \sin\left(l_y \pi \frac{y}{b}\right) \sin\left(l_z \pi \frac{z}{c}\right)$$

describing an electron trapped in a three-dimensional rectangular potential well of dimensions a, b, c, and infinite depth, is $A = (8/a b c)^{1/2}$. Here, l_x, l_y, and l_z are positive integers. [From Pain 1999.]

11.11 Derive Equation (11.89).

11.12 Consider a particle trapped in the finite potential well whose potential is given by Equation (11.75). Demonstrate that for a totally-symmetric state the ratio of the probability finding the particle outside to the probability of finding the particle inside the well is

$$\frac{P_{\text{out}}}{P_{\text{in}}} = \frac{\cos^3 y}{\sin y \,(y + \sin y \,\cos y)},$$

where $(\lambda - y^2)^{1/2} = y \,\tan y$, and $\lambda = V/E_0$. Hence, demonstrate that for a shallow well (i.e., $\lambda \ll 1$) $P_{\text{out}} \simeq 1 - 2\,\lambda$, whereas for a deep well (i.e., $\lambda \gg 1$) $P_{\text{out}} \simeq (\pi^2/4)/\lambda^{3/2}$ (assume that the particle is in the ground state).

11.13 Derive expression (11.105) from Equations (11.101)–(11.104).

11.14 Derive expression (11.114) from Equations (11.110)–(11.113).

11.15 The probability of a particle of mass m penetrating a distance x into a classically forbidden region is proportional to $e^{-2\alpha x}$, where

$$\alpha^2 = 2\,m\,(V - E)/\hbar^2.$$

If $x = 2 \times 10^{-10}$ m and $V - E = 1$ eV show that $e^{-2\alpha x}$ is approximately equal to 10^{-1} an electron, and 10^{-38} for a proton. [Modified from Pain 1999.]

11.16 A stream of particles of mass m and energy $E > 0$ encounter a potential step of height $W (< E)$; that is, $U(x) = 0$ for $x < 0$, and $U(x) = W$ for $x > 0$, with the particles incident from $-\infty$. Show that the fraction reflected is

$$R = \left(\frac{k - q}{k + q}\right)^2,$$

where $k^2 = (2\,m/\hbar^2)\,E$ and $q^2 = (2\,m/\hbar^2)\,(E - W)$.

11.17 Consider the half-infinite potential well

$$U(x) = \begin{cases} \infty & x \le 0 \\ -V_0 & 0 < x < L \\ 0 & x \ge L \end{cases},$$

where $V_0 > 0$. Demonstrate that the bound-states of a particle of mass m and energy $-V$ $E < 0$ satisfy

$$\tan\left(\sqrt{2\,m\,(V_0 + E)}\;L/\hbar\right) = -\sqrt{(V_0 + E)/(-E)}.$$

11.18 Given that the number density of free electrons in copper is 8.5×10^{28} m^{-3}, calculate Fermi energy in electron volts, and the velocity of an electron whose kinetic energy is equal to the Fermi energy.

11.19 Obtain an expression for the Fermi energy (in eV) of an electron in a white-dwarf star a function of the stellar mass (in solar masses). At what mass does the Fermi energy equal the rest mass energy?

Bibliography

Abramowitz, M. (ed.), and Stegun, I. (ed.) 1965. *Handbook of Mathematical Functions: with Formulas, Graphs, and Mathematical Tables*. Dover.

Batchelor, G.K. 2000. *An Introduction to Fluid Dynamics*. Cambridge.

Born, M., and Wolf, E. 1980. *Principles of Optics: Electromagnetic Theory of Propagation Interference and Diffraction of Light*, 6th Edition. Pergamon.

Crawford, F.S. 1968. *Waves*. McGraw-Hill.

Dirac, P.A.M. 1982. *The Principles of Quantum Mechanics*, 4th Edition. Oxford.

Fitzpatrick, R. 2008. *Maxwell's Equations and the Principles of Electromagnetism*. Jones & Bartlett.

Fitzpatrick, R. 2012. *An Introduction to Celestial Mechanics*. Cambridge.

Fitzpatrick, R. 2015. *Plasma Physics: An Introduction*. CRC Press.

Fowles, G.R., and Cassiday, G.L. 2005. *Analytic Mechanics*, 7th Edition. Thomson Brooks/Cole.

French, A.P. 1971. *Vibrations and Waves*, 1st Edition. W.W. Norton & Co.

Gasiorowicz, S. 1996. *Quantum Physics*, 2nd Edition. John Wiley & Sons.

Goldstein, H., Poole, C., and Safko, J. 2002. *Classical Mechanics*, 3rd Edition. Addison Wesley.

Gradshteyn, I.S. and Ryzhik, I.M. 1980. *Table of Integrals, Series, and Products*, Corrected and Enlarged Edition. Academic Press.

Grant, I.S., and Phillips, W.R. 1975. *Electromagnetism*. J. Wiley & Sons.

Hazeltine, R.D., and Waelbroeck, F.L. 2004. *The Framework of Plasma Physics*. Westview.

Haynes, W.M. (ed.), and Lide, D.R. (ed.) 2011a. *CRC Handbook of Chemistry and Physics*, 92nd Edition. CRC Press, Taylor & Francis Group. Section 1.

Haynes, W.M. (ed.), and Lide, D.R. (ed.) 2011b. *CRC Handbook of Chemistry and Physics*, 92nd Edition. CRC Press, Taylor & Francis Group. Section 6.

Haynes, W.M. (ed.), and Lide, D.R. (ed.) 2011c. *CRC Handbook of Chemistry and Physics*, 92nd Edition. CRC Press, Taylor & Francis Group. Section 14.

Hecht, E. 1974. *Schaum's Outline of Optics*. McGraw-Hill.

Hecht, E., and Zajac, A. 1974. *Optics*. Addison-Wesley.

Held, M.A., and Marion, J.B. 1995. *Classical Electromagnetic Radiation*, 3rd Edition. Saunders College Publishing.

Ingard, K.U. 1988. *Fundamentals of Waves and Oscillations*. Cambridge.

Jackson, J.D. 1975. *Classical Electrodynamics*, 2nd Edition. J. Wiley & Sons.

Lamb, H. 1932. *Hydrodynamics*, 6th Edition. Cambridge.

Landau, L.D., and Lifshitz, E.M. 1959. *Fluid Mechanics*, Course of Theoretical Physics, Volume 6. Pergamon.

Lighthill, J. 1978. *Waves in Fluids*. Cambridge.

Longair, M.S. 2011. *High Energy Astrophysics*, 3rd Edition. Cambridge.

Love, A.E.H. 1944. *A Treatise on the Mathematical Theory of Elasticity*, 4th Edition. Dover.

Pain, H.J. 1999. *The Physics of Vibrations and Waves*, 5th Edition. J. Wiley & Sons.

Press, W.H., Teukolsky, S.A., Vetterling, W.T., and Flannery, B.P. 1992. *Numerical Recipes in C: The Art of Scientific Computing*, 2nd Edition. Cambridge.

Park, D. 1974. *Introduction to the Quantum Theory*, 2nd Edition. McGraw-Hill.

Reif, F. 2008. *Fundamentals of Statistical and Thermal Physics*. Waveland.

Rindler, W. 1997. *Essential Relativity: Special, General, and Cosmological*, 2nd Edition. Springer-Verlag.

Riley, K.F. 1974. *Mathematical Methods for the Physical Sciences*. Cambridge.

Schiff, L.I. 1955. *Quantum Mechanics*. McGraw-Hill.

Spiegel, M.R., and Liu, J. 1999. *Mathematical Handbook of Formulas and Tables*, 2nd Edition. Shaum's Outline Series, McGraw-Hill.

Stix, T.H. 1962. *The Theory of Plasma Waves*, 1st Edition. McGraw-Hill.

Storey, L.R.O. 1953. *An Investigation of Whistling Atmospherics*. Philosophical Transactions of the Royal Society A **246**, 113.

Swanson, D.G. 2003. *Plasma Waves*, 2nd Edition. Taylor & Francis.

Watson, G.N. 1962. *A Treatise on the Theory of Bessel Functions*, 2nd Edition. Cambridge.

Wikipedia contributors 2018. *Wikipedia: The Free Encyclopedia.* http://en.wikipedia.org

Zygmund, A. 1955. *Trigonometrical Series*. Dover.

Index